Fossils
in the Making

PREHISTORIC ARCHEOLOGY AND ECOLOGY

A Series Edited by Karl W. Butzer and Leslie G. Freeman

Fossils
in the Making

Vertebrate Taphonomy and Paleoecology

Edited by
Anna K. Behrensmeyer
and Andrew P. Hill

The University of Chicago Press

Chicago and London

ANNA K. BEHRENSMEYER is a paleoecologist currently working as a postdoctoral fellow in Yale's Department of Anthropology. She has done extensive on-site research in Kenya, western North America, and Pakistan, and has contributed numerous articles to scholarly journals in the field.

ANDREW P. HILL, a research fellow of the International Louis Leakey Memorial Institute for African Prehistory in Nairobi, carries out research in East Africa and Pakistan, and is at present writing a book on taphonomy.

The University of Chicago Press, Chicago 60637
The University of Chicago Press, Ltd., London

© 1980 by The University of Chicago
All rights reserved. Published 1980
Printed in the United States of America
84 83 82 81 80   54321

Library of Congress Cataloging in Publication Data

Main entry under title:

Fossils in the making.

       (Prehistoric archeology and ecology)
       Papers presented at a symposium, July 1976,
sponsored by the Wenner-Gren Foundation for
Anthropological Research held at the Burg-
Wartenstein castle in Austria.
       Includes bibliographical references and index.
       1. Vertebrates, Fossil--Congresses.
2. Paleoecology--Congresses. I. Behrensmeyer,
Anna K. II. Hill, Andrew P. III. Wenner-Gren
Foundation for Anthropological Research, New York.
IV. Series
QE841.F66   566   79-19879
ISBN 0-226-04169-7
ISBN 0-226-04168-9 pbk.

Bill Bishop provided inspiration to all who knew
him as a scientist and a friend.
This book is dedicated to him.

Walter William Bishop
1931--77

CONTENTS

FOREWORD

During the past ten years, the ecological interpretation of fossil "death assemblages" has emerged as a new research perspective in human and vertebrate paleontology
as well as archeology. Under the designation *taphonomy*, such studies have begun to il-
luminate the conditions of bone accumulation in several key hominid and archeological
sites. Critical questions include the extent of postdepositional disturbance or distortion of bone assemblages, and whether human or carnivore activities are chiefly responsible for them. Equally pertinent to human origins is the potential of taphonomic work
to elucidate the ecological preferences of early hominids.

Much of the recent taphonomic work exploring the evolution of human behavior has been
associated with Kay Behrensmeyer in East Africa, and with C. K. Brain and Richard Klein
in South Africa. The present volume attempts to transcend the technical limits of
taphonomy, to bring together representative recent efforts that are all relevant to human
and vertebrate paleoecology. The result is a first general work that provides a coherent
conceptual framework, as well as a general methodology, for vertebrate paleoecology. A
second and equally important synthetic contribution is the information marshaled on
vertebrate community evolution in Africa, including that of the hominid line. In these
areas *Fossils in the Making* makes an original, significant contribution.

As we have suggested, taphonomy is a fresh and expanding field, so that it is not
surprising to find that research efforts are still to a large degree experimental.
Empirical results are still few in comparison to the need for accumulating data, and thus
far only a handful of large paleontological samples—unbiased in the process of removal—
have been analyzed. Many more large-scale, exceedingly careful excavations of fossil
assemblages—in a variety of sedimentary and paleotopographic contexts—will be required
before interassemblage comparisons can provide more useful and more interesting answers.
Statistical expertise adequate to deal with large samples does not necessarily follow
from elegant mathematical models, but must be generated and tested empirically within the
actual research framework in order to achieve sound results from a finite body of
imperfect data. *Fossils in the Making* does not aspire to be a definitive work; it does
provide an exciting assessment of the state of the art and of the problems that confront
it, and in so doing, suggests the ways in which future work should be practiced and the
many field and laboratory approaches integrated.

It is our belief that, in putting together this volume, Behrensmeyer and Hill have
done a great professional service, towards both the education of a wide spectrum of
interested readers and the crystallization of an increasingly effective research enter-
prise of growing importance to human paleontology and archeology. We hope that it will
stimulate many more, patient, competent, and ground-breaking studies in Africa and on
other continents.

Karl W. Butzer

Leslie G. Freeman

ix

PREFACE

Fossils are the remains of single organisms, but they are also pieces of much larger, more complex, and fascinating puzzles: the ecosystems of the past. In paleoecology we attempt to reconstruct these ecosystems and their animal and plant communities, using evidence from fossils and environments of burial. Because the fossil and geological records are incomplete, and because what samples we have may result from any number of processes which bias the original ecological record, taphonomic analysis of fossils is an essential prerequisite for paleoecology. Taphonomy means "the laws of burial," and it is a field of study that serves to focus all the varied interests that describe and analyze how organisms become fossils. The taphonomy and paleoecology of vertebrates, including humans and their ancestors, are the subjects of this book.

Interest in paleoecology, and recognition of the importance of taphonomy, have grown rapidly in the past several years. Researchers in diverse fields have more or less independently pursued a variety of taphonomic problems, without the benefit of mutual interchange concerning the common areas of study. The Wenner-Gren Foundation for Anthropological Research in New York, well known for its interest in fostering new or multidisciplinary areas of study, sponsored a symposium held in July of 1976 to bring together representatives of the many disciplines involved in taphonomy and paleoecology. The symposium was entitled "Taphonomy and Vertebrate Paleoecology, with Special Reference to the Late Cenozoic of Sub-Saharan Africa" (Burg Wartenstein Symposium No. 69). Participants included scientists with expertise in anthropology, archeology, recent ecology, functional anatomy, geochemistry (of bone trace elements and amino acids), geology, geomorphology, hydrodynamics, mathematics, and vertebrate paleontology.

The confrontation of these disciplines and the varied approaches they employ provided the basis for fruitful discussions of the central problems of taphonomy and paleoecology. The diversity of the papers presented in this volume corresponds to the diversity of approach. They are unified by the essential problems that lead toward understanding of the ecology of animals in the past. As is customary at Wenner-Gren symposia, the papers were prepared as preprints and circulated before the meeting. They have been revised by the authors in the light of discussions at the conference, and thus provide up-to-date information on the present status of paleoecological research. Some of the papers review and synthesize existing methods and concepts. Most of them present previously unpublished data and new approaches to problems in paleoecology and taphonomy.

We believe that the papers are relevant to the study of terrestrial fossil vertebrates of any place or time period, but we have chosen to illustrate the questions and problems by reference mainly to sub-Saharan Africa during the late Cenozoic. There are a number of reasons for this. Africa has many modern communities of large land vertebrates, and they have been studied in some detail by ecologists. Also, much of the work in recent taphonomy, the study of the remains of modern animals soon after death, has been done there. Finally, in this area, during the late Cenozoic, the human species is believed to have evolved. Since it is obvious that man has had some significant effects as a member of the mammal community in Africa during the past million years, the study of the interaction of evolving man and changing faunas and environments is of special interest. Human paleoecology in Africa has helped to stimulate a great deal of research in taphonomic and paleoecological problems, and has also provided fertile ground for the application of new methods and ideas which have grown from this research.

The conference was interesting not only at the level of basic data and theory in paleoecology, but also because it revealed differences in attitude among the participants concerning the very nature of the science and the way it should be done. To polarize these shades of opinion somewhat, it seems that, from one point of view, paleoecology, paleontology, and archeology are regarded as "historical sciences," implying that workers in these fields should simply tell the story of what happened in the past. The historical character of the subject matter limits the possibility or desirability of looking for "laws" or for general principles whereby fossil or archeological occurrences might be better explained. Every occurrence of fossils is assumed to be unique and must be approached as a completely isolated case. The contrasting viewpoint regards this narrative version of paleontological and paleoanthropological activity as inadequate: without the active attempt to generalize about the nature of fossil and archeological occurrences, fields such as paleontology and anthropology can hardly claim to be sciences at all.

Another fascinating feature of the conference concerned discrepant opinions as to what is acceptable as evidence for a theory. One viewpoint appears to be satisfied with simple coherence of fact and hypothesis; another wishes to explore alternative explanations that might lead to new observations or to new kinds of information that would attempt to falsify presently accepted hypotheses.

Without the Wenner-Gren Foundation a conference of this kind would not have been possible. We are very grateful to them. Their conference center in Austria, Burg Wartenstein, provides a unique opportunity for enforced seclusion and concentration upon a particular set of topics which makes it a most valuable asset to anthropology and related sciences. In particular, we thank Lita Osmundsen, the Director of Research, for her unfailing help and good advice at all stages in the development and progress of the meeting. She and her staff are to be thanked for producing at Burg Wartenstein the perfect ambience in which an intensive ten day meeting of this kind could occur. Throughout the conference and its preparation we received the help and advice of our two coorganizers, Bob Brain and Alan Walker, and we are very grateful for their assistance and tolerance. Of course we thank all the participants contributing to such a memorable and useful occasion. Dorothy Dechant Boaz was the rapporteuse for the meeting and helped enormously by taking extensive notes of the discussions and assisting with the editing of this volume. We thank John Barry for help in correcting the manuscripts, and Marie Pietrandrea and Liz Kyburg for their excellent typing.

A major influence on us before and during the conference was Bill Bishop. His energetic personality and wide-ranging interests led him to associate with all aspects of Cenozoic studies in sub-Saharan Africa, and to most he was not only a colleague but also a personal friend. His death at the age of 46, only six months after this meeting, came as a great shock and is a loss to us all. We are dedicating this volume to him.

INTRODUCTION

This is a time when serious ecological questions can be asked about the near and distant past of the vertebrates. During the century since Darwin's *Origin of Species*, fossil bones and teeth have provided us with abundant documentation of the origin, evolution, and extinction of an impressive array of backboned animals. In connection with the new methods and insights introduced in this volume, fossils and their enclosing rocks can now also tell us something about how these organisms related to one another as members of ecosystems and animal communities, how they were distributed in the ancient environments, and even something about their behavior.

Paleoecology is relatively new as a field of concentrated scientific investigation, although paleontologists have long been intrigued by ecological questions concerning such animals as the first land vertebrates, the dinosaurs, and our own human ancestors. With the recent increase in understanding of modern ecosystems and vertebrate communities, theories and models necessary for the understanding of past ecology are at last available. We are now at the stage of defining what we can and cannot learn about paleoecology from the fossil record. And we are also beginning to see exciting possibilities--that we may be able to perceive evolutionary changes against a background of consistency, or consistencies in a setting of change, which recur in vertebrate communities of very different ages, and with very different species filling the recognizable ecological roles.

The ideals of what we would like to know are relatively easy to formulate. However, the reality of vertebrate paleoecology is that it is a difficult subject for sicentific study, so difficult that significant progress has only begun during the past fifteen years. There are too many variables, too many unknowns, and a general lack of understanding of how bones become fossils. Nevertheless, there have been exemplary studies, imaginative, perceptive, and hopeful works of research which show that there is valuable paleoecological information tied up in fossils and their geological contexts. These studies underline the fact that something potentially important is being missed every time a fossil bone or tooth is collected without regard for its associations with other fossils or with the sediments that buried it.

What is being missed? Why is it potentially important? Confronted with these questions, the vertebrate paleoecologist often has had to depend on his own specific experiences or intuition for answers, answers which may need more substance to convince the skeptic. There has been no centralized source of information and no generally accepted body of data to draw upon.

1

We believe that this book will serve as a first important step in the organization of facts and ideas needed to put vertebrate paleoecology on a solid scientific foundation. It provides many of the answers to "What is being missed?" and "Why is it important?" reinforced with original data and thoughtful analysis which exemplify the new paleo-ecological approach to vertebrate fossils. It also provides the first comprehensive bibliography of research in vertebrate taphonomy and paleoecology.

What is the role of taphonomy in vertebrate paleoecology? While studies in modern ecology can give us the essential guidelines for important characteristics of vertebrate communities, in pursuing these ecological variables back through time we are soon confronted by a massive stumbling block: fossil assemblages are not fossil com-munities; they are only samples, ghosts of the original communities. And we know very little about the sampling processes. Thus we must turn to the realm of taphonomy, where we can hope to find the means for reconstructing animals in their community context through studies of the samples of fragmentary fossils. In taphonomic studies we come to understand the sampling processes, the biasing filters which have given us our record of the past. Without taphonomy, there can be no valid paleoecology. In this book are many examples of the recent, innovative work which has been directed toward understanding of taphonomic processes through theory, experiment, and studies of death and burial in modern ecosystems.

The papers in this volume present information at three different but complementary levels. First, they provide a variety of examples of what paleoecologists and taphonomists do, how they tackle the problems presented by the fossil record. Second, they give a wealth of new information concerning specific regions, time periods, and fossil localities that have served as testing grounds for paleoecological research. Third, they illustrate the present state of theory, hypothesis, and method in taphonomy and paleoecology, and point the way to future development in these fields.

Diversity of expertise and perspective characterize the authors of the papers as well as their contributions. As fields of study, taphonomy and paleoecology are very much in a phase of exploration in which the asking of new questions has touched off a search for relevant data and methodologies in many realms of science. It is too early for new syntheses, and the material presented in this book provides much more by way of ideas and directions for the future than answers or clear-cut prescriptions for paleoecological research. We feel that this is appropriate and realistic. We hope that the reader will be motivated to take up the threads from these works and combine them with his own ideas and knowledge to further increase our understanding of taphonomy and paleoecology, and ultimately, of the history of life.

Anna K. Behrensmeyer
Andrew P. Hill

Part 1

HISTORY AND BACKGROUND

Olson introduces the subject of taphonomy in terms of its origins, which perhaps surprisingly lie in dialectical materialism, and traces its historical development. He shows that initially the search for "laws of burial" preceded any compilation of a significant body of facts or attempts at empiricism. This may help to explain why it has taken over thirty years since the original definition of taphonomy for it to begin to mature as a field of scientific research. Olson's perspective, which stems from his many years of experience with taphonomic and paleoecological problems, allows him to outline the most general and important aspects of paleocommunity analysis and the state of research at present, and where it may go in the future. He provides essential definitions of paleocommunities and discusses how they may be conceived of in geological time. His paper ties the diverse subjects of other contributions in this volume to the common theoretical background of paleoecology and taphonomy.

Bishop's viewpoint is no less extensive. He takes a global perspective in order to analyze the nature of the samples of fossil vertebrates that are available to us for study. What major kinds of environments are preserved in the fossil record? What kinds of geomorphological and geological circumstances have led to their preservation? We learn that most depositional environments where vertebrates can be safely assumed to have lived do not preserve their remains, and Bishop provides reasons for this. His organization of the facts and problems will be of particular use to readers who may not be familiar with the vertebrate fossil record. His general theoretical treatment of the major features of the vertebrate record is followed by specific examples which illustrate some major points. It is apparent that we cannot assume that the record provides us with a truly representative sample of life in the past. However, imperfections in the fossil record, once confronted and understood, need not block the way to paleoecological reconstruction.

1. TAPHONOMY: ITS HISTORY AND ROLE IN COMMUNITY EVOLUTION

Everett C. Olson

Introduction

Taphonomy literally means "the laws of burial" and less formally concerns all aspects of the passage of organisms from the biosphere to the lithosphere (Efremov 1940, 1950, 1953). Severe losses of biological information occur during the taphonomic interval (see, for example, Olson 1957, 1971; Clark, Beerbower, and Kietzke 1967; Lawrence 1971). If the purpose of a study of samples from fossil assemblages involves the biology of the organisms and/or the populations of the parent assemblages, some major part of this loss must be retrieved.

Any such studies, of course, must begin with the data obtained from the fossil assemblage of interest and from the associated sediments, including the relationships between the organic and inorganic parts of the system. Taken by themselves, these data provide for inferences concerning the deposits with utilization of only the most general sort of biological and physical uniformitarian principles. Beyond this type of preliminary inference, modes of study fall into two broad categories. One involves direct recourse to modern analogues, either actual or models based on data derived from extant populations. At the species level, for example, this may involve comparison of size-frequency distributions of measurements from the two sources (for example, Boucot 1953; Olson 1957; Craig and Oertel 1966). At more complex levels samples of the full fossil assemblage may be compared in a variety of ways with living interspecies populations or with full ecological complexes of appropriate kinds (for example, Johnson 1972; Valentine 1972).

The other approach involves "working back" from the fossil assemblage to the composition, structure, and dynamics of the parent population(s) using evidence in the assemblage and the enclosing sediments and analysis of the history of their formation. This type of analysis is critical for an understanding of the ecology and community structure and evolution of extinct populations. Basically, of course, this is the method of taphonomy, and it was Efremov's hope that continued studies of the formative processes might eventually produce a set of principles generally applicable to taphonomic analyses, these being the "laws" he sought in his initial studies.

Taphonomic work, of course, cannot be independent of information on biological and physical processes determined from studies of modern circumstances, but the application of such information is general rather than specific to particular instances. The conservatism inherent in the alternative approach, involving the use of analogies based on modern situations, is minimized.

5

During the last two decades many papers have consciously used taphonomic approaches, and several of these have included short analyses of the history of the subject. Prior to that time, and going well back before 1900, efforts had been made to reconstruct the ways that fossil deposits came into being (see, for example, Abel 1912, 1914; Barrell 1916; Case 1919, 1926; Lull 1915; Matthews 1915; Richter 1928, 1929; Walther 1930; Weigelt 1927; Wapfer 1916). Formulations that have recognized the field of taphonomy as an entity of interest in its own right and have made full use of the methods have largely postdated the formulation of Efremov (1950). In the following section, I will attempt to bring together the principal aspects of the history of taphonomy and to analyze some of the contributions to the principles of study of the field. This will be followed by special attention to the problems of community evolution which, of course, involves extensive use of taphonomy.

The History of Taphonomy

Efremov and Taphonomy

The term taphonomy was defined and briefly described by I. A. Efremov in 1940. It carries somewhat the same meaning as biostratonomy (Weigelt 1927), but has sometimes been expanded to include all postmortem events affecting fossils and sediments, including diagenesis (see Lawrence 1971). In 1950 Efremov published an extended study, intended to be the first part of a more complete work which did not eventuate. The 1950 study included an analysis of the basic propositions of taphonomy and applications of these to various geological circumstances in different parts of the geological column. The compilation was actually completed in 1943, but publication was delayed by World War II and its aftermath.

In his book published in 1950, Efremov outlined what he termed "laws" of taphonomy as well as the "laws" of the opposite process, the formation of sedimentary rocks (*litholeimonomy* of Efremov). He presented in considerable detail, with examples, methods of analysis of the processes of destruction and preservation of continental sediments through time, and illustrations of how such analyses could lead to understanding the meaning of deposits of fossil organisms. The examples illustrated concrete methods of analyzing the transference of biological materials from living populations to their incorporation into sedimentary deposits. This is the heart of taphonomic study.

The 1940 paper of Efremov is in English and the 1950 paper is in Russian, and was translated into French in 1953. Some aspects of taphonomy had been treated by Efremov during the 1930s, and he published a few later studies directed specifically to taphonomy (1957, 1958, 1961).

Some aspects of Efremov's concepts of taphonomy have not received much attention but are useful in understanding the scope and methodology of his work.[1] First, it is important to recognize that Efremov was a vertebrate paleontologist with a wealth of experience in geology. His interests were largely with terrestrial vertebrates and to a lesser extent with terrestrial plants. He paid much less attention to marine deposits and marine invertebrates, and developed only rather broad methods appropriate to their taphonomic histories. Approaches are necessarily quite different, as shown by the

---

[1]The following comments are drawn in part from the Russian volume on taphonomy (1950), which is faithfully translated in the French treatment (1953), and partly from my close association with Professor Efremov from 1959 to the time of his death in 1972.

partially independent development of terrestrial and marine studies of taphonomy in the Western world. Beyond this, an unnecessary gap of understanding and credibility has existed between the marine and terrestrial biostratigraphers and biostratonomers. Efremov frequently expressed his displeasure with the studies of marine paleontologists and stratigraphers.

Professor Efremov thought dialectically, with a pervasive recognition of opposites. It was this point of view that first turned his attention to the erosional areas where terrestrial life existed and depositional areas where the remains accumulated. The linkage of these opposites emerged as taphonomic processes. His thinking was not specifically in the historical dialectical mode of Hegel or the socio-historical-scientific vein of Marx-Lenin-Engels. Rather, it was largely a simple, commonsense concept that questions and processes have two sides, usually opposites, and that neither should be neglected if full explanation is to be attained. In his philosophy as expressed in some of his literary writings, Efremov extended beyond this level, but these extensions do not appear in his science.

The dialectical approach is evident throughout the 1950 (and 1953) treatment, but it is not clear in the 1940 summary. The use of dialectical thought was made more explicit by Efremov (1958, p. 91) as follows:

> The use of the dialectical method will make application of biological
> data in paleozoology more fruitful. I wish to dwell on it because this
> method, which will in the future be substituted for "monolineal" formal
> logic, is but little developed. The essence of dialectical analysis
> of biological and paleontological phenomena, as well as phenomena in
> general, lies first of all in its revelation of the duality and opposi-
> tion of phenomena and their development. The analysis of the develop-
> ment of contradictions and the unity of opposites is certain to produce
> good results when combined with the inevitable historical approach of
> paleontology.[2]

A further aspect of Efremov's analysis, stemming from his predilection for dialectical reasoning and rejection of "monolineal" logic, is the absence of the use of abstractive mathematical approaches and stochastic models. He was well aware of such approaches, which had become prominent in the late 1950s and 1960s, but regarded them as sterile and misleading. Tendencies to use such methods in taphonomy, which are increasing rapidly, were an anathema to Efremov, although to many of us, at least, they are enriching the field materially.

Trends during the 1950s

Efremov's term taphonomy was paid little heed during the 1950s, and he frequently expressed his regret that this was so. Some studies were conducted, but largely in ignorance of his efforts to formalize the field. In 1961 Simpson summarized the status of taphonomy as follows:

> Sampling problems, already briefly mentioned, are here acute, and
> there is great need for better understanding of the factors that
> act between the living fauna and the preservation of part of it in
> the fossil state, as well as factors involved in the formation of fossil
> deposits in general. Study of such factors has been called "bio-
> stratonomy" by Weigelt and "taphonomy" by Efremov, although it may be
> a little premature to designate as distinct sciences fields in which,
> unfortunately, there is as yet little concrete accomplishment.

---

[2]The passage has been somewhat modified from the precise wording of Efremov (1958 in English) to eliminate some of the "Russianisms" from it. To do this without altering the meaning, translation of the Russian of the same passage in Efremov (1961) has been used as a basis for the minor changes.

Although this estimate was probably a little on the conservative side, it serves to emphasize, when read today, the great strides that have been made over the last fifteen years.

Some work paralleling that outlined in Efremov's taphonomy developed during the 1950s, and trends that began to emerge set patterns for more extensive later studies. Lithology as a source of basic evidence, supplemented by repeated samplings of distinctive, recurrent assemblages, was used to infer community composition and structure among terrestrial vertebrates by Olson and his colleagues (Olson 1952, 1958; Olson and Beerbower 1953). The 1958 paper, the last of a series, was noted as "true taphonomy" by Efremov in a personal communication. Fager (1957) also used studies of recurrent assemblages, and, of course, recurrence had been recognized and used earlier by paleobotanists and invertebrate paleontologists.

Shotwell (1955, 1958) employed a different approach, recognizing and characterizing communities on the basis of very large samples of mammals to which he applied the method of proximity, on the assumption that the animals living nearest to the sample site, other things being equal, would be the most abundant in the sample. This approach, as well, had been used by paleobotanists (see, for example, Axelrod 1950).

A trend which developed and which became increasingly intense during the 1960s was the separation of work, and for the most part of cross-references, by paleontologists working either with the marine or with the terrestrial records. Lines of taphonomic study tended to proceed independently. The problems faced and the specific nature of taphonomic processes do differ sharply, but the lack of cross-communication has resulted in considerable duplication of effort and failure to apply information from one field to the other even where appropriate.

This trend has been particularly evident in the development of mathematics and statistics in the area of taphonomy, something which, as noted, was quite contrary to the thinking of Efremov. *Quantitative Zoology* by Simpson and Roe (1939) and *Tempo and Mode in Evolution* by Simpson (1944) gave an impetus to use of these tools, especially in vertebrate paleontology, and increasing quantitative analyses of sedimentary processes provided the needed complement. Various analyses, such as Kurtén's studies of population dynamics and genetics (1953) or "Morphological Integration" (Olson and Miller 1958) are examples of early stochastic studies which, although not strictly taphonomic, do involve mathematical modeling applied to complex problems of fossils, including those of extinct communities. The emphasis in these studies was on terrestrial vertebrates, but somewhat parallel studies also developed in areas of invertebrate paleontology. It is of more than passing interest that studies of size-frequency distributions, such as those of Boucot (1953), Olson (1957), Rigby (1958), and Craig and Oertel (1966), based on their reference lists and contents, seem to have been carried out with relatively little cross-reference between fields of major interest. Two fairly recent books, *Models in Paleobiology* (Schopf 1972) and *Introduction to Quantitative Paleoecology* (Reyment 1971), include materials broadly applicable over much of the area of paleoecology but draw their substance very largely from invertebrates, with only passing references to parallel work in the fields of terrestrial plants and vertebrates. Much the same applies to most studies of vertebrates and plants. The dichotomy is partly due to the differences in the nature of problems and partly due to the bifurcation of disciplines based on different materials and interests; the results, perhaps unavoidably, provide less than a full synthesis of the work in taphonomy, paleoecology, and community evolution.

The status of paleoecology near the end of the 1950s was reviewed in a critical article by Cloud (1959) which drew data from both invertebrate and vertebrate studies, and this may be taken as an authoritative statement of the status of study at that time. It is indicative of the extent of clearly identified taphonomic work that in Cloud's review this area of study was not treated as distinct from other paleoecological endeavors, and the term taphonomy was not used.

Work from 1960 to 1976

An exponential increase in the number of studies involving taphonomic analyses has taken place during this interval, in part realizing the hopes of Efremov expressed in his efforts to establish taphonomy as a science with its own "laws." In the bibliography of representative papers given at the end of this book, drawn largely from the United States, may be found titles of many of the major studies made during this decade and a half.

One aspect of the hopes of the 1950s has been realized, for taphonomy assuredly has assumed an important role in the analyses of ancient ecologies. Many, although not all, of the papers listed include explanations of methods of research, and some, such as Lawrence (1971), are explicit in presenting a basis for systematic codification of the studies. At the theoretical level his analysis is very useful, especially in suggesting close links with modern ecological studies. Experimental work has been featured in a number of papers, for example, Voorhies (1969) and Behrensmeyer (1975), and the utilization of models of various types has become increasingly popular (see Schopf 1972). Some papers of this period considered the paleoecology of specific sites, others made comparisons between sites, and still others used the paleoecology of multiple sites as bases for biogeographical and evolutionary interpretations (for example, Bakker 1975a; Olson 1966).

What emerges most clearly from the studies made during these sixteen years is that we are far from the establishment of a universally applicable array of principles or "laws" of taphonomy. It is possible to group similar kinds of sites into suites and to determine for each suite some common properties that may be similarly analyzed. But the wide variety of circumstances, both with regard to the original living populations and to modes of death and deposition, make it difficult to generalize. This probably is inevitable and certainly is one aspect of the difficulty of relating studies of marine invertebrates and terrestrial vertebrates.

Transportation of bones of large animals has been studied in detail, for example by Voorhies (1969) and Behrensmeyer (1975), and their results can be applied to analyses of transportation and deposition of many kinds of large vertebrates. Transportation of shells has been studied by various persons, for example Johnson (1957). There is little in the results of these two types of studies, or similar ones, that can be applied over the wide spectrum of organisms in the fossil record. Size-frequency distributions supply the bases for models for many different groups of organisms, but each is applicable only to a limited spectrum of types of organisms because of differences in morphology, population structure, and dynamics. Models are nicely applicable within limited areas, but rarely can be useful over a broad suite of organisms and populations.

What these studies show, as logically they should, is that each is treating a singular historical event (see Simpson 1963) and that each event has unique characteristics that require a particular approach. Where we may infer reasonable similarities we may also, with caution, generalize approaches, but we cannot, it would seem, hope to find principles that will be useful across the board. Each analysis is to some extent an ad hoc experience.

Variety, then, currently seems to be the dominant aspect of paleoecological situations, with some grouping possible on the basis of constituents and modes of formation of deposits. But dependable grouping can be made only after analyses during which the initial inferences may be substantiated or falsified. A significant practical effect follows—that usually it is necessary to revisit and resample deposits two or more times, often in different ways, before the samples become appropriate for study of the site. We know that our studies must consider transportation of materials, the associations of organic remains, the relationships of organic and inorganic elements of the deposits, sedimentation and relationships of sedimentary types, the biological nature of materials, and so on, but also it is evident that useful analyses must select carefully from among these variables and apply only those that are pertinent to the unique situation.

Among the most challenging efforts of taphonomic research of the last twenty-five years or so have been attempts to study the evolution of complex ecological systems, usually called studies of community evolution. Such investigations have been conducted for a number of reasons, among them the following: to understand the evolution of a particular community itself; to use an ecological framework in the study of particular phyletic lineages; and to analyze the processes of evolution at work in an ecological matrix through time. The remainder of this paper is devoted to a consideration of the problems of community evolution and taphonomic processes.

Community Evolution

General Statement

The changes of ecological complexes through time have been called faunal evolution, chronofaunal evolution, community evolution, and ecosystem evolution, depending in part upon which aspect of the biologico-physical system is being treated. Studies of community evolution, regardless of their scope, have focused in one way or another on changing ecological systems, and this type of work has been carried out on marine invertebrate assemblages and upon both terrestrial vertebrate and terrestrial plant communities. As in the case of paleoecology, the material and aims in these various areas have been different and the work has diverged notably. Because our interests are mainly with terrestrial vertebrates, what follows applies primarily to their communities. References at the end of this book include some of the more comprehensive studies on marine invertebrates.

A survey of the literature on fossil communities and community evolution at once shows that terminology and concepts have been varied and that a coherence in methods and scope has yet to emerge. This circumstance makes it difficult to generalize with reference to ecological complexes. Along with this problem, and the ever-present matter of the scope and composition of the assemblages treated, is the dual conception of a community as a substantive entity—such as a pine forest in the Black Hills—and as an abstraction—for example, a mountain pine forest. It seems necessary before we go farther to clarify definitions and concepts. In the following sections are framed definitions that are applied in this paper and that, it is hoped, will prove more generally useful.

Some Definitions

In what follows, the difference between community, the substantive entity, and community type, a conceptual extension, is maintained. A *community* is defined in the ecological sense as a *group of organisms living together within a definite locality*. Further, living together is assumed to imply interaction of the constituent individuals either directly or by influence of organisms upon each other through their impacts on the common environment. The integrated structure of interactions of the elements of the system (e.g., trophic structure) defines the limits of the system both geographically and temporally. When the structure becomes seriously altered, the community as such ceases to exist. Although the last aspect, as well as the areal and compositional limits, is necessarily imprecise, the definition as a whole serves reasonably well as a working base for studies of fossil communities and eliminates at least part of the ambiguities of multiple definitions.

Evolution of a community, of course, refers to change of some aspects of the community through time. The phrase "community evolution," however, has been an elusive one which has many interpretations. Some of the problems relative to terrestrial vertebrate communities have been considered by Shotwell (1964) and Olson (1966). Sloan (1970), in his analysis of the late Cretaceous and early Cenozoic mammals, uses the term community partly in a substantive and partly in a conceptual sense. "Late Cretaceous" community, for example, appears to be conceptual, whereas much of the evolution reported deals with one or another existing assemblage. His paper was one of a series presented in a symposium published under the editorship of Yochelson in 1970. The papers in this series, all on community evolution, were criticized by Durham (1971) both because of the vagueness of definition of community and community evolution and because of the general paucity of evidence of the actual nature of the communities. Each of the papers treated some part of the problem of community evolution and, in so doing, revealed the wide range of concepts and the many difficulties that arise in treating the evolution of extinct communities.

To give some coherency to this area of study, the following suggestions are made. Community evolution should be defined to encompass only those changes that occur within integrated complexes of organisms that maintain direct continuity through time by persistence of their *basic ecological structure*. The definition of community is that given earlier in this section. Usually, but not necessarily, some major part of the phyletic lineages of the community will persist throughout the life of the community. In essence this definition refers to the evolution of a "resident community" in the sense of Shotwell (1964) or of the chronofauna of Olson (1952). Any change within such a complex, beyond mere short-term fluctuations, constitutes community evolution. Such evolution may be considered to persist for as long as the defined basic ecological structure persists in recognizable form.

If this concept of community evolution is used, many of the phenomena that have been so designated, for example in "Community Evolution and the Origin of Mammals" (Olson 1966), may best be considered as community succession, much in the sense of Shotwell (1964). Within the framework of community succession, each evolving community is a temporally extended unit in a series of successive communities. Such succession—not of course to be confused with ecological succession within a community—is one of the most common patterns observed during analyses of the changes of terrestrial vertebrates through time. Distinguishing community evolution and community succession in this way will, I believe, reduce confusion and provide a way for coherent use of taphonomic procedures as means to understanding the basic nature of communities through time.

Aspects of Community Evolution

1. The Data   The definitions of community evolution and community succession introduced in the preceding section have several consequences. The concepts of community as applied to fossil and living populations are brought into conformity, although the levels of observation are somewhat different. The losses of information during the

formation of fossil assemblages are not directly recoverable, but the close conceptual relationship to living assemblages opens the way to precise use of information from modern
analogues in interpretations of extinct communities and to inferences about the properties
of their original populations through taphonomic analysis.

Localities and areas in the fossil record where analyses of community evolution of
terrestrial vertebrates can be carried out are limited, but it appears that there are
enough scattered through the geological record that a great deal can be learned. Currently,
however, the data for such analyses are restricted, and the focuses of interest of the
majority of studies have not produced the required kinds of information. This can be
remedied, but usually only if a problem is attacked in a taphonomic context.

Data used in the following section, in which efforts are made to extract some of
the general aspects of community evolution, have come in large part from the following
studies: Clark, Beerbower, and Kietzke (1967); Olson (1952, 1958, 1962, 1966, 1976);
Olson and Beerbower (1953); and Shotwell (1955, 1958, 1964). In addition, information
has come from the work of Clapham (1970), Gingerich (1976), Gould (1970), and Van Valen
(1970), plus material from Bakker (1971, 1975a) as well as from conversations with him in
the course of seminars.

2. Interpretations   Whenever it has been possible to follow a community through
time in the geological record, with minimal interruptions, the changes or evolution that it
has undergone appear to be very conservative. How general this conservatism is and how
much of it may be a product of the coarseness of observations are questions clearly in
need of extensive study. Within the community over a short period of time--a season or
a few years--we may assume that the population dynamics fluctuated in ways comparable to
those observed in living communities. Most of the fluctuations, such as changes in
relative and absolute number of individuals of species, seasonal modifications of trophic
circumstances, changes in genotypic composition of species and infraspecies populations
in successive generations, and even the development of sibling species by local isolation,
cannot be determined with confidence for most fossil assemblages. Such changes, however,
are part of the basis for long-term evolutionary changes according to generally accepted
concepts. If this is the case, the small modifications become manifest in the fossil
record only in the grosser modifications from which they may be inferred. In very recent
deposits of fossils, some of the small changes may become apparent, but for most of the
record this is not the case. The strong tendency for stability, even including the rarity
of evident speciation in an evolving community, suggests that the impact of fluctuations
such as noted above is generally minimal. As will be discussed under the last section of
this paper, which treats community succession, the contrary seems to apply during the
events of succession.

From observations of the rather limited known examples of evolving communities,
in the sense defined above, the following generalizations can be made:

1. Many phyletic lineages persist with little or no morphological changes over
the life of a community. The principal modifications, where change does occur, are increases in size and accompanying allometric changes in proportions. Size increase may
occur in diverse lines including omnivores, herbivores, and carnivores. The largest
animals of a lineage tend to occur in the terminal phases of the evolution of a community.
Examples are to be found in Olson (1958), Clark, Beerbower, and Kietzke (1967), and
Kurtén (1963). The same trend has been recorded for the lower part of the Permian by
Bakker in as yet unpublished material. There are exceptions. Kurtén (1960), for example,

has given examples of decrease in size, and there are various unanalyzed cases where diminution in size clearly has occurred in well-established lineages. Gingerich's study (1976) of early Tertiary mammals indicates a general tendency for size increase but also some instances of decrease in closely graded samples. His lineages are not integrated into community units but appear to be phyletic lines evolving in continuing community systems.

The slow, directional pattern of phyletic changes raises, as Gingerich has pointed out, the matter of phyletic speciation, in contrast to the usual speciation by geographic (or sometimes niche) isolation. Gingerich has contrasted his results with the concept of punctuated equilibrium of Eldridge and Gould (1972). If the linear changes are to be called speciation, or phyletic speciation, then two patterns of species change must be recognized.

Just how these two aspects of change are to be related to evolution is, of course, closely related to the question of the scope of evolutionary impact of the changes that occur within communities. Relatively little speciation by isolation, other than temporal, appears to occur in evolving communities, although there are exceptions as noted below. The relationships of changes within evolving communities and the evolutionary phenomena observed in the fossil record will be explored further under the heading of community succession.

Some lineages within evolving communities show little or no change over long periods of time, not even appreciable changes in size. The most stable and persistent elements of communities appear to be the large carnivores, especially those at the peak of the trophic pyramid. Increase in size over time among them is not uncommon, but this is far from a universal trait. The capacity of these large carnivores to survive a wide range of modifications in climate and composition of the evolving community probably is a function of their capacity to utilize a great variety of foodstuffs, both animal and vegetable.

2. Species extinctions occur frequently within evolving communities. This is evident in the dropping out of species from the record and may, of course, be a local phenomenon, not one that applies to the species in its entirety. In all cases that have been studied, climatic change occurs during the life of a community and is often an important factor in species extinctions. Losses of species are irregular in time, but the greatest numbers occur at periods when there is physical evidence of an increase in environmental stress. Evolving communities generally survive moderate losses of species, unless such species are so critical that their disappearance severely disrupts the trophic structure. Massive losses tend to result in the demise of the community.

3. The extinction of species to some extent alters the trophic structure, but if it is not severe, a return to stable conditions has been inferred to take place in one of several ways: (a) One or more species of the system may expand to fill the role of the lost species in the event that the roles of the two or more species have been more or less alike in the economy of the system. (b) A new species from outside of the community system may enter and assume the role of the extinct species, or assume at least an approximately similar role. (c) Speciation within the evolving community may produce a new species to fill a vacated role in the event that one of the existing species is pre-adapted to this role. This appears to be a rather unusual event that occurs when the loss of a species creates the opportunity for isolation of some part of an existing species. The usual pattern appears to be that environmental stress, as physical conditions change, results in the extinction of a species. Subsequent return to earlier

physical environmental conditions leads to a redevelopment of the lost or reduced habitat. Adaptive penetration of some members of a persisting species into this redeveloped habit may result in their isolation and the formation of a new species. An early Permian instance was cited by Olson (1952b) and one from the Pleistocene by Kurtén (1963). There is evidence of such change, although not explicitly noted, in the study of Clark, Beerbower, and Kietzke (1967) and also in studies of Shotwell (1955, 1956, 1964). (d) The trophic structure of an evolving community is very stable and may, as noted, continue in spite of losses of species, additions of species from outside of the system, and rather strong fluctuations of the physical environment. Even with severe climatic change the structure tends to endure, if only as a reduced skeleton. As the community becomes increasingly depauperate in species, the final outcome appears to be extinction of the community rather than significant adaptive adjustment to the changes of environment.

In summary, it appears that major evolutionary changes which appear significant in the long course of evolution do not themselves occur within evolving communities. The stage for such changes may be set, but they occur during the complex of events that take place during what we have here termed community succession.

Community Interaction and Succession

General Problems

Whereas the evolution of particular fossil terrestrial communities, especially those of limited scope, can be followed in the most favorable cases, the interactions between adjacent contemporary and successive communities pose very different problems of analysis. First, of course, thorough taphonomic analysis is essential, as in all cases of paleoecology, to determine whether the fossil assemblages in the area of collection were derived from one or more than one living community. Then, if more than a single source is detected, it is necessary to sort out the elements from the contributing sources and to establish as far as possible the structure and population dynamics of each of the contributing communities. Thereupon it may become possible to make some estimates to the extent to which the living communities may or may not have overlapped and impinged upon each other, that is, to what extent they interacted during their existence. If, as is sometimes possible, the contributing communities can be identified in separate deposits, the task may be facilitated. Detailed analyses of this sort of interaction can be found in Shotwell (1964 and earlier), Kurtén (1960, 1963) and Olson (1974, 1975, 1976).

On a much broader and less definitive scale, faunal interactions and successions are the objects of most paleobiogeographical analyses, even though single groups or organisms may be the focus of attention. The interaction between late Tertiary and Quaternary mammals of North and South America is a case in point. The effects of the apparent Oligocene incursion of rodents into the mixed marsupial-placental communities of South America is another. Analyses of the puzzling distributions and history of the marsupials cannot stand apart from an understanding of the roles of these animals in communities or of the impact of the interaction of placental- and marsupial-dominated communities as they came into contact. To some extent the analyses of Sloan (1970) involve this broad recognition of community interaction without detailed analysis of the nature of the structure in particular communities.

All such studies are important in biogeography and the history of vertebrate evolution. Here, however, our aims are more limited, and comments will be directed to cases in

which discrete communities can be recognized and analyzed. First the matter of interaction
will be considered, followed by a short discussion of some aspects of community succession.

Community Interactions

Information upon the interactions of adjacent communities as defined in this paper
is quite restricted in the fossil record and, of course, interactions can only be inferred
after the fact. The most commonly observed pheonomenon is the appearance of a new
species (or often a new genus) in what has been a stable community over time. For such
an event to be recognized, sampling over a stratigraphic sequence must have been fairly
extensive, enough so that the integrity of the community is established and so that the
"invader" can be seen to have become a part of the community; multiple samples are essential. Such new species first appear as "erratics" in the assemblages (see, for example,
Olson 1974; Shotwell 1964). Only if the new species appears consistently in temporally
successive samples after the first appearance can it be judged to have become assimilated
into the community structure, rather than merely having been introduced into an assemblage
by physical processes.

Such information, viewed in the perspective of a single community, as often must
be done, gives the impression of representing an introgressive phenomenon. If this apparent introgression is sufficiently massive, involving a number of species, it may alter
the character of the evolving community to the extent that it loses its structural
identity, producing a new community which succeeds the older one but maintains at least
some continuity with it at the transition.

Apparent introgression may, of course, be but one side of the mutual interaction
between two communities, but as a rule it is too much to hope that records could be
complete enough to reveal this except over long periods of time involving massive actions
of many interacting and succeeding communities. The late Tertiary interactions between
North and South America show such mutual interchanges and interactions, but the long
time spread and the multiple units involved tend to lead to shaky inferences on the
details of the interchanges, as evident in the strongly divergent interpretations of
Patterson and Pascual (1968) and Hershkovitz (1966).

Community interactions considered to this point have involved somewhat similar
adjacent community types of terrestrial vertebrates. In these cases the species invading
a community are more or less equivalent in behavior, physiology, and morphology to those
in the invaded community. Incorporation and replacement of the extant species may
proceed with relatively little disruption, and the evolutionary course of the invaded
community may not be seriously altered. Another broad spectrum of associated community
systems may exist, and has existed many times in the past. In such systems two or more
adjacent communities may be very different in character. Various instances come readily
to mind; the juxtaposition of freshwater aquatic communities with crossopterygian fishes
and the terrestrial invertebrate-plant communities of the Devonian Period; terrestrial-
reptilian communities of the lowlands and the more upland invertebrate-plant communities
somewhat later in the Paleozoic; or the small mammal-insect-plant communities and the
dominant reptilian communities during the last half of the Mesozoic.

When, for any of many possible reasons, ecological balances change and introgression
of one such community type by the other takes place, rapid establishment of a radically
new community type, different from either of the progenitors, can occur. Intensified
selective pressures during the establishment of this new community type and rapid evolution following an evolutionary release of both the introgressing and introgressed species,

appear to be critical in this process. The circumstances are rather closely analogous
to the evolutionary phenomena concomitant with the invasions of islands as developed by
MacArthur and Wilson (1967). Under these conditions rapid speciation may occur (see, for
example, Zimmerman 1960), and, it would appear, rapid adaptation to new circumstances
might produce new types of organisms in a short period of time. The period of time involved in the founding of new communities by the introgression of one community by members
of a very different community is undoubtedly short relative to the usual life span of
communities, and appears to take place by means of events that favor rapid evolution, as
discussed in the following section.

Community Succession

Evolution necessarily takes place within the framework of communities and during
establishment of new communities. Much of the evolutionary change observed in the fossil
record may be profitably viewed in the context of community succession, irrespective of
the aims of any single study. The bases for the more apparent, major changes are laid
within communities as they change with time, with the development of organisms preadapted
to conditions other than those of the community in which they exist. As indicated
earlier, the changes within evolving communities are small and relatively few, often involving only phyletic speciation. Changes during successions of communities are in some
instances seen to be slow as well, but in other cases they appear to have been rapid, sometimes almost revolutionary. This occurs at times of the development of radically new
types of communities.

Studies of community succession have in general been of two types. Most have dealt
with succession within a limited geographical area, for example the Big Badlands of South
Dakota (Clark, Beerbower, and Kietzke 1967), and of the Northern Great Basin of the United
States (Shotwell 1964). Others have had no geographic restrictions but have recognized
community continuity only by the composition and structure of the successive communities—
for example "Community Evolution and the Origin of Mammals" (Olson 1966). As far as the
geological record shows, these are "noncontiguous" communities, and they will be identified
by this term hereafter. The first type (contiguous) allows careful and critical analysis
of closely graded events and reveals in these only moderate changes, related to environmental modifications. The second is necessarily less critical but seems to reveal much
greater evolutionary change relating to development of distinctly new community types, often
under rather sharply differing environmental circumstances. The differences between the
two, of course, are neither truly discrete nor incompatible. They are, however, in part
real, representing two extremes of a spectrum, and so are treated separately in the
following sections.

1. Noncontiguous Communities   For noncontiguous communities to be recognized
as truly successive they must, of course, have appropriate stratigraphic relationships and
include a reasonably large increment of phylogenetically related organisms throughout the
series. The presence of new organisms (species, usually representing new genera or even
families), change in trophic structure, and, in some cases, changes in population
dynamics are indications of extensive evolutionary differences in the successive communities. These are the kinds of changes found, for example, in the shift in composition and
trophic structure in the derivation of therapsid-carrying communities from those with
pelycosaurs (Olson 1966), and the shifts in mammalian groups in the late Cretaceous and
early Cenozoic (Sloan 1970). It may be supposed that most of the major events of evolutionary history took place as such new communities formed during rapid modifications of

the environments and the habitat zones available to parts of the previously existing communities. Soon after the formation of a new, complex community, it is probable that a stable, slowly evolving unit would be formed.

Validation of these generalizations poses a serious problem, one that may not be entirely solvable. We are faced with a series of "stopped-action" sections of a process for which the dynamics are evident only when traced through time. Whether we are treating a modern community or an extinct one, the possible transitional nature of the complex cannot be recognized from data derived from the particular community alone. The living community, however, can also be studied in the context of adjacent existing communities and by use of a wide range of types of information. A similar complex of adjacent communities of demonstrably equivalent age is at best a rarity in the fossil record. What we generally find as the result of analysis of a series of samples is the stable, evolving community noted earlier. The lack of evidence of rapid changes has been variously explained as due to small samples, to very short times of change, to accidents of sampling, and so forth. Taphonomic considerations, however, suggest that it may well be as much the difficulty of recognition as the absence of preservation that is involved. A real question exists as to whether such rapid transitions, stopped in time, *can* be so identified, that is, whether the types of data available make recognition possible.

Sampling of a given fossil assemblage must be extensive to determine that the materials came from a single, living interspecies population. Unless this can be established, which generally requires validation by repetition of suites of species in multiple samples, there can be no assurance that what may appear to be a new mix of species, relative to earlier ones, is not the result of taphonomic mixing. The difficulties of recognizing an unusual assemblage as a new community type are evident. The short time or existence of the formative processes of a community and its probable, limited areal distribution makes multiple sampling difficult if not impossible. The use of this critical procedure is largely denied by the circumstances. What is seen in the record is something which is in existence during a short period and which, alone, cannot indicate that it is a formative stage leading to a more mature ecological complex.

This difficulty is evident in one of the basic processes in the formation of radically new communities, the origin of new species. Even within modern communities, using all available evidence, the process of development of new species is difficult to detect. Among extinct communities, except in the case of phyletic speciation, it is virtually impossible, except by accepting a considerable inferential jump (see, for example, Olson 1952b for the case of speciation in *Diplocaulus*).

What is seen, then, in the fossil record of noncontiguous, successive communities is the result of processes of change which may well be partially preserved but which, viewed at any moment in time, appear as static. If the evolutionary modifications occur rapidly during the establishment of new communities, as comparisons of successive communities suggest they must, then even in temporally closely spaced samples, changes of composition and structure will usually not be sufficiently great that successional relationships between samples will be evident. The short-term, dynamic processes, even though faithfully recorded by the participating organisms, may not be interpretable in view of the loss of dynamics that inevitably accompanies the course of taphonomy.

2. Geographically and Temporally Contiguous Community Succession   Relatively few cases of succession of this type have been studied carefully, but there are exceptions, among them Shotwell (1955, 1958, 1964), Kurtén (1960, 1963), Clark, Beerbower, and

Kietzke (1967), and Olson (1952, 1958, 1976). In each case there was climatic change through time in the area covered. What is witnessed in each instance is relatively slow evolution, with persistence of some phyletic lines, disappearance of species, and introduction of new species. Successive communities show moderately close resemblances. In such successions, as within communities, new species are rarely seen to arise, but rather seem to appear full-blown, presumably coming from other communities. All of these studies vacillate somewhat on the matter of just what the community of concern is: whether, for example, it is just a part of a larger complex—say a riparian assemblage which is part of a larger complex, including grasslands, woodlands, and so on—or whether the unit is the full complex, the different habitat zones to be considered as subcommunities. (Definition, of course, is necessary for treatment of succession both in specific and general cases, but primarily is a matter requiring clarification in the context of any particular study.)

In the succession of contiguous communities one might hope to see the dynamics of community formation in an interpretable form. What appears, however, is a depositional succession during which physical conditions are modified, either in a directional or fluctuating manner, and a series of faunas which change moderately in response to the physical modifications. The principal biological changes are in the different percentages of species from different subenvironmental zones and the introduction and loss of species. Shotwell (1964), in his consideration of this problem, noted the masking effects of migration relative to the development of new communities. He, like others, has stressed the apparent conservatism in community evolution and succession.

Kurtén's work on the Villafranchian (1963) and Pleistocene (1960) shows the effects of fluctuating climates during the Cenozoic and their impact on mammalian faunas in the succession of communities. He stressed, in the 1960 paper, the effects of extinction, evolution, and migration on the changes in faunal composition. From extensive data he estimated rates of origination and extinction of species and their roles in temporal changes. However, here as elsewhere, species appear and disappear, but by the very nature of the observations and the data, the kinds of information needed to view events relative to the initiation of distinctly new kinds of communities are not at hand.

Only when a strong environmental stress had taken place in the course of development of geographically and temporally contiguous community successions does there seem to be a hope of "catching" the formation of a truly distinctive community "in the act." The closest approximation to this that I know is seen in the transition from the Arroyo to Vale Formations in the Permian of Texas. At this time many old species failed to survive, and after the transition many new ones are present. The trophic structure of the community, well known from the Arroyo, was seriously disrupted, but after strong environmental stresses at the transition, it was reformulated with several new species in the trophic roles formerly occupied by some that had become extinct. The additions, the new species, however, cannot be traced with any certainty to known sources. The possibility of an adjacent community as a source has been studied (Olson 1975), but the results are inconclusive. The species in the reconstituted community seem to appear *de novo*, and whether they developed rapidly from phylogenetic predecessors in the earlier part of the succession, were derived from an adjacent community, or arose from members of this adjacent community by adaptive speciation as they moved in, cannot be determined. Here again, we are faced with the dilemma of inferring the actions that led to results seen in samples in which only morphology, biological associations, and physical conditions supply the data. Much of what we might like to know about transitions and interactions in the course of community successions and origins seems to escape even the most carefully planned and executed taphonomic research.

It was suggested above that this may be the inevitable consequence of the nature of paleontological data. Perhaps this is true, but hopefully it is not. A combination of new ideas for taphonomic research (and there have been many during the last decade) and development of directive models, particularly of the type arising in marine paleoecology, may show new ways to get at these very old problems which lie at the heart of understanding major aspects of evolution as viewed in the fossil record.

2. PALEOGEOMORPHOLOGY AND CONTINENTAL TAPHONOMY

Walter William Bishop

Preamble

In 1971 a highly respected professor of geology suggested that one solution to the global problem of disposal of macrogarbage was to feed it into a deep sea trench. Here, if the site was selected with care, the rubbish would proceed into a subduction zone. Having been subducted, it would be recycled on a major scale, to be served up much later, reconstituted and possibly on an entirely different plate.

The suggestion is perhaps fanciful. However, it seems appropriate to look at those natural processes operating on a continental or even a global scale, stopping short of metamorphism, which aid selected structural elements of living organisms to be preserved and become part of the fossil record. In this paper I seek to delineate continental areas where the dominant physical processes favor retention of exact organic morphology, by preserving hard parts with their mineral structure augmented, altered, replaced, or removed. Such a view of environments that lead to preferential preservation of fossils on a global scale should provide insight into the broad sampling biases that dictate the nature of the terrestrial fossil record.

Potential areas where the first stages towards fossilization can be studied exist in various present-day environments. However, in order to relate such studies to the whole range of vertebrate evolution it is necessary to refer to the stratigraphic record of Phanerozoic time. This paper has three parts. It sets out guidelines for a systematic geological approach to taphonomy; it attempts a classification of fossiliferous environments in continental regimes; finally, it illustrates selected paleogeomorphological settings by outlining four case histories.

Guidelines for a Geological Approach to Taphonomy

Three categories of information are involved: rock, space, and time (fig. 2.1).

I. Rock perspective. This leads to paleoenvironmental data but also provides the raw material for dating and establishes the pedigree of individual dating-specimens or fossils. Recording and description may be classified under four headings (after Selley 1971).

Professor Bishop died suddenly on 20 February 1977 before he was able to revise his contribution to the symposium. The volume editors are responsible for the final form of the work and any errors it may contain.

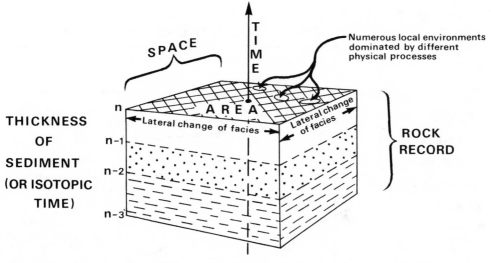

Figure 2.1. Diagram illustrating the relationships between rock, space, and time perspectives.

A. Geometry. Thickness, attitude of strata; e.g., planar bodies, shoestrings.

B. Lithology. Composition and fabric; grain and/or crystal size, shape, chemistry; fabric; diagenetic factors; by inference, permeability, etc.

C. Sedimentary structures. From these former flow regimes, their probable directions, and other depositional factors can be inferred.

D. Biological content. Fossils considered as an integral part of the rock; as indicators of former living organisms or communities; as peculiarly shaped sedimentary particles. Data recorded on their nature, shape, size, taxon, proportional representation, etc.

    —Taphonomic implications. A to D above allow reconstruction of environment of deposition (i.e., burial environment) of the contained fossils.

II. Space perspective. Location, distribution, and areal extent of strata are fundamental parameters. *Mapping*, by accurate surveys establishing precise areal coordinates and height above datum plus interpretation of air photograph or satellite imagery, *is essential to all geological investigations*. It establishes the raw material of stratigraphy and also the local configuration of a fossil locality or archeological "site."

    —Taphonomic implications.

    1. Extent of the deposit and hence some approximation of scale of the former environment.

    2. Configuration of area where fossils occur: possibility of a catchment to allow secondary concentration of eroding fossils.

    3. The basis for regional paleogeomorphological reconstruction.

III. Time perspectives. Four categories are suggested.

    A. Long-time-base sequences. Stability of geomorphological process results in sedimentation which is independent of short-term environmental change. Hence, thickness of sediment (even if the pattern is cyclic) is related to time lapsed, which in turn provides a time base against which biological data can be set; examples are Texas Permian and Pakistan Siwalik channel and overbank (although in a large-scale tectonic framework). Refer to figure 2.2.

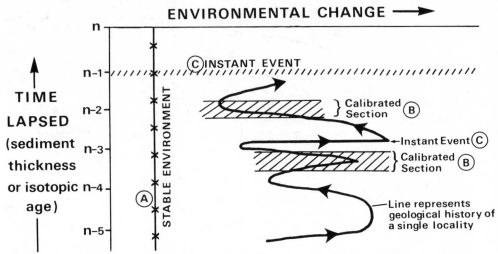

Figure 2.2. Diagram illustrating the relationships between (A) long-time-base sequences in stable environments; (B) short-time-base sequences in calibrated successions; (C) instant events producing "frozen" instant environments.

—Taphonomic implications.

1. Such sequences imply a probability that the overall environment of life of individuals contributing to fossil faunas was similar to the death environment (e.g., the Siwalik fossiliferous sediments all imply a broad setting of river bank and floodplain), even if its detailed location was not identical in each case (see fig. 2.1).

2. As thickness of sediment is equivalent to time lapsed, evolutionary bursts or periods of biological quiescence should be revealed against a consistent environmental background.

B. Short-time-base sequences. Although rapid changes in environment occur and are recorded in the sedimentary record, good time resolution (^{14}C; ^{40}K-^{40}Ar, paleomagnetic time scale, etc.) allows change of large amplitude to be calibrated. Examples are East African Rift Valley late Cenozoic sequences ("stop-go" deposition involving many gaps with lava flows and sedimentary units interbedded); Holocene lacustrine sediments in East African lakes related to rapid climatic change giving alternating arid and "pluvial" conditions; intermontane basins of western U.S.A.: fluctuations of Lakes Bonneville and Lahontan.

—Taphonomic implications.

1. Establishes rate of environmental change.

2. Provides an accurate timescale.

3. Probably samples several paleoenvironments.

4. May reveal faunal shifts due to ecology or evolution.

C. "Instant" environments. Short-term events enable moments in geological time to be frozen and studied in detail by reference to single, more or less catastrophic happenings or by study of deposits on, or related to, isochronous surfaces. Examples are archeological occurrences: occupation surfaces, cave hearths, butchering sites, etc.; surfaces buried by tephra; paleosols; deposits resulting from catastrophic flood conditions.

—Taphonomic implications.

1. Different environments which are broadly contemporary can be sampled and compared, with minimal problems resulting from time-averaging.

D. Others.

1. Chaotic sequences: very complex sedimentary, volcanic, or other regimes without clear or persistent pattern; areas deformed by tectonic processes; metamorphic areas. All background noise and no clear signal. Examples are parts of California coast ranges.

2. Areas with minimal or nil fossil content: so few fossils survive that burial laws are difficult to establish.

—Taphonomic implications. Better start work somewhere else! However, conditions and parameters leading to lack of preservation of fossils can also be of interest to the taphonomist.

Note that rates of operation of sedimentary or evolutionary processes may be by reference to *either* radiometric (radiocarbon, potassium-argon) years, *or* may be only relative and derived by reference to some modern analogue of sediment accumulation (under ideal conditions annual or other rhythms may be recognized in sediments, e.g., varves or "blooming" of algae). It should be noted, however, that sedimentation rates are *not* the same as average rates of sediment accumulation when deriving a time base from sediment thickness.

<u>Classification of Fossiliferous Paleoenvironments
in Continental Regimes</u>

In 1961 I attempted what now seems a very naive classification of "suitable environments for death" based upon late Cenozoic fossil mammal localities in East Africa (Bishop 1963). At that time I recognized only four categories: volcanic, lacustrine, fluviatile, and cave. If we widen the horizon to encompass continental deposits over the time span of vertebrate evolution a possible first-order classification might be as follows.

A. Tectonic situations inducing rapid and frequently cyclic sedimentation

1. Molasse and fan deposits in mountain girt basins and on the flanks of growing mountain chains, e.g., North American Tertiary intermontane basins, Siwalik formations of India and Pakistan (see Case History 1 below, page 25).

2. Rift Valley sumps and grabens, e.g., Karroo grabens of Africa (see Case History 2 below, page 26), Gregory Rift Valley late Tertiary.

B. Volcanic fields yielding tephra of particular petrological composition

3. Primary carbonate fallout onto land surfaces, e.g., Miocene carbonatite— nephelinite volcanic suites of East Africa (see Case History 3 below, page 30).

4. Secondary deposits derived from volcanic sources

   a. Calcareous tuffs/steppe limestones

   b. Trachytic pumices: pumice as a natural form of expanded polystyrene packing round fossil bones in channel lag situations

    C. Continental margins

        5. Distal ends of rivers, especially those flowing from semiarid hinterlands, e.g., Old Red Sandstone of Welsh Borders in Britain

        6. Shallow water estuarine, deltaic, and lagoonal deposits, especially those associated with marine transgression

            a. Bay bars, e.g., Langebaanweg, South Africa (see Case History 4 below, page 35)

            b. Bone beds, e.g., Rhaetic and Ludlow bone beds in England

            c. Fissure fills, e.g., Rhaetic fissures into Carboniferous limestone in Mendip Hills, England

            d. Sea caves

    D. Inland basins and traps: temporary burial sites

        7. Lacustrine, fluvio-lacustrine, swamp, and spring-eye margins

            a. Deltas, e.g., the Omo River, Ethiopia

            b. Channel lag ("dropout" situations)

            c. Peat bogs and tar pits, e.g., "Bog people" from Tollund, Denmark, Rancho La Brea tar pits, California

            d. Springs (petrification)

        8. Karstic settings: favorable chemical burial environments—e.g., South African australopithecine sites (pitfalls and cave occupations)—calcrete and also silcrete, ferricrete, or phoscrete

        9. Impedence and sedimentation on planation surfaces yielding concentrations of surface sweepings, e.g., Miocene deposits underlying volanics at Moroto Mountain, Uganda

      10. Deserts

            a. Hot desert dune fields: playas with salt effects, calcification, etc.

            b. Fringing deserts: coastal dunes, e.g., Elandsfontein, Saldanha Bay, South Africa

            c. Cold deserts, permafrost: deep-freeze situations preserving mammoth and rhinoceros

      As many of the above can only be isolated in an artificial classification, I have chosen four case histories, for areas where I have carried out fieldwork, in preference to a systematic review of the voluminous bibliography. The North American intermontane basins, South African australopithecine caves, the Old Red Sandstone in Britain, and various other topics cited above are not dealt with in detail here. As so much work has been published recently on the Gregory Rift Valley of East Africa (Isaac and McCown 1976; Coppens et al. 1976; Bishop 1978) I make no further reference to work on Cenozoic rift valley paleoenvironments.

      The case histories selected are as follows:

    1. Siwalik formations of the Potwar Plateau, Campbellpore District, Pakistan (A1);

    2. Karroo grabens of the Ruhuhu Valley, southern Tanzania (A2);

    3. Miocene carbonatite volcanoes of Kenya and Uganda (B3);

    4. Pliocene bay bar, Langebaanweg, Cape, South Africa (C6).

Letter-numbers in parentheses refer to items in preceding outline.

Selected Case Histories

Case History 1. Siwalik Formations of the
Potwar Plateau, Pakistan
(Example of tectonically controlled long-
time-base sequence.)

Lithological Units

The Potwar Plateau is a dissected area of some 20,000 km^2, with an upper surface
about 500 meters above sea level bounded to the south by the Salt Range, and to the north
by the Kala Chitta and Margala Hills, and to east and west by the Jhelum and Indus Rivers
(Anderson 1927; Gill 1952a,b). During and after deposition of the Siwalik Group the region
was subjected to folding and faulting. Many prolific fossil mammal localities lie on the
limbs of the Soan Syncline in the south of the plateau. The structure strikes approximately
east-west, lies south of the Soan River, and is broadly parallel to it. To the north of
the syncline is an anticlinal structure where the strata have also yielded fossils
(Pilgrim 1910, 1913; Cotter 1933).

Siwalik Group sediments span a period from mid-Miocene to Pleistocene in age. They
consist of alternating beds of gray-green sandstone and fine-grained "red bed" deposits.
The cyclothems vary from being predominantly sandstone to mainly clays or silts, and
locally conglomerates are common (Anderson 1927; Gill 1952a, Cotter 1933). The group is
composed of four formations, many of which are fossiliferous, with some being much richer
than others. The deposits are interpreted as fluvial in origin, deposited on steadily
subsiding floodplains, probably by several major streams. The headwaters and sediment
sources for this Siwalik molasse were on the emergent Himalayas. These were later elevated
more strongly, and their structures encroached upon the Potwar Plateau in the north
and east.

Type sections of the formations of the Siwalik Group have been defined by the
Pakistan Geological Survey as the Chinji, Nagri, and Dhok Pathan Formations, named from
localities on the Potwar Plateau (see Pilgrim 1913; Lewis 1937; Hussain 1971; Fatmi 1973).
The original Tatrot Formation also had its type section on the Plateau, but the Pinjor
Formation was based on a section in the Simla Hills of India. These two units have been
considered upon very slender evidence to be the equivalent of the Soan Formation
(Kravtchenko 1974). The type locality of the Soan Formation, near the village of Mujahad
in the Potwar Plateau, is poorly defined and lacks good faunal evidence. Until the Soan
Formation type area has been mapped and studied in detail in relation to its surroundings
little can be said of its age or correlatives. The current grouping of Siwalik lithological units is as follows:

> SIWALIK   Soan Formation (considered to
> GROUP   be equivalent to the Tatrot
>   and Pinjor Formations)
>
> Dhok Pathan Formation
>
> Nagri Formation
>
> Chinji Formation

Only the Chinji, Nagri, and Dhok Pathan Formations are included in the following
comments. Sedimentation throughout these three Siwalik formations is cyclic, with each
cyclothem consisting of a sandstone resting with sharp contact upon underlying clays and
silts. The sandstones are interpreted as representing the deposits of migrating point
bars of major rivers. They fine upwards into silts and clays having an overbank relationship to the active channels at their time of deposition.

Paleoenvironments

Early attempts were made to establish Siwalik depositional environments by means of
sedimentological and petrographical laboratory techniques (Krynine 1937). Fossil plants,
invertebrates, and vertebrates have also provided valuable information (Tattersall 1969a,b;
Prasad 1971; Vokes 1935). However, since virtually no taphonomic work has yet been carried
out, the vertebrates must be used with circumspection.

Siwalik deposition in the Potwar Plateau area occurred in a subsiding basin close
to the developing Himalayas. Deposition was dominantly fluvial with regimes ranging from
sediments of major rivers to ephemeral streams and temporary floodplain lakes. Sedimentation rhythms appear to have been fairly regular, and each cyclothem was probably of
relatively brief duration; thus, on the order of sixty to seventy cyclic units are found
in the combined Chinji, Nagri, and Dhok Pathan Formations, spanning a period of time
probably not greater than 10 million years. Estimating the ages of the base of the Chinji
Formation at thirteen to fifteen m.y. and the top of the Dhok Pathan Formation at five to
eight m.y., each major cycle would average between about 75,000 and 150,000 years.

The sandstones, which are sometimes multistory units, appear to have been
deposited in broad channels on point bars and as bank undercuts. The clays and silts
represent lower energy environments, and are interpreted as floodplain overbank or
temporary lacustrine deposits of various kinds. They exhibit generally deep weathering
and soil formation, with small-scale local channel sands, conglomerates, and crevasse splays.

Bone concentrations usually occur in association with a channel. The channel
deposits consist of medium-grained, moderately well sorted sands, cross-bedded in places,
and shallow lenses of intraformational, pellet rock conglomerates whose clasts include
silt or clay, calcium carbonate concentrations, and ferruginous pisoliths in a matrix of
poorly sorted sediment. Figure 2.3 illustrates a locality in the upper part of the Nagri
Formation where outcrops of three such pellet rocks (intraformational conglomerate),
are tentatively interpreted as a meander splay deposited after crevassing of a levee.
They have yielded numerous mammalian fossils.

Case History 2. Karroo Grabens of Southern
Tanzania
(Example of rift valley graben setting,
short-time-base sequences.)

Grabens are likely to preserve, within their fault margins, down-thrown strata
which may be lacking outside their confines owing to removal by erosion. Much of the
surface of Africa is underlain by crystalline rocks of pre-Carboniferous age across which
successive flights of planation surfaces are traditionally preserved. Profiles showing
deep lateritic weathering are characteristically associated with these surfaces throughout
areas having tropical rainfall regimes. On the surface of such soils bones and teeth
are seldom preserved, even if they are rapidly covered and buried. It is principally in
tectonically controlled basins of sedimentation that they have survived as part of a
coherent fossil record.

In Southern Tanzania the Songea Series of Karroo age survives within a graben
110 km long having a northeast to southwest trend and with boundary faults some 30 km
apart.

The Karroo rocks are being differentially eroded by the present drainage system.
This is controlled by the base level of Lake Nyasa (500 m [1,568 ft] above sea level)
which lies as a north-south-trending fault-bounded lake within the later rift system.
Exhumation of the older graben has progressed sufficiently far for boundary escarpments

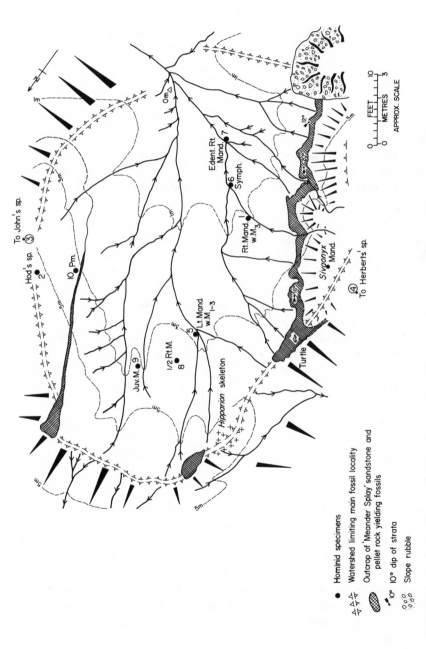

SKETCH MAP-LOCALITY 182-Upper Nagri Formation

• Hominid specimens

△△△ Watershed limiting main fossil locality

Outcrop of 'Meander Splay' sandstone and
pellet rock yielding fossils

10° dip of strata

Slope rubble

Figure 2.3. Geological sketch map showing the location of surface fossil finds at Locality 182, Upper Nagri Formation, Potwar Plateau, Pakistan. Mandible: 1 + 5 + 6 + 7 + 8 = one individual; specimens 2, 3, 4, 9, 10 may represent five other individuals.

once again to dominate the sedimentary basins containing Karroo stages K_1 to K_8 and to etch out fault blocks and resistant strata.

The exposed sequence in the area ranges from the Ruhuhu Formation (K5) to the Manda Formation (K8) of Stockley (1932) as follows:

Approximate Thickness

K8 MANDA FORMATION	2,150 meters	(6500 ft)
K7 KINGORI SANDSTONES	750 meters	(2300 ft)
K6 KAWINGA FORMATION	300 meters	(900 ft)
K5 RUHUHU FORMATION	580 meters	(1750 ft)
K5 to K8 Total	3,780 meters	(11,450 ft)

K5 Ruhuhu Formation

The sediments are typically green to gray in color, consisting of mudstones and cherty siltstones with interbedded sandstones. In the Ruhuhu Valley fossil reptiles were discovered, and tree stumps in position of growth were found at several horizons, suggesting inundation of temporary land surfaces by shallow lake waters. Many of the trees exhibit growth-ring structures, indicating a seasonal and possibly temperate climate.

K6 Kawinga Formation

The Ruhuhu Formation is overlain with at least local unconformity by the Kawinga Formation (Charig 1963; Lower Bone Bed of Stockley and Oates 1931a,b). A persistent cuesta is formed by a basal arkosic grit some 65 meters (200 ft) thick, which dips at 14° towards the boundary scarp. The abundance of fresh feldspar and the occurrence of current bedding in the arkose suggest rapid erosion of fresh rock together with strong current action in shallow water. Shale and mudstone fragments, derived from pre-existing sediments, may indicate intra-Karroo fault movements in the basin. Although no fossils were seen in the basal grit, the overlying beds include brown mudstone with articulated fossil bones, thin sandstones, and a dolomitic limestone conglomerate containing rolled bones. Tree stumps in position of growth also occur. The deposits are only sparsely fossiliferous, although the formation is rich in reptiles at other localities in the Ruhuhu basin.

K7 Kingori Sandstones

The Kingori Sandstones overlie the Kawinga Formation, probably unconformably. The massive sandstone beds invariably give rise to pronounced escarpments with good exposure, but no fossils were seen.

K8 Manda Formation

The Kingori Sandstones appear to grade up into the Manda Formation. This is typically a series of purple-to-chocolate mudstones interbedded with gray-to-whitish sandstones and grits of variable thickness and persistence. The Manda Formation contains the richest fossiliferous strata in the area. The mudstones and sandstones yielded reptilian material at a total of twenty-seven localities. In the mudstones the remains of anomodonts, rhynchosaurs, and archosaurs were frequently found, occurring as parts of associated and sometimes articulated skeletons. In addition to such fossil bones found in the Manda mudstones, excellently preserved by siliceous or ferruginous replacement in a fine-grained matrix, cynodont remains were found associated with surface concentrations of small nodular calcareous concentrations (kunkar). The fossil cynodonts often have a thin calcareous encrustation. The specimens themselves, although shattered and deformed, are held together by the secondary limestone cement and are still recognizable. Perhaps the finest preservation was that seen when the bones had become the nucleus of an ironstone nodule, found frequently in a fine sandy matrix. Many of the best skulls obtained, apart from some of the smaller cynodonts, were from such nodules. The distribution of identified fossils through the Manda Formation is indicated diagramatically in table 2.1.

Table 2.1

Number of Fossils Found at Different Levels in the Manda Formation

Thickness in Feet	Positions of Fossil Levels		Cynodonts	Anomodonts	Rhynchosaurs	Archosaurs	Indeterminate Reptiles	Total	
6000'	———				1		1	2	
	———		1				3	4	
5000'	———		3	6		1		10	
4000'	———	Main Fossil Levels	22	3	5	3	11	44 }	300' thick
	———		15	11	19	14	17	76 }	65% fauna
	———		7	1	2	4	5	19	
3000'	———		1	1	2		1	5	
2000'	———			5		2	2	9	
	———		2	2	1?		4	9	
1000'	———		2	2		1	1	6	
0		Totals	52	31	29	27	45	184	

This confirms that the fossil-bearing horizons are never in any sense a single "bone bed," far less several "bone beds." A pronounced concentration of finds, however, occurs within a small thickness of the upper part of the Manda Formation; 65% of all the specimens found in that formation were obtained from 100 meters (300 ft) of strata.   The fauna represented by these finds was not strictly a contemporaneous assemblage as several different horizons were spanned. However, their stratigraphical concentration implies that there was a period of time during which conditions more suitable for fossilization were widespread throughout the area. It is not yet clear what factors contributed to these more favorable conditions.

Paleoecological Conclusions

Shallow freshwater conditions and repeated emergence seem indicated by assemblages of tree stumps in growth position, by current bedding, and by the presence of land-dwelling reptiles. Recurrent minor tectonic movements, intensifying early faults, may be implied by the influx of fresh feldspar and by the presence of derived fragments of earlier sedimentary rocks. Most of the fossils of large reptiles appear to represent complete, or largely complete, carcasses stratified in mudstone, found where the animals were buried after grounding in lake shallows or on flood plains at the margins of muddy basins. It seems unlikely that a carcass would drift for great distances before becoming lodged or sinking following a bloat phase.

Cynodonts are best preserved where they lay on or immediately under the surface and where secondary limestone formation (calcrete) prevented the breakdown of the bone chemistry. However, the shattered and distorted nature of many of these fossils suggests exposure to physical and chemical weathering on land surfaces and lake flats prior to stratification.

Rolled bone fragments and isolated large specimens in sandstone may have been washed in by streams and rivers. Such washing-in was frequently followed by the secondary precipitation of iron around nuclei provided by the bones themselves.

In essence the story is one of the gradual filling of an unstable tectonic trough in which sedimentation appears to have been periodically balanced locally to compensate the deepening of the basin of deposition. This resulted in the formation of shallow bodies of water. The detritus carried into these was generally fine, but occasional incursions of coarser sediment (including rolled bones) also occurred. Sometimes the coarser influx may have resulted from minor earth movements.

Conditions for fossilization seem to have been most suitable for a short period during later Manda times representing 100 meters (300 ft) of sediment (table 2.1) out of a total thickness of 1500 meters (5,000 ft) for the Manda Formation. Some fossils occur intermittently throughout the series from the upper Ruhuhu Formation (K5) to the upper Manda (K8).

Case History 3. Miocene Carbonatite Volcanoes
of Kenya and Uganda
(Example of volcanoes with primary carbonatite
fallout, short-time-base sequences.)

The volcanic environments contain fossils preserved in pyroclastic rocks or their derivatives. Fully lacustrine environments occur at some localities in Kenya at Karungu, Mfwanganu, and Rusinga and in Uganda near Bukwa on Mount Elgon. These sediments are frequently dominated by tuffaceous material derived from the flanks of nearby volcanic cones. The Mammalia are well preserved, and long faunal lists have been assembled which include both large and small mammals and are similar to the assemblages obtained from primary tuffs. Crocodile teeth and scutes are abundant, while the fully lacustrine conditions are indicated by the presence of fish and freshwater Mollusca and also by horizontal bedding and the evidence of current action.

The greatest number of fossiliferous localities with well-preserved specimens occur in deposits of fine-grained subaerial primary tuffs as at the following localities: Uganda: Napak I, IV, V, and IX; Kenya: Koru and Songhor. At these localities the occurrence of paleosols suggests intermittent pyroclastic activity with long periods of quiescence and weathering punctuated by subaerial fallout of fine ash carried downwind from the source vents.

The abundance of species of small mammals at Napak, including rodents, primates, and insectivores, is also typical of Songhor, Koru, Mfwanganu, and Rusinga. At Napak I, IV, V, and IX the various large mammals (e.g., *Deinotherium*, *Brachyodus*) which are typical of fluviatile localities appear to be absent. *Trilophodon* is present only at two of the four tuff localities at Napak and is represented by only a few individuals. Fossil invertebrates include the shells of land snails which, together with casts of their eggs, are abundant.

Crocodiles are absent except for two fragments of scutes of juveniles, but vertebrae, often associated in series of up to seven or eight, and other fragments of snake skeletons are important elements. Lizard remains also occur. Bird bone is present but chelonian fragments are virtually absent. Fossil fruits occur at certain levels, rootlet casts and fossil wood are abundant on many horizons, and casts of leaves, frequently rolled and transgressing the bedding, are common at some levels. At Napak X, which is also in the subaerial tuff sequence, the preservation is in calcite and, in addition to fruits and gastropods, fossil millipedes also occur. All these paleontological data support a subaerial environment of deposition which is confirmed by the presence of paleosols, by the haphazard distribution of mica flakes at all angles to the bedding, and by the presence of rain spotting and accretionary lapilli. Leaves from Napak IV indicate a tropical vegetation of trees and shrubs. The lack of broad-leaved monocotyledons and of ferns shows that the climate was not very moist, while the absence of microphylls suggests that it was not very dry. The lack of narrow-leaved savannah grasses indicates a closed forest or thicket.

More detailed analyses of the total fossil mammal collections from the four subaerial tuff localities at Napak are helpful in assessing the nature of the assemblages (table 2.2). The features which can be noted are as follows:

Table 2.2

Percentages of Fossil Elements at Four Napak Localities

	Napak I	Napak IV	Napak V	Napak IX	Grand Total All Locations
Total Number	2008	4176	1291	401	7876
Percent Indeterminate	62.7	73.1	68.6	71.3	69.6
Percent Better Bones	23.3	12.2	179.9	16.7	16.2
Percent Skulls, Teeth, and Jaws	14.0	14.7	13.5	12.0	14.2
Analysis of Skulls, Teeth, and Jaws					
Percent Rodent	42.2	81.8	46.3	29.2	64.0
Percent Proboscidea	35.1	-	-	45.8	10.9
Percent Artiodactyl	6.4	4.9	29.1	10.4	9.3
Percent Primate	3.5	6.6	9.1	2.1	6.0
Percent Others	12.8	6.7	15.5	12.5	9.8

a. *The high proportion of crania, teeth, and jaws* in relation to the total.

b. *The dominance of rodents* within the total of crania, teeth, and jaws. Many of the rodent finds are complete or nearly complete skulls.

    c. *The high proportion of primates*. This is particularly striking when it is
remembered that primates are normally "shy" candidates for fossilization.

    d. One further peculiarity is that the localities Napak I, IV, V, and IX are
discrete patches where bone is abundant. The localities preserve virtually contemporaneous
assemblages at points within one mile of each other on the flanks of a growing volcanic
cone (fig. 2.4). There is almost continuous outcrop of lithologically identical strata
which represent the same sequence of ash falls. Despite this fact, diligent searching has
failed to find fossil material eroding out on the surface between the main fossiliferous
patches.

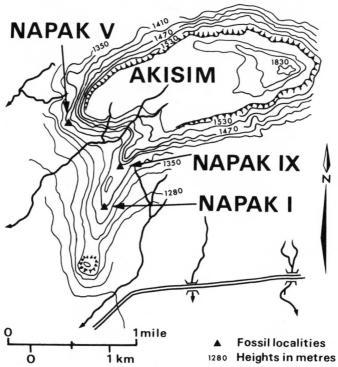

Figure 2.4.   Location of localities yielding Miocene fossil mammals, Akisim, Napak, Uganda.

    A threefold coincidence seems required to account for these peculiar patches, as
follows:

    i) The occurrence of calcareous tuffs which gave rise to chernozem-like calcite-
rich soils. These were gradually colonized by mammals following in the wake of vegetation
which provided only incomplete ground cover at first. Water courses, supplied by precipitation as a result of the local climate induced on its higher slopes by the volcano itself, gave gallery forest with lianas as suggested by some of the fossil wood and fruits.
Natural deaths allowed gradual accumulation of potential fossils over all the slopes of
the volcano as bone became scuffed into and covered by the calcareous soils. The next
layer of wind-borne ash effectively buried the bone and allowed mineralization to take
place. The chemistry of the environment of burial seems to have been crucial, and it is
likely that the fine matrix was rich in calcite as the ash came from a carbonatitic center.
The primary calcareous nature of the tuff is difficult to prove owing to recrystallization
of the calcite. However, trace elements help to confirm this (see table 2.3).

Table 2.3
Analyses of Calcareous Tuffs and Limestones from Kenya

Loc. and Lab. No.	Description and Field No.	La	Ba	Nb	Y	Zr	Sr	Rb	MnO	TiO_2
SONGHOR:										
4901 A	Pink, marly tuff. 30/5	.01	.06	.004	.005	.01	.04	trace	.05	.1
4901 B	Flaggy "limestone". 30/6A	.02	.2	.02	nd	nd	.2	nd	.15	nd
4901 C	Flaggy tuff. 30/6F	.01	.07	.03	.005	nd	.14	nd	.15	nd
4901 D	Calcareous tuff. 30/6G	.03	.12	.02	nd	nd	.22	nd	.1	nd
4901 E	Gray medium tuff. 30/6I	.01	.12	.03	.005	nd	.25	nd	.1	nd
KORU:										
4903 A	Pink fossiliferous tuff. 24/2	nd	.09	.004	.005	nd	.18	.004	.06	.4
4903 B	Gray fossiliferous tuff. 26/2E	.04	.3	.03	.01	nd	.3	nd	.3	nd
FORT TERNAN:										
4902 A	Coarse (Carbonatitic?) tuff. 25/4	.01	.06	.006	.005	.02	.04	.002	.08	1.0
4902 B	Coarse tuff (Bed 4). 25/9	nd	.3	.008	trace	.03	.015	.002	.06	2.0

Approximate X-ray Fluorescence Analyses by Dr. A. Livingstone. nd = not detected. All figures expressed as percentages.

ii) The second stage required seems to be a later concretionary concentration of calcium carbonate to strengthen the calcified fossils and allow them to hold together after reerosion. Many of the fossils are seen to be riddled with small cracks but are held together as coherent specimens by concretionary carbonate. At Napak IV a spring-eye, which seems to have been active during the later Pleistocene but before erosion lowered the water table to its present level, is centered above the cone of fossil-bearing strata. The amount of concretionary calcium carbonate (kunkar) falls off very rapidly away from the center marked by the travertine of the former spring-eye.

iii) The third requirement is a catchment area in the form of a shallow valley (as at Napak IV) or a flat (as at Napak I) to allow the slow and natural accumulation under present-day conditions of a remanié concentrate of calcified and reinforced fossils (a similar catchment seems to have played an important role at the Siwalik fossil locality 182; see fig. 2.3).

Additional information relevant to the role of carbonatite tuff in the Miocene environment of fossilization can be obtained from study of the late Pleistocene to recent carbonatitic centers. The following paragraphs give examples of the importance of analyzing the chemical background in taphonomic studies.

Dawson (1964a) showed, on the basis of both field relationships and trace elements, that calcite-rich tuffs in northern Tanzania with calcium carbonate content up to 70% of the total rock have volcanic rather than sedimentary affinities and can be traced into carbonate tuff rings and cones at carbonatitic centers of explosive activity.

In other studies Dawson (1962, 1964b) has investigated the eruptive history of Oldoinyo Lengai and commented upon carbonatitic ashes from this center and at other vents in northern Tanzania. The field relationships inferred for the Miocene volcanoes which dominated the fossil localities at Napak, Songhor, Koru, Rusinga, Mfwanganu, and Fort Ternan show that they are all closely associated with intrusive carbonatitic centers.

To test the similarity of the Miocene calcite-rich tuffs containing fossil mammals to modern carbonatite ashes and to undoubted carbonatites, a series of specimens was collected from Kenya and Uganda Miocene fossil mammal localities. The details of investigations of samples from three localities are shown in table 2.3.

Of the five samples from Songhor, one is from the "Red Clay" unit of Kent (1944) and the remaining four are from different levels within the "Limestone" sequence of Kent (1944) and Shackleton (1951). Many of these "limestones" can be seen from detailed investigations to be calcified tuffs. In all but the most dense calcretes ghost crystal outlines of the original tuff can be seen in thin sections. The overlying red marl is a red weathering erosion product which was also dominantly tuffaceous in origin. These two lithological units form the main source of the Songhor fossil mammal fauna.

At Koru two deposits were sampled. One is a pink marl, which yielded the bulk of small mammals collected by Hopwood (1933). The other is a calcified tuff from one of a series of shallow quarries in surface limestone near the carbonatite center to the northeast of Legetet Hill. These limestones are also calcified tuffs and contain fossil land snails. The final two specimens of calcite-rich tuff were collected from horizons in the Fort Ternan fossil mammal excavation.

The results in table 2.3 can be compared with trace element percentages recorded by Dawson (1964a) for recent carbonate tuffs and surface limestones from northern Tanzania and also with trace elements from Indian carbonatites (Deans and Powell 1968). In each case a broadly similar pattern can be seen which contrasts sharply with the trace element spectrum from normal, freshwater, limestones.

Dawson (1964a) comments that the carbonatitic affinities of the Tanzania tuffs and limestones were shown in all cases by high trace amounts of barium, strontium, lanthanum, and titanium, accompanied in some of the analyses by yttrium, zirconium, and niobium. Deans and Powell (1964a) state that, of the Indian carbonatites, "the high abundance of strontium and phosphorus is perhaps most striking but the presence of lanthanum and cerium is equally significant." Although they note that "barium and niobium are rather capricious in their abundance," yet "collectively enrichment in a majority of six elements can usually be regarded as diagnostic."

The analyses from Songhor, Koru, and Fort Ternan all have some features which confirm their carbonatitic affinities, and they contrast sharply with analyses of normal

limestones. The two analyses from Koru are remarkably similar to those from the carbonatite center northeast of Legetet Hill, Koru. Further work is necessary, based upon suites
of carefully selected samples from a range of Miocene limestones and tuffs yielding mammalian faunas, to ascertain the areal importance of the carbonatite influence. However,
from this preliminary trace element evidence, primary falls of carbonatitic ash seem
likely to have been an important factor in the preservation of bone at several fossiliferous localities. The analyses also confirm that the Songhor limestones are calcretes
developed within and upon tuffs and resulted from diagenetic redistribution of calcite
under processes of subaerial weathering. The trace elements help to establish that there
are broad continental domains where the background chemistry is likely to favor the
preservation of fossils. The presence of carbonatitic volcanic centers is certainly one
of the factors of fossil preservation peculiar to parts of the East African rift system.

Case History 4. Langebaanweg Pliocene Bay Bar,
South Africa
(Example of a continental margin, short-time-
base sequence.)

     Langebaanweg is located some 110 km northwest of Capetown (32° 58' south, 18° 9'
east) in southwestern Cape Province, South Africa. The site lies about 15 km inland
from Saldanha Bay.

     Various early accounts of the phosphatic deposits near Langebaanweg (Langebaan Road)
and elsewhere in South Africa have been published (DuToit 1917; Haughton 1932), and a
large amount of unpublished information lies in mining company records. The finding of
new fossil-bearing horizons (Singer and Hooijer 1968; Singer 1961; Boné and Singer 1965),
coupled with increasingly extensive open-cast mining activities over the last fifteen
years by companies quarrying the phosphatic deposits, has resulted in better exposure and
bore-hole coverage in the area of occurrence of the fossiliferous strata.

     The preliminary descriptions of Langebaanweg that have been published have been
concerned with theoretical environmental interpretation (in terms of sea level change or
variations in the course of the Great Berg River) rather than with the definition or
mapping of the distribution of lithological units exposed by quarrying activities and
confirmed in bore holes. The relationship of the fossils to the sediments has
received little attention. Tentative conclusions concerning environments of deposition
can only follow from the interpretation of a firmly established litho/stratigraphic framework. It is unfortunate that attempts to date the Langebaanweg mammalian assemblage have
been made in the light of assumed maximum heights for former marine sediments in the
Langebaanweg area in relation to Pleistocene strandlines elsewhere in South Africa. For
instance, correlations have been attempted with sequences recorded by Richards and
Fairbridge (1965), Carrington and Kensley (1969), and even with the "classical" Mediterranean sea levels of Zeuner (1959).

     The data recorded here were obtained from my brief study of "E" Quarry, in 1967,
amplified by bore hole records kindly supplied by AMCOR. The earlier worked Baard's and
"C" Quarries (see Boné and Singer 1965, fig. 2; Hendey 1970a, fig. 2, p. 79) are not discussed as they had ceased to be worked and were no longer well exposed in 1967. The complex
relationships of facies changes in the zone of inferred interplay between littoral, possibly fluviatile, and aeolian sediments make attempts at extrapolation and correlation
highly speculative over even such short distances as the 2000 meters which separate
Baard's and "E" Quarry. The assemblage from "E" Quarry is much larger than that from "C"
or Baard's (tables in Hendey 1969, 1970a,b). Virtually all the taxa listed for Baard's are

known from "E" Quarry (the exceptions being cf. *Mellivora capensis, Hyaena* cf. *brunnea, Redunca* cf. *ancystrocera,* and cf. *Raphicerus* sp. which occur only at Baard's). However, the slight importance of these apparent differences is further diminished when one considers that in the family Mustelidae, an *incertae sedis* is noted from "E" Quarry while the Baard's specimen is not definitely diagnosed. Again, *Hyaena* cf. *brunnea* is noted at Baard's, while *Hyaena brunnea* is found at "C" Quarry and *Hyaena* sp. is known from "E" Quarry. As the Baard's area was almost worked out when paleontological studies commenced in 1958, gaps in the fossil assemblage from Baard's may result from a combination of lateral change of facies and collection failure. It would be dangerous to infer a major difference in time of deposition, although this remains a third possible factor.

The "E" Quarry sediments are interpreted as comprising the Varswater Cyclothem. This commences with a phosphatized calcrete which has been converted to phoscrete (0.5 m-1.0 m). It is succeeded by marine beach gravels represented by cobbles and pebbles of phoscrete surviving in hollows and grooves on a wave-eroded surface together with marine molluscs and sharks' teeth. This passes up into a sequence of sandy silts succeeded by richly phosphatic sands (up to 2 m). The surviving upper part of the Varswater Cyclothem comprises less phosphatic sands whose former thickness is not known as they are terminated by an erosional unconformity above (up to 7 or 8 m survive). The succeeding formation exhibits an erosional, unconformable relationship and is of windborne calcareous sands with calcrete horizons containing *Dorcasia* sp. The calcretes have been patchily converted to silcrete or ferricrete in some areas.

The Varswater Cyclothem is interpreted as representing a single phase of marine transgression and regression of Pliocene age during which the sea level rose, on evidence from the vicinity of "E" Quarry, from below 12 meters (40 ft) O.D. to at least 36 meters (130 ft) O.D. Movement of an offshore bar which progressively contained a lagoon was suggested by the form of the sand body encountered in bore holes. An extension to the quarry on its west side has confirmed this and revealed carbonaceous lagoon sediments. The original conclusions are in keeping with the results obtained by Butzer (1973). More recent detailed research by Tankard (1974) has established details of the lagoon and its relationship to the formation of pelletal phosphorite and the preservation of fossil bone.

A Pliocene age for the transgression which initiated the Varswater Cyclothem is in accord not only with the mammalian fossil evidence but also with recent studies of the sedimentary history of the continental shelf off Saldanha Bay (Dingle 1971, 1973). The last major marine transgression (T_4) is shown by Dingle to be of probable Pliocene age.

Conclusion

From the case studies very little emerges in the way of simple laws which govern burial and survival of bone and which might assist in the search for possible sites where fossils are being naturally exhumed. In a broad sense each of the groups A to D provides conditions which facilitate the survival of potential candidates for fossilization. However, it is probably best to use these in a negative way in that, if an area does not fulfill one or more of the conditions in categories A to D (pp. 23-24), then it is not worthwhile wasting time in searching. Thus in a sense the categories do assist by suggesting where not to look.

However, if one proceeds to one of the areas A to D, there is no guarantee of success within these macro-groupings. It is still essential to consider one or more micro-factors in order to decide where a search should be localized. That is, the nature of conditions favoring meander splay deposits within the Siwalik succession, or the localization within a graben succession of the situation that dictates what narrow span of the 1600 m thick Manda Formation will produce abundant fossils, are still the questions which it remains for taphonomy to answer.

Part 2

MODERN ECOLOGY AND MODELS FOR THE PAST

Paleoecology must rely heavily upon work in modern ecology for basic information concerning the nature of ecosystems and for models that can be related to the fossil record. What inferences about important ecological variables can be made using information from fossils and their environments of burial? Ecologists who have seriously considered the evidence of the past in the light of modern ecological theory are beginning to provide possible answers to such questions.

Western, who has studied the dynamics of animal and plant communities in Africa, shows us what paleoecologists should look for in reconstructing ecological history. He discusses the standard characterizations of community structure such as species composition and relative abundance, and also points out that features such as the body weight of species may hold important clues for paleoecology. Weight can be correlated very closely with birth rate as well as other life-history parameters in mammals, opening up a broad and potentially exciting basis for interpreting a great deal about extinct species and their roles in paleocommunities. Western also focuses on biological factors such as death rate which affect the bone sample derived from a living community. He points out a number of ecological factors that may bias the composition of such a bone assemblage and which need to be carefully analyzed before paleoecological interpretations are made. Regarding the role of ecologists in paleoecology, Western feels that they have much to contribute, but that they may profit as well from an exchange of information, since the fossil record can contribute a missing element in most ecological research--time depth.

The nature of animal communities in Africa and their relationship to vegetation and climate are described by Coe. He provides us with an understanding of what variables are thought to be most important in regulating community structure. Of particular interest is the relationship between rainfall and primary productivity. Through Coe's discussions, we can begin to comprehend the present limits of knowledge in modern ecology, and this provides a helpful perspective regarding what we can hope to know in paleoecology. He also relates some of his findings about mortality in vertebrates and the decomposition of their carcasses. Evidence from his work on elephants shows that the rate and style of decomposition varies strongly with substrate and climate. This has important consequences for interpreting the taphonomic history of animal remains in paleontological situations.

3. LINKING THE ECOLOGY OF PAST AND PRESENT MAMMAL COMMUNITIES

David Western

Introduction

An interpretation of past faunas and their environments from the fossil record is limited by our knowledge of both ecology and taphonomy. Given a perfectly preserved assemblage of animals we can reconstruct the past no better than we can reconstruct the ecology of contemporary communities from dead animals and carcasses. However, from the time an animal dies and passes into the inorganic world processes are at work which determine what traces remain intact at a given point in time and, therefore, how precise our reconstruction can be. Taphonomy is the study of these processes, the biases which influence what is preserved as fossil material. It should also include the study of sampling biases involved in estimating living populations with which a death assemblage is being compared. The more we know of such natural biases, the more contemporary ecology can be applied to paleoecology by either correcting or accounting for them.

Paleoecologists necessarily take ecological determinism to its limits; they must become skilled "ecosleuths" to do so. The form and function approach in ecology (Thompson 1944), which is increasingly related to physiology, energetics, and engineering (Kleiber 1961; Pennycuick 1972; Moen 1973), has already contributed a great deal to an interpretation of the past (Shotwell 1955; Bakker 1971, 1975b). At the level of the individual species this approach has proven merits. In fact, implicit in the niche theory (Hutchinson 1967), the postulate that each species is functionally adapted to a particular structural niche, is the reverse assumption, that an assessment of the environment of an animal can be made from its morphology. Examples are the inferences made about the structural niches of birds from their bill shapes (Cody 1974). However, a single species only exploits a limited component of the environment, and insights into the entire system will be limited accordingly. The more species that are considered, the greater will be the potential information on the ecosystem, and it is at this level that ecologists are beginning to make rapid progress, for example by relating species diversity to environmental stability and structure (Elton 1958; Margalef 1968). It is on this level that I feel an interchange between ecologists and paleontologists will benefit most, and on which I propose to concentrate.

To date, the inferences made by paleontologists about past environments from fossil assemblages have been made almost entirely without allowance for natural sampling biases, such as the death rates of animals, which are likely to vary with different-sized animals

41

(Hill 1975). I want to examine some of the ecological processes involved as an animal passes from a living population to fossil deposits, and to consider what biases are introduced in terms of sampling a large mammal fauna. Finally, I will give a few examples of how biases can be corrected or allowed for in an interpretation of fossil assemblages.

It is neither feasible nor desirable to separate the ecological and taphonomic processes that operate after an animal dies, and only those which are largely ecological will be considered in detail. Most of the information used will be confined to and drawn from East African large mammal communities which probably approximate natural assemblages that existed prior to man's massive impact of the last few thousand years.

Sampling the Living and the Dead

We need to know what factors bias an estimate of a living population, the "control" against which a dead assemblage is being compared. Counting living populations of animals accurately is a formidable task. Rarely can populations be totally counted since they normally cover areas too large and heterogenous to census at reasonable cost. In most circumstances, particularly in large mammal communities, sampling techniques must be adopted. The conditions required for valid estimates of a population using various sampling techniques can be found in Cochran (1963) and Seber (1973). In most cases the biologist is also interested in the distribution and movement of populations in relation to environmental resources, and further difficulties arise in deriving unbiased data of this sort. Most samples of animal distributions assume that the estimate is based on an instantaneous sample, an impossible objective except in a localized population. However, provided the extent of movement over the sample period is minimal and random with respect to the sampling strategy, reasonable estimates of animal density distributions can be made. To estimate the use of habitats and resources through time, a successive series of counts must be made over a number of seasons. Various large mammal ecosystems have been studied in this way (Lamprey 1964; Western 1973).

Despite the refinement of sampling techniques in recent years, particularly in aerial counting, biases still exist. A count is affected by the observer's search image, the size and color of the animals being counted, their mean group size, and behavior, the nature of the terrain, habitat, etc. (Graham and Bell 1969; Pennycuick and Western 1972). To date there has been no complete assessment of the range utilization and density distributions of an entire large mammal community. Many species are omitted, particularly the nocturnal animals. Most emphasis is given to the numerically dominant species and those most easily sampled—the diurnal animals that prefer open habitats.

The age composition of a species population is also extremely difficult to assess from the living population, there being no ready way of aging animals visually in the field. Ecologists generally infer the structure of the living population from dead animals which can be more easily aged by such criteria as tooth wear (Laws 1966).

Many of the problems that confound accurate estimates of the parameters of living animals also apply to counting dead animals. There will inevitably be a bias in favor of counting large bones in open terrain. However, counting bones of nocturnal animals should be no more difficult than for diurnal species, and for the zoologist this method does offer one way of complementing counts on live populations. Another difference is that the time duration of a sample is no longer of overriding concern since it can be assumed that the distribution of skeletons will reflect the average distribution of the population multiplied by the site-specific death rate. To obtain accurate estimates of the numbers rather than distributions of the living populations would, however, demand that the rate of bone accumulation be known. Accumulation rates can be measured through time, but so far cannot be inferred from a spot sample since no accurate technique exists for giving a precise age of a bone to within, say, a month, the resolution needed in such remote sampling. The time dimension becomes even less definable in looking at fossil accumulations, and its resolution using various radioisotope techniques diminishes inversely with the period since death.

We know too little about the relationship between the density distribution of living animals and their carcasses to infer confidently very much about a community from carcass data. Even if every animal that died did so in situ, one could still not infer the average distribution of the living population exactly since the death rate varies with location and season. A locale where animals spend the harshest season will tend to have a higher than expected density of carcasses. The same may apply to areas in which animals give birth since the death rate in young mammals is characteristically high, irrespective of their mode of death (Caughley 1966). We need, therefore, to distinguish between death ages, sites, and seasons to apply suitable bias corrections to carcass data if we intend to derive density patterns of the living population from them.

Before we can confidently extrapolate from fossil assemblages to their environment we need to know more about the relationship between carcasses and the populations from which they accumulated. To my knowledge the only attempt to do this so far is Behrensmeyer's study (this volume) in the Amboseli basin of southern Kenya.

Changes in Composition after Death

With contemporary faunas one cannot calculate the size of the living population from carcasses without data on the durability of the specimen, the death rate, and the area used by the population, that is, its range. The average population that uses a given area can be calculated given the durability of bones and the area-specific death rate, but these variables have to be established initially by reference to the living population and are liable to vary somewhat in time.

For more remote populations that can only be inferred from fossil remnants, no estimate of the original population size can be given since the area of use (or the area-specific death rate), the time over which the samples accumulated, the ratio of fossils to original carcasses, and the species-specific death rate are all unknown. Even if these variables were known, the problems of excavating a large enough area of a given time horizon in order to gain a respectable sample size would be prohibitive.

The most that paleoecologists can hope to do is to reduce the time period over which they are sampling and concentrate on the relative proportions of species within it. Many attempts have been made to do so (Shotwell 1955, 1958; Bakker 1975; Behrensmeyer 1975). However, this method is liable to considerable bias because of varying death rates between species (Hill 1975). The proportions in a fossil assemblage must be a function of the original population size of each species, multiplied by its death rate. Other factors discounted for the moment, an elephant shrew with a death rate twenty-five times that of an elephant should appear twenty-five times more commonly in the fossil record for the same number of animals in the living population.

From the literature, I have accumulated data for twenty-one species of mammals studied under natural conditions, and have given birth rates for each in table 3.1. The populations I included were, as far as I could determine, more or less stable. In this event the death rate is equal to the birth rate. A highly significant relationship is

Table 3.1

Annual Birth Rate and Body Weight for Various African Mammals

Number	Species Name	Weight (kg)	Annual Birth Rate (%)
1	Elephant, *Loxodonta africana* (Blumenbach)	2575	9
2	Hippo, *Hippopotamus amphibius* (L.)	1000	11
3	Rhino, *Diceros bicornis* (L.)	816	11
4	Giraffe, *Giraffa cameleopardalis* (L.)	750	16
5	Buffalo, *Syncerus caffer* (Sparmann)	450	17
6	Zebra, *Equus burchelli* (Gray)	200	22
7	Wildebeest, *Connochaetes taurinus* (Burchell)	165	25
8	Waterbuck, *Kobus ellipsiprymnus* (Ogilby)	160	23
9	Kongoni, *Alcelaphus buselaphus* (Pallas)	125	26
10	Lion, *Panthera leo* (L.)	120	34
11	Lechwe, *Kobus leche* (Gray)	72	32
12	Uganda kob, *Adenota kob* (Schlater)	58	30
13	Hyena, *Crocuta crocuta* (Erxleben)	55	37
14	Warthog, *Phacochoerus aethiopicus* (Pallas)	45	35
15	Grant's gazelle, *Gazella granti* (Brooke)	40	38
16	Impala, *Aepyceros melampus* (Lichenstein)	40	36
17	Thomson's gazelle, *Gazella thomsoni* (Guntter)	15	59
18	Dik-dik, *Madoqua kirki* Thomas	6	74
19	Mongoose, *Mungos mungo* (Gmelin)	1.47	126
20	Elephant shrew, *Rhynchocyon chrysopygus* Gunther	0.54	156
21	Elephant shrew, *Elephantulus rufescens* Peters	0.07	232

Note: Data are taken from the literature and include species with more or less stable populations, so that birth rate and death rate are equal.   A full analysis of the data will be published elsewhere.

found between the birth rate (or death rate) and the body weight of a species (fig. 3.1). The data can be used as an estimate of the death rate of animals based on their size. For fossils, an anatomical correlate could be substituted for body weight, from which the death rate could be estimated. The proportions of species in the living population from which they were derived can then be calculated by the number of individuals (measured by, for example, mandibles) divided by the death rate.

For heterotherms a similar size-dependent birth and death rate will exist, and Fenchel (1974) has shown from empirical evidence that the intrinsic rate of increase (r_m) of a species can be calculated from body weight. The slope of the regression that calculates r_m from body weight is similar for heterotherms and homeotherms, though lower in the former, and is a function of the metabolic rate. The realized birth rate is 65% of the intrinsic rate in mammals, with a similar slope to r_m. A comparable ratio of actual to potential birth rate might apply to the heterotherms, but this assumption needs testing. The data on birth rates do include species with multiple births such as hyena, lion, and mongoose, but additional information on small mammals would be needed to test their approximation to the regression equation.

Bone size and density will probably be the most significant variables determining fossilization, and they depend on the size and age of animals. The effect of bone size on

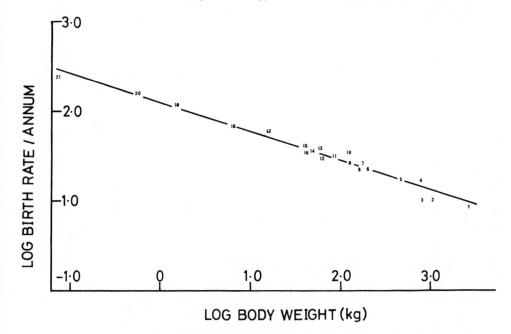

Figure 3.1. Relationship between birth rate and body weight for various African mammals.
Data from table 3.1. Log y = 2.096-0.327 log x, r = 0.989.

durability can be indirectly tested from data collected in Amboseli (table 3.2). I have
assumed that bone size will be related to animal body weight. Figure 3.2 shows the rela-

Table 3.2

Number and Proportions of Animals and Carcasses in the Amboseli Ecosystem

	No. in Population (N)	Proportions (Np)	Carcasses in Sample (C)	Proportions (Cp)	NxDeath Rate (ND)	Proportions (NDp)	C_p/N_p	C_p/ND_p
Zebra	2200	0.27	243	0.28	484	0.20	1.04	1.40
Wildebeest	2473	0.30	336	0.38	618	0.26	1.27	1.46
Thomson's gazelle	788	0.10	36	0.04	465	0.19	0.40	0.02
Grant's gazelle	1230	0.15	77	0.09	468	0.19	0.60	0.15
Impala	700	0.09	38	0.04	252	0.10	0.44	0.10
Buffalo	457	0.06	49	0.06	78	0.03	1.00	2.00
Elephant	218	0.03	28	0.03	20	0.008	1.00	3.75
Giraffe	115	0.01	38	0.04	18	0.008	4.00	5.00
Rhino	40	0.001	30	0.03	12	0.005	3.00	6.00

Note: The expected proportions of bones are calculated both with and without allowance for
the species-specific death rate. Animal data from Western (1975), bone data from Behrensmeyer, collected 1975. Rhino death rates were derived from observations since the population has been declining in recent years. Elephant carcasses are probably underrepresented
in the study area since poaching levels elsewhere in their range is high. They have been
excluded from figure 3.2.

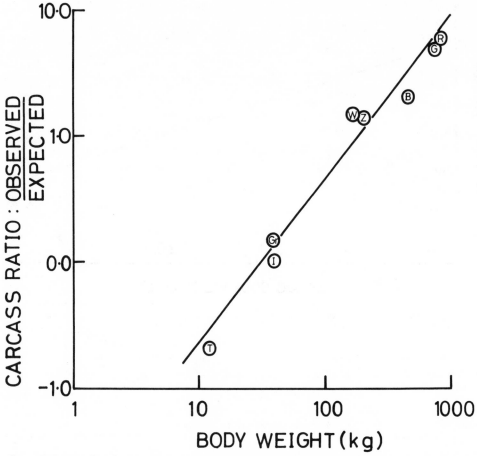

Figure 3.2.   Ratio of skeletal proportions found to those expected (population number x death rate) for animals of varying body size in the Amboseli ecosystem. Data from table 3.2.   R = rhino; G = giraffe; B = buffalo; W = wildebeest; Z = zebra; Gr = Grant's gazelle; I = impala; T = Thomson's gazelle. Regression equation, log y = -2.105 + 1.351 log x, r = 0.982. When the species-specific death rate is ignored, r = 0.643. Skeletal data (preliminary) from Behrensmeyer, personal communication.

tionship between bone durability (observed divided by expected numbers) and body weight (an approximation to bone size).   In the case where the death rate of each species is ignored, a low correlation exists between bone size and bone durability, but when death rates are incorporated, a highly significant correlation is found.   Small bones are retained in the bone assemblage in lower proportions than expected, while those larger than 100 kg are retained in higher proportions than expected.   Such differential retention in contemporary bone assemblages would affect the fossilization ratios in a similar manner.

Similar sampling procedures can be applied to other environments where the durability of bones may be different, depending on the physical agents of weathering and the predator-to-prey ratios.   The slopes of the regressions established under such conditions are not likely to differ so much as the intercepts.   Data for other areas could be collected relatively easily.   The only figures needed are the population size for each species, the death rate, and the number of bones per unit area of ground surface.

A number of localities have figures available on population numbers (Coe et al. 1976), and death rates could be calculated from figure 3.2, leaving only the surface bone-density data to be established. Obviously data are needed from many sites if we are to establish the pervasiveness of size bias and its reliability in reconstructing fossil communities.

    Within a species the age at death will affect a bone's chances of fossilization because young animals will be more completely consumed by predators and rapidly broken down by weathering. The durability of bones expressed on the basis of animal body size (fig. 3.2) will apply in some approximate fashion to bone sizes within a species and hence to the age at death. There is adequate evidence to show that young animals are in fact grossly underrepresented in both contemporary bone remnants and fossil assemblages (Caughley 1966; Spinage 1972; Olson 1957). A fossil record will therefore give a highly biased reconstruction of the age distribution at death. Fortunately, among large mammals there does not appear to be much difference in the age-specific mortality (expressed as a percentage of a total life span) among species. This is demonstrated by data I have as-sembled for African herbivores, for which reasonably good life-table data exist (table 3.3 and fig. 3.3). The curve in mammals is generally a Type I curve (Deevey 1947), indicating

Table 3.3

Survivorship, e, of Various African Large Herbivores Expressed as a Percentage of Total Life Span

Species	0	10	20	30	40	50	60	70	80	90	100	e_o (%)	References
Loxodonta africana	1000	490	417	355	263	170	120	93	62	22	0	30	Laws (1966)
Hippopotamus amphibius	1000	440	350	300	240	200	170	125	60	20	0	29	Laws (1968)
Diceros bicornis	1000	550	407	309	214	138	93	56	32	12	0	28	Goddard (1970)
Syncerus caffer	1000	513	468	445	380	252	195	102	28	6	0	34	Spinage (1972), Sinclair (1974)
Equus burchelli	1000	741	661	475	513	407	275	132	46	15	0	44	Peterson and Casebeer (1972)
Connochaetes taurinus	1000	643	600	539	375	250	165	202	63	31	0	39	Andere (1975)
Kobus deffasa	1000	585	485	460	370	290	270	180	70	40	0	38	Spinage (1974)
Alcelaphus buselaphus	1000	525	437	417	347	269	209	132	87	26	0	35	Gosling (1975)
Kobus leche	1000	676	575	513	417	339	269	191	115	50	0	41	Sayer and Van Lavieren (1975)
Adenota kob	1000	550	490	470	440	360	250	160	70	30	0	38	Modha and Eltringham (1976)
Aepyceros melampus	1000	490	331	302	269	214	166	58	35	24	0	29	Spinage (1972)
Tregelephus scriptus	1000	691	589	479	389	316	234	145	93	55	0	40	Allsopp (1971)
Phacochoerus aethiopicus	1000	436	245	66	40	35	28	20	13	10	0	19	Spinage (1972)
Mean	1000	564	466	402	328	252	188	123	60	26.2		34	

Note: Survivorship is expressed as the number remaining from the initial cohort of 1000 at a given percentage of lifespan. Life expectation at birth (e_o) is given as a percentage of total life span in the final column.

that the patterns of mortality are similar, irrespective of the mode of death (Caughley 1966). The implication is that no corrections are needed in comparing the age at death of various species, apart from the size factor, or at least that such variations as do exist among larger mammals are impossible to generalize about. With invertebrates in particular,

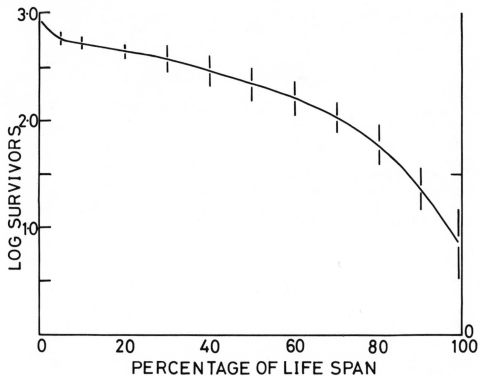

Figure 3.3. Generalized survivorship of various African large herbivores (table 3.3).
Vertical lines are standard errors.

a Type II or Type III curve is more common, and a smaller fraction of the total population
can be expected to fossilize since more animals will die when they are young compared to
species with a Type I curve.

Other biotic factors that affect the composition of a bone collection prior to
fossilization are predator-to-prey ratios, the season of death, and an animal's ranging
patterns.

The higher the predator-to-prey ratio in the living population, the less likely
are bones to survive. It follows that the rate and type of death will influence predator-
to-prey ratios. The analysis can be reduced to a supply and demand approach, where the
greater the supply of carcasses relative to the predator demand, the greater will be
the chances of bones being fossilized, ignoring durability. In much the same way that
there is a lag between the population increase of predators and that of their prey
(Huffaker 1958), there will also be a delay between an increased supply of carcasses and
a carnivore demand sufficient to balance it. In general, the more peaked and seasonal the
death rate, the less the carnivores will be able to consume the carcasses before they dry
and become unpalatable. Shipman (1975) suggested that unpredictable events such as
droughts and floods produce an oversupply of carcasses so that the chances of fossilization are increased. However, such events are infrequent and may in fact contribute substantially fewer carcasses to the fossil record than accumulations due to normal annual
mortality. For example, droughts in Nairobi Park that led to observable deaths through
malnutrition occurred more than a decade apart, in 1961 and 1973/74. The latter drought,
for which data were recorded (Hillman and Hillman 1977), resulted in a total mortality in

wildebeest of only 1.34 times the annual average of 27.7%. One must therefore conclude that, over a ten-year period, the noncatastrophic mortality was greater than catastrophic deaths by nearly an order of magnitude. This is not atypical, to judge from records available elsewhere in East Africa. Normal mortality will in most cases probably overcompensate for the surfeit of bones that are left untouched by carnivores after catastrophic events.

For herbivores with highly mobile foraging patterns it is difficult for predators to keep up and to exert the pressure they would on less mobile prey (Kruuk 1972). Coupled with the greater saturation due to a highly seasonal birth and death rate, there will be a distinct bias to oversampling grazers which tend to be both mobile and seasonal breeders (Western 1975).

Detailed discussions on the changes in bone distribution that occur after death can be found in Behrensmeyer (1975) and Hill (1975).

Sampling Biases in Fossilization and Recovery

Biases in the site of fossilization are likely to be due in part to the association of species with sedimentary environments. Many of the volcanic deposits presently yielding East African fossil material are likely to be biased towards grassland habitats since, even today, extensive grasslands, apart from floodplains, are associated with volcanic soils, e.g., Serengeti, Amboseli, the Athi Plains, and the Eastern Rift Valley.

In recovering fossils, a large number of biases are obvious. The same search image that occurs in counting animals applies in locating fossils—especially the bias in counting large bones and large concentrations. Sedimentary environments are selected because of the probability of finding fossils and the high return on investment. The resultant bias is in favor of concentrations of fossils and unusual mortality events. There is also an enormous predilection for hominid material, introducing a clear choice of fossils associated with environments favored by early man. Search effort is additionally concentrated where the rate of erosion is high, to minimize excavation. High erosion rates are largely confined to low rainfall zones (Langbein and Schumm 1958), and even though climates have changed through time, they are likely to have done so in a spatially correlated manner. Sampling bias would therefore seem to be in favor of lower rainfall areas.

An appreciation of the sorts of sampling bias illustrated above is needed, and an examination of contemporary environments should help in an interpretation of what situations are underrepresented.

Interpretation

An analysis of the biases associated with a collection of fossils should be central to the process of interpreting past environments. Comparisons of past and present faunas without allowance for the more obvious sampling errors will give dubious results at best. I have suggested that various biases can be recognized, particularly those relating to size, and will give a few examples of how they can be utilized in interpreting the ecology of past faunas.

A major dichotomy arises from fossils preserved at the site of death (autochthonous) compared to those transported prior to preservation (allochthonous). In the latter case it is difficult to know the area from which the fossils were derived since small bones in particular can be transported considerable distances (Hill 1975). Behrensmeyer (1975) has suggested that most emphasis be placed on autochthonous faunas as the least biased samples of the original community. Most of the following remarks are concerned with this category of fossils.

There will be a difference in the age structure of a fossil assemblage, depending on whether animals died ordinarily or catastrophically. A catastrophic event will reveal a life table that reflects the living population (progressively fewer animals in each age class) provided the event was widespread and instantaneous. Noncatastrophic agents of death will typically show a U-shaped distribution of age classes. Bearing in mind the small sample of the original population that stochastic events such as droughts and floods account for over time, the data produced can be extremely useful. Where the event was pervasive and followed by rapid burial, a range of different-sized animals will be preserved, giving an insight into size frequency of species in a community that would not be readily obtainable from other modes of death. Even though young animals are seldom preserved (Olson 1957), the size frequency of the older age classes should still be helpful in establishing whether the cause of death in a species was unselective of age—e.g., a widespread catastrophic event.

I have suggested that a size-specific correction factor of the sort given in figure 3.2 must be applicable to fossil assemblages. However, because the bones of small animals of, say, 15 kg have such a slight chance of fossilization, comparisons of fauna from different areas can best be made by restricting the analysis to larger animals, perhaps of 100 kg and larger. Such decisions must be based on the relative preservation rates observed for each collection, which can be gauged from the age or size composition of each species. Restricting the comparison to animals above a certain size is no more than ecologists routinely do in looking at the diversity of communities; an entomologist may look only at the diversity of moths (Williams 1964), while a mammalogist may restrict his analysis of diversity to animals of, say, 15 kg and larger (Stanley-Price 1974). Provided the cutoff point is the same in faunas being compared, the relative proportions can be very revealing.

The size frequency of animals above a selected minimum weight may relate to the productivity and mass of vegetation. Data are only now being assembled on this. Figure 3.4 shows a summary of data I have collected for Amboseli from which the distribution of animal weights has been plotted against the standing mass of vegetation on which they preferentially feed. The selection of denser vegetation by larger animals is apparent. The relationship is perhaps simply derived from the fact that there is a minimum quantity of food per unit area needed to support a species of a given body weight. Data on these aspects should soon be available for a variety of East African mammals.

It is widely held as a theoretical premise that stable environments favor animals with a low birth rate (K-strategists), while variable environments favor species with a high birth rate (r-strategists)(MacArthur and Wilson 1967; Pianka 1970). Animals in variable environments are typically colonizers which reproduce rapidly following a sudden increase in resource availability. It is argued that, with stability of environment, more effort will be put into maintenance and competition and less into reproduction since the uncertainty is less and pressure for limited resources greater (Pianka 1974). Reference to figure 3.2 suggests that, in mammals at least, most of the differences in reproductive performance can be accounted for on the basis of body size and therefore of metabolic rate. Most species seem to be reproducing at a rate which is a constant function of their maximum potential. If this is so, then it should follow that more stable environments will favor large-bodied animals which have lower birth rates and comparatively more energy contained in standing mass.

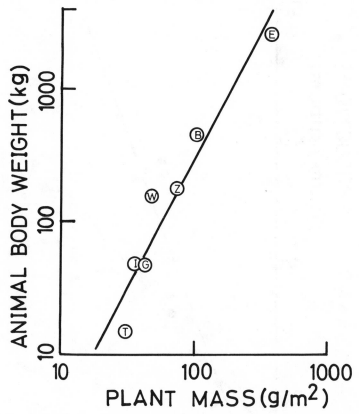

Figure 3.4. Relationship of an animal's body weight to the average vegetation mass on which it feeds. Data from the Amboseli ecosystem (Western 1973). E = elephant; B = buffalo, Z = zebra; W = wildebeest; Gr = Grant's gazelle; I = impala; T = Thomson's gazelle. Log y = 1.48 + 1.96 log x, r = 0.961.

I have assembled data to test this hypothesis for both plants and animals and have assumed that a gradient of low to high rainfall will provide a trend of decreasing climatic (and therefore resource) variability, a relationship that has been found to apply fairly universally (Hershfield 1962). It has been recognized for some time that primary production increases with rainfall (Whittaker 1970; Walter 1973), and Coe et al. (1976) have recently shown that a similar relationship exists between large-mammal biomass and rainfall. Data on biomass and production for both plants and animals is plotted against rainfall (figs. 3.5 and 3.6), and for both, the ratio of mass to production increases with total precipitation. It seems evident that with more stable conditions and greater biological productivity there is a shift to larger-sized organisms which have a lower maintenance cost per unit body weight. In the case of plants, most of the mass becomes fixed in large forest trees at higher levels of rainfall, while in large mammals there is a shift to large-bodied species such as elephant, hippopotamus, and buffalo. If more refined data can validate this point, there is every reason to suspect that an analysis of size-frequency distributions within taxa will prove equally profitable in interpreting past climates, vegetation, and successional changes.

Body size and reproductive rates may, for example, be relevant to an interpretation

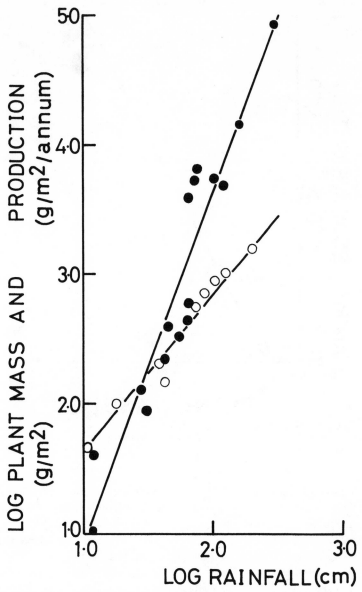

Figure 3.5. Relationship of plant biomass and production to rainfall. Biomass data from Langbein and Schumm (1958), production data from Whittaker (1970). Regression equation from biomass to rainfall, log y = -1.80 + 2.72 log x, r = 0.944. Regression equation for production to rainfall, log y = 0.44 + 1.21 log x, r = 0.0978.

of phyletic size increase (Newell 1949), the tendency in a large number of taxa of increasing body size in the course of phylogeny (Cope 1885). The converse tendency of dwarfism is found, but it is not as common. From a correlation of body size and reproductive rates (fig. 3.2), it might be expected that communities will show a shift to larger-bodied animals (K-strategists) with a moistening of climate. With a transition to a drier climatic regime, large-bodied animals will be at a competitive disadvantage compared to small-bodied species which have a higher reproductive rate (r-strategists) and are

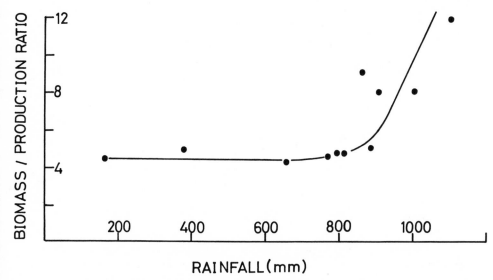

Figure 3.6.   Relationship of large mammal biomass/production to rainfall.   Data from Coe et al. (1976).   The points represent values for various African large mammal ecosystems, and those included are the least affected by recent human impact.

selected for in unstable environments. Both species diversity and niche breadth (Cody 1974) are likely to be lower in drier, less stable environments. The combined effects would lead to an exclusion of various species, particularly the slow-reproducing, large-bodied species. They would be unable to change their reproductive rates rapidly compared with small-bodied species with short generation times. In effect the reduction of body size is a move against a competition gradient since there are many more small than large animal species (Pianka 1970).

An analysis of size-frequency distributions in various taxa offers other possible interpretations of animal and plant communities, past and present. Size is a key variable that influences age at maturity, longevity, mean generation time, birth rate, and the potential rate of increase (Fenchel 1974; Western, unpublished). Each can be related to the metabolic rate, an outcome of the size of an animal through its heat exchange with the environment (Kleiber 1961). The energy expenditure on reproduction relates closely to an animal's metabolic rate, and most of the variation between species will revolve around when and how to make the investment to ensure maximum survival of young. Size will predominantly influence an animal's reproductive characteristics and potential, and individual strategies will influence the way in which that energy is dispersed.

It is surprising, therefore, that so little attention has been paid to the significance of size to environmental production and variability. A shift in body size could account for most of the r and K correlates listed by Pianka (1974), and similarly the ecological and energetic characteristics found in succession communities (Odum 1969) and maturing ecosystems (Margalef 1968). Succession and maturation are perhaps largely a consequence of changes in size-frequency distributions and the associated energetic properties.

Bakker (1975) has shown how size and metabolic rate can be used to interpret much about the ecology of past communities. Much more could be done to examine the significance of changes in animal size through time, an approach that could be complemented by spatial studies on contemporary communities. Data such as those of Goulden (1966) and Tsukada (1967) on lake microfossils have already contributed significantly to testing the hypothesis that changes in environmental stability will be reflected in changes in species diversity.

Further generalizations may be possible, but would need testing on contemporary communities. It has been shown that the range used by an animal is proportional to its size when a larger number of species are considered (MacNab 1963). It is also likely that mobile animals will show a greater departure in the distribution of the living and the dead than less mobile species. In this event small-bodied species may be a better habitat guide than large species, at least in autochthonous faunas. Water-dependent species may be preferentially preserved because of the high predation rate around water holes, which are also good sites for fossilization. Finally, generalist feeders may show a greater variability in the site of death than specialist feeders, and may therefore be less useful as site-specific indicators.

The few examples given are only intended to illustrate various ways of linking the ecology of past faunas with interpretations that can be made from contemporary communities. However, ecology can greatly benefit from paleoecology, and will profit increasingly as the methodological sophistication of taphonomy refines our knowledge of the processes and biases that occur between an animal's death and its recovery as a fossil.

Acknowledgments

I am grateful to Kay Behrensmeyer, Andrew Hill, Galen Rathbun, Jim Allaway, and Nancy Hurxthal for the various comments and suggestions that they contributed.

4. THE ROLE OF MODERN ECOLOGICAL STUDIES IN THE RECONSTRUCTION OF

PALEOENVIRONMENTS IN SUB-SAHARAN AFRICA

Malcolm Coe

Introduction

When examining a contemporary environment in sub-Saharan Africa we may experience great difficulties in interpreting what we observe in functional ecological terms. This is not in the least surprising, for we are dealing with an incredibly complex system comprising a variety of biotic and abiotic parameters, each of which may influence the other to a greater or lesser degree.

Since the time when biologists first began to recognize that a group of living organisms interacted with their environment as a dynamic entity which they called an ecosystem (Tansley 1935), ecologists have been looking for common denominators that will tell them what a particular environment is capable of sustaining in terms of species composition. Thus a community is comprised of floral and faunal elements that have evolved by processes of natural selection so that each may fulfill a particular role within the system. Every one of these biotic units has its own life-style which will ensure that its full reproductive potential is realized in terms of the number of its genes that may be passed on to the next generation.

Wallace (1878) was aware of the fact that animal and plant species tended to be more abundant and varied in the tropics than in other parts of the world. Clearly, the reasons for this greater diversity are complex and may involve the time available for species to evolve, the heterogeneity of the environment, and biotic factors such as competition, predation, and productivity (Pianka 1966).

If we examine the mammalian fauna existing in sub-Saharan Africa at the present, we find that species or groups of species are characteristic of particular environments in terms of both their structure, physiology, and behavior. The ungulates of the rain forest are dominated by the family Cephalophinae (the duikers), which are characterized by relatively small size, rounded backs, thick, rather coarse fur, backward-sloping horns in the males, and generally small family units. By contrast, the Antilopinae (the antelopes) are larger, slender-limbed, mobile animals, with more erect horns, whose behavior is often characterized by a more complex herd structure. Recently, Jarman (1974) has drawn attention to the way in which the behavior of the ungulates may be viewed in relation to environmental factors. He has drawn attention to the fact that size is related to feeding strategy and that particular families are characteristic of particular vegetation types.

The African Mammal Fauna

The mammalian fauna of sub-Saharan Africa represents the richest and most diverse assemblage of mammal species existing in any of the main zoogeographical regions. To a large degree we may divide these species into three broad groups that have either narrow, intermediate, or broad habitat requirements. The factors which determine an animal's habitat range are many and varied, but we can make a significant primary division on the basis of food requirements. Insectivores (e.g., shrews) may be observed to have a wide habitat preference, for, although they may be considered specialized mammals, their prey is available from the rain forest to the edge of true desert. However, the comparative lack of seasonality in the forest will favor a greater species range than the more arid regions. The same may be said of the largely insect-eating bats (Microchiroptera), which also have a broad habitat range, with the largest number of species being again found in forest. By contrast, though, the fruit-eating bats (Megachiroptera) are restricted to environments where the availability of their main food is a prime limiting factor to their distribution. Largely arboreal creatures such as the squirrels (Sciuridae) are severely limited by the horizontal and vertical diversity of vegetation, so that woodland and semiarid environments are unlikely to have more than one or two species of lowland bush squirrels and probably a single species of ground squirrel.

Sadly, the fossil record is noticeably lacking in an adequate spectrum of small mammal species which would allow the prediction of conditions pertaining in paleoenvironments. These circumstances leave the paleoecologist little alternative but to use the species composition of larger mammals to interpret these remains in a climate-habitat context.

Before we examine the evidence available from the fossil record, we should consider the factors that govern the distribution and abundance of modern large mammal species. In terms of the trophic levels they occupy, primary consumers will always be more abundant than secondary consumers. I shall restrict my comments to the large African herbivores.

When an ecologist attempts to study a community of large mammals, one of the first aims is usually to determine by what means a diverse range of species can coexist in the same area. Such coexistence may be at the species level (so-called sympatric species) or the family level. In order to carry out such a study it is necessary first to carry out an accurate census to determine the number of individuals of each species present and to record at the same time the habitats in which they are located. Clearly, in regions with marked seasons such studies need to take into account the whole seasonal range of each species.

The first attempt in East Africa to demonstrate habitat separation in large ungulate populations was that of Lamprey (1963), working in the Tarangire Game Reserve in Tanzania. This study illustrated the importance of wet- and dry-season movements of these species, and their differing degrees of water dependence. Since this time more sophisticated attempts have been made to more precisely delimit species habitat requirements in both spatial and temporal terms (Western 1973; Bell 1970; Sinclair 1974).

In considering the woodland and grassland environments of East Africa, which have for a long period probably been the predominant vegetation types over much of the area where fossiliferous deposits have been studied in detail (Livingstone 1971), we now have an adequate knowledge of the ecology of most major ungulate groups to be able to look at past species assemblages from an ecological viewpoint.

The ungulate species occurring in modern habitats may be divided into two broad groups, those feeding mainly on grass (grazers) and those feeding mainly on woody material (browsers). Clearly, there are a number of species which fall between these two groups, but to a large degree the buccal and internal anatomy of each group is quite distinctive. From the paleontologist's viewpoint it is the jaws of mammals that are most valuable for making inferences regarding the dietary habits of fossil species and hence, perhaps, the percentage of browse and grass available at the time. Since the amounts of silica and other hard materials differ between grass and woody vegetation, micro-patterns of wear on fossil teeth might be used to distinguish fossil grazing and browsing elements of the fauna (see Walker, this volume). Vrba (this volume) has discussed the importance of bovid morphology and has shown how valuable these features may be in predicting the diet and hence the environment type occupied by paleofaunas.

It is, however, necessary here to add a note of caution, for while dental characters may indicate a species' position in the grazing-browsing habit of herbivorous mammals, they do not indicate any more than a broad association with habitats dominated by grass or woody vegetation. The differences between browse-dominated habitats in East Africa and Central Africa at the present time are a striking indication of the caution that is required in such attempts at environmental prediction.

The Climate and Food Availability

Climatologists have explored the relationship between rainfall and temperature and have described a series of climatic regions, each of which may be expected to sustain a characteristic vegetation type. Since it is to be expected that climatic variations, especially in precipitation, affect the production of plant material, it might also be predicted that the same factors have a direct influence on the carrying capacity of an ecosystem.

Walter (1954) and Whittaker (1970) have demonstrated that rainfall may bear a predictable relationship to primary production, while Rosenzweig (1968) has described a similar predictive correlation between evapotranspiration and primary production. Workers in East Africa have demonstrated an association between annual rainfall and large African herbivore biomass (Watson 1972; Leuthold and Leuthold 1973; Sinclair 1974; Western 1975). In a detailed study of large African herbivores Coe, Cumming, and Phillipson (1976) have shown that for areas receiving rainfall of less than 700 mm y^{-1} (per year), a high degree of correlation exists between wildlife standing crop biomass, mean annual precipitation, and predicted above-ground primary production.

Clearly, if it were possible to predict the standing-crop biomass of a fossil assemblage of large herbivores, this should at the same time enable us to make valid inferences regarding the climate and vegetation at that time. The chances, however, of a fossil mammal sample being truly representative of the standing-crop biomass at some point in the past are extremely remote, and this means that assumptions drawn from such assemblages regarding either climate or vegetation would be fraught with difficulty.

We might here ask ourselves if there are any features of modern large herbivore faunas that might prove to be a useful indicator of environmental conditions pertaining to faunas in the past. It has already been pointed out that certain groups of large mammals characteristically perform a grazing or browsing role. Clearly, since the percentages of grass or woody materials will differ between different climate and vegetation zones, it should be possible to make broad environmental predictions even when faced with a sample of modern mammals whose collection site is not known.

Mortality in Modern Ecosystems

When we study the population dynamics of a large-mammal community we are basically concerned with natality and mortality. Although at the same time we must take into account immigration to and emigration from the population under study, it is reasonable to presume that in a natural system considered over a long period of time these two factors should be equal. Under ideal conditions we may presume that in a population which is in equilibrium with its surroundings natality will equal mortality. Since, however, the environment will never be entirely constant, the climate (e.g., rainfall), through its direct influence on food availability for herbivores, may influence both birth and death, so that in consequence the population size will also vary. Additionally, since the requirements of each species also vary, the degree to which each will be affected by such environmental fluctuations will also vary so that the relative species composition may change. Under some circumstances such changes may be cyclical.

Many attempts to demonstrate short-term climatic cyclicity have met with limited success, but Phillipson (1975), in a study of rainfall in the Tsavo National Park, Kenya, finds strong evidence that drought may occur in these regions of Eastern Kenya on 5, 10, and 40 to 50 year cycles, the last-named being the period at which severe drought conditions might be expected. Cobb (1976) has also analyzed the Tsavo rainfall and has concluded that the strongest cycles occur at 5 and 35 to 40 year periods, and suggests that an apparent 10 year rhythm may be a cumulative effect from 5 year peaks.

The degree of aridity of an area must itself have an important influence on the biomass that the area can carry. Clearly, short periods of high precipitation will have a more dramatic effect in an arid region than they will in forest, even though the amplitude of the climatic oscillations may be the same. Over two-thirds of the land surface of Africa experiences periods of drought, and whether they are cyclic or not will place a potential nutritional stress on the herbivore population which may result, depending on its relative severity, in mass mortality. Phillipson (1975) has shown that for the elephant population of Tsavo, the danger point comes when precipitation can be expected to produce above-ground primary production of less than 200 g $m^{-2} y^{-1}$.

We must remember, however, that the effect of such periodic if not cyclic conditions on individual species will differ greatly. Water-dependent species such as water buck, buffalo, and zebra will lose condition during a drought much faster than oryx and Grant's gazelle, which are virtually independent of water. In addition, the size of the animal may also have an important influence on potential mortality, in open woodland and grassland habitats, due to the inability of the larger species to seek shade in the face of thermoregulatory stress. Indeed, this factor alone may be an important reason why open habitats tend to support animals with smaller body sizes.

In the semiarid regions of Africa where drought is responsible for large-scale mortality in wild and domestic mammals, we can also see that large animals will be affected more severly than small animals. If we consider that the generation time of an animal is related to size, we can see that creatures such as elephants with slow development to puberty and long generation times will be quite incapable of taking advantage of short-term periods of climatically induced abundance. This is especially true in arid regions which are characterized by variable climatic regimes. Additionally, of course, following large-scale mortality such populations can be expected to recover more slowly. Phillipson (1973), in constructing a tentative energy-flow model for Serengeti, has pointed out that small mammals and invertebrates can be expected to take advantage of such periods of abundance by rapidly increasing their numbers.

Comparatively few really good life tables are available for large African herbivores, but it is possible to use estimates of secondary production to estimate the rate of "turnover" of such populations. Such estimates can be made by using the limited data on production and biomass that are available in the literature for different size classes of these mammals. Production to biomass (P/B) ratios have been calculated by dividing communities into large (>800 kg), intermediate (100-750 kg), and small (5-90 kg) herbivores (Coe, Cumming, and Phillipson 1976). P/B ratios of 0.05, 0.20, and 0.35 were obtained for these three size classes, indicating the rough rate of turnover that might be expected and hence the number of mammals that might be expected to be potentially available for fossilization. It is therefore quite clear in this context that the smaller species represented by these P/B ratios would be more numerous in a contemporary bone assemblage than large species.

During any one year a varying percentage of each species' population may be expected to die and pass to the decomposer chain either via predators, scavengers, or invertebrates and microorganisms. Foster and Coe (1968) calculated that up to 15.5% of the total biomass of large herbivores in the Nairobi National Park might be expected to be removed by lions and cheetahs each year. Under these circumstances the predators, together with the scavengers (vultures, hyenas and jackals), consume and/or scatter most of the bones, except for the skulls belonging to the larger species. Thus this steady turnover might provide little material that would be available for fossilization in a complete state, so we have to ask ourselves under what circumstances such material might be produced.

It has become clear in recent years that catastrophic death of large numbers of mammals is not as unusual a phenomenon as had often been supposed. The great migrations of wildebeest across the Serengeti plains of Tanzania have been shown to be characterized by massive mortality of young animals through trampling and drowning, especially at river crossings (Talbot and Talbot 1963; Watson 1967). It would not seem unreasonable to presume that many of these bodies could be buried quite rapidly in silt, although it is clear that such deaths would provide material quite unrepresentative of the size classes available in the population at large.

Drought conditions have also led to massive mortality of large herbivores in Kenya since 1960. The population of the Nairobi National Park was halved in 1961 due to drought and floods (Foster and Coe 1968). Later in 1973-74 the population of the same area, together with the adjacent Athi Kapiti plains, suffered a 75% reduction in numbers due to drought (Casebeer and Mbai 1974). Under such circumstances of catastrophic death the amount of carrion available completely swamps the scavengers, which in consequence leave a large percentage of the carcasses to be decomposed by invertebrates and microorganisms. At such times a large number of entire skeletons could become fossilized if they were buried. In the instances noted above the carcasses were well distributed over the plains, with large aggregations of material around water holes and other water sources.

In 1960-61 the drought that affected areas in central Kenya also affected the Tsavo East National Park, where up to 300 rhinoceroses died along the Tsavo River. A later and more severe drought occurred in Tsavo in 1971-72, and on this occasion the main area influenced was quite different. As a result of these conditions up to 7000 elephants died and, as in the Nairobi National Park, the majority of carrion produced was virtually unaffected by scavengers. Corfield (1974) has studied this mortality and has demonstrated that the main area affected coincided with the 10" (25 mm): 10% rainfall probability isohyet.

Additionally, this author has shown that a large percentage of the carcasses were located along the permanent water courses of the Galana and Tiva rivers. It would again seem probable that such a catastrophic event could potentially provide material for fossilization. It is interesting here to note that Phillipson (1975) has shown that the area in which maximum mortality occurred coincided with that in which we would expect to find insufficient food to sustain the population.

The Decomposition of Carrion

During the course of this (1970-71) drought the author (Coe 1976) located and observed the decomposition of elephant carcasses. Four bodies were located and observed in detail while up to 100 others have been the subject of irregular observation from the time of death to the present (fig. 4.1). It was found that, in the absence of scavengers, a large carcass would pass through all the wet stages of decomposition (that due to the loss of soft tissues) in two weeks, while in the wet season a further 21 days would see the skin and sinews removed by invertebrates (*Dermestes* beetles) at the astonishing rate of 8 kg per day. During the dry season the skin remains intact for many months since the lack of water acts as a severe limiting factor to the activity of *Dermestes* beetles.

The fate of the skeletal remains is of great significance to the paleontologist, and it is surprising that comparatively few detailed studies have ever been carried out on this important if aesthetically distasteful topic. In open habitats in Tsavo where the surfaces of bones may be expected to experience a diurnal temperature range of at least 35-40° C, they begin to flake and crack within five weeks of death. These changes may be slightly delayed during the dry season when the skin remains intact for an extended period. It is only, however, for the larger mammals (e.g., elephant, buffalo, rhino, and giraffe) which possess a particularly thick and keratinous skin that this covering will afford such protection, and even then it is unlikely to do so for more than a few months. In addition to the bones, the teeth will also develop vertical cracks and the outer layers will begin to flake within the same time scale.

Considerable microclimatic amelioration can occur even in sparse thickets compared with open situations. Therefore, animals which die in riverine habitats and thick scrub may be expected to suffer slower exfoliation of bones and teeth than those in the open. Temperature measurements made by the author in Tsavo and South Turkana have shown that the diurnal range recorded at ground level in thickets may be less than 20° C, compared with twice this figure in the open less than 2 meters away.

During observations of the decomposition of elephant carcasses in Tsavo, the majority of the bodies were located in the open on red lateritic soils where exfoliation was apparent within five weeks. A single body was located on porous gravel close to the Aruba Dam where during rain the nitrogenous putrefaction liquids were rapidly leached away. In this situation rhizomatous grasses rapidly recolonized the surface of the ground, and within two months of death the whole skeleton was covered with vegetation except for a small portion of the skull. Subsequent observation of this carcass has demonstrated that even a shallow covering of this type is sufficient to reduce the degree of bone flaking. In addition to the creation of a less extreme microclimate around the skeleton, the accumulation of litter at the soil surface completely buried the smaller vertebrae, ribs, and other small bones within two years (fig. 4.2). In this situation bony material was stained brown by humic materials and was consequently less bleached than that in the open. Such vegetation cover is not permanent, though, and a further severe drought in 1975 once more exposed all the larger skeletal remains.

Figure 4.1*a* (top). Elephant corpse three days after death.
Figure 4.1*b* (bottom). Elephant corpse three weeks after death.

When such large amounts of carrion are produced and initial scavenger attack thus
reduced, it might be expected that the skeleton would remain intact. Elephant carcasses
observed over five years were scattered up to 50 meters from the site of death by trampling, scavenger attack, and elephants, which have been shown by Douglas-Hamilton (1972) and
Trevor (personal communication) to carry tusks and bones for considerable distances.
The observation of such large-scale mortality poses the interesting problem of how

Figure 4.2*a* (top). Skeletal remains of elephant carcass one year after death.
Figure 4.2*b* (bottom). Elephant skeletal material partly buried two years after death.
Note the fissuring and exfoliation of the surface.

high the chances are of any of this material being fossilized. In the case of the massive
elephant mortality in Tsavo it appears unlikely that, except for bodies being buried in
gullies by subsequent rainfall or being transported and redeposited in silt by one of the
main rivers, much if any of this material will be mineralized. The rapidity with which
skeletal remains are fragmented by heating and cooling suggests that fossilized material
which is recovered without appreciable signs of weathering must have been rapidly buried

after death either as a result of rainfall, eolian deposits, or other catastrophic
phenomena such as volcanic ash. Indeed, since damage is much delayed under conditions of
microclimate amelioration, observations of the degree of weathering of fossil bones
might be used as a measure of the degree of ground cover by woody vegetation pertaining at
the time of death or the speed with which the remains were interred. For material that
can be located and examined "in place," a detailed study of the surrounding soil structure
might well also indicate the conditions under which the carcass was buried, a field now
being explored by the taphonomist.

Reconstructing the Paleoenvironment

The biggest question we have to ask in relation to reconstructing paleoecology
is to what degree a fossil assemblage of large herbivores reflects the structure of the
community at large at that time. Additionally we must also ask if it is possible or permis-
sible to extrapolate such information to predict the environmental conditions under
which the large mammal community existed.

If we take a contemporary large herbivore community, we immediately observe that
the turnover time of each species bears a relationship to the animals' size, so that in
any one time period we might expect larger numbers of small animals to be potentially
available for fossilization than the larger forms. Smaller species and the juveniles of
larger species will be broken up and fragmented much faster than the remains of large
species, though it must be admitted that they might at the same time have a greater chance
of being interred. The decomposition of an aardwolf (*Proteles cristatus*) carcass was observed in November 1975 in Tsavo East National Park, and the soft tissue from this
body was completely removed by dipterous larvae in 3 days, leaving an intact skeleton and
hair remaining on the soil surface. Torrential rain 2 days later partly buried the long
bones, while the smaller structures were washed down a gentle slope up to 2 feet (61 cm)
from the site of death. Bourn and Coe (1977), working on the giant tortoises (*Testudo
(Geochelone) gigantea*) of the Aldabra Atoll in the Indian Ocean, have noted that the
carapace remains of small individuals are fragmented very rapidly so that estimates of
age-specific mortality from skeletal remains are greatly biased in favor of the more
persistent remains of adult animals. Such considerations are, however, only pertinent to
the student of population dynamics who wishes to obtain data relevant to the age structure
of a species population, and are not of great importance to the assessment of the species
structure of a community. Since this information tells us that smaller specimens of a
single species will be underrepresented in a sample, it clearly follows that small species
are likely to be subject to a similar bias in attempting to interpret a fossil assemblage
in terms of a whole community. While such problems must be recognized, the interpretation
of community structure and related environmental correlates can be carried out by limiting
the observations to the material falling in the larger size classes.

It has already been noted that water-independent species are not likely to be affected by even fairly severe drought conditions so that they would not tend to be commonly
included in fossil material derived from catastrophic climatic phenomena. Unless these
species are considered to be particularly indicative of a specific climatic-vegetation
zone, their exclusion may not be serious. When, however, we consider that mammals such
as gerenuk (*Litocranius walleri* Brook) and the lesser kudu (*Tragelaphus imberberis* Blyth)
are good indicators of *Acacia—Terminalia—Commiphora* "Nyika" habitats, their absence from
an assemblage might well be serious. Clearly, we must remember that in dealing with

paleoenvironments the degree of discrimination that is acceptable must be, of necessity, much less sensitive than that which is possible when attempting to predict the composition of a modern community from samples of contemporary skeletal remains.

I have already indicated that we can predict primary production and the potential carrying capacity of an environment from measures of the integrated rainfall of an ecosystem. If we examine figures derived from a series of East and Central African wildlife areas, we can compare the biomass and density of large mammals in a variety of climatic regions. These data are derived from counts which also indicate the species structure of each community. These different communities of large mammals are characterized not only by a variety of species but a varying spectrum of size classes, as if each habitat has its own particular range of mammals. This is certainly not a new observation, for it has long been recognized that species of the genus *Oryx* are to be found in very open country while sable and roan antelope are more characteristic of dense woody environments. If we accept that the climate largely determines the type of vegetation that exists, then, since each plant community has a definite physiognomy, mammal species of particular size groups may also be selected as the best herbivore strategy for utilizing it in the most efficient way. One must accept that, since some genera and species have broader environmental requirements than others, detailed studies using environmental association indices should reveal the best indicator groups for comparison with fossil material.

An examination of the data for wildlife areas in East and Central Africa does show different size group percentages in different areas. In order to test these observations the biomass and density figures have been reduced to a mean weight (kg) for each habitat. Thus high figures indicate environments dominated by large animals, while low figures represent those where smaller animals are dominant (table 4.1). In order to test the assumption that such information might be correlated with climate, these mean weights have been plotted against rainfall for each area. Although the results show considerable scatter, they do indicate that a broad categorization of size classes might be used as an indication of the climatic-vegetation type in which the community occurs (fig. 4.3). Western (this volume) has explored the size relationships of African herbivores in a similar way and has underlined their importance in relation to generation time and recruitment.

Using this information for interpreting fossil assemblages would clearly require that the investigator be able to estimate a mean weight for the group under consideration. Within the errors implicit in studying such material, it is quite possible that skull or long bone characteristics could be employed for such estimates so that a broad prediction of the paleoenvironment could be attempted. The sensitivity of such determinations could not and perhaps should not be expected to yield a division into more than three or four broad vegetation types.

The modern ecologist can go into the field and describe or, still better, measure the biota present, while the paleoecologist is left at best with a gross species bias in his samples. Whether or not the material available can assist in interpreting paleoenvironmental conditions will depend on the sensitivity demanded. If we accept that we are dealing with no more than a general picture of conditions in the time period under consideration, we cannot expect anything more than a general answer to our question, however sophisticated the numerical techniques employed may appear. There is indeed a great danger in demanding sensitive interpretations from necessarily insensitive data.

With the assistance of the geologist, geochemist, palynologist, and biochemist as a starting point for paleoecological investigations, the interpretation of fossil

Table 4.1. Biomass, Density, and Mean Weight Data for African Wildlife Ecosystems in Relation to their Prevailing Rainfall Regimes

	Biomass (kg km⁻²)	Density (no km⁻²)	Mean Weight (kg)	Rainfall (mm)	Authority
1. Rwindi Plain, Virunga National Park, Zaire	17,448	34.74	502.25	863	Bourlière and Verschuren 1960
2. Rwenzori National Park, Uganda	19,928	66.37	300.26	1010	Field and Laws 1970
3. Bunyoro North, Uganda	13,261	12.8 (- hippo) 32.0 (+ hippo)	414.00	1150	Laws, Parker, and Johnstone 1975
4. Manyara National Park, Tanzania	19,189	30.71	624.85	915	Watson and Turner 1965
5. Ngorongoro Crater, Tanzania	7,561	46.83	161.46	893	Turner and Watson 1964
6. Lake Nakuru National Park, Kenya	6,688	64.20	104.17	878	Kutilek 1974
7. Amboseli Game Reserve, Kenya	4,848	70.83	68.45	350	Western 1973
8. Lochinvar Ranch, Zambia	7,568	63.17	119.80	813	Robinette 1964
9. Lake Turkana (East), Kenya	405	2.40	160.75	165	Stewart 1963
10. Samburu, Isiolo, Kenya	2,018	25.00	80.72	375	Barkham and Ridding 1974
11. Nairobi National Park, Kenya	4,824	44.29	108.92	844	Foster and Coe 1968
12. Tsavo East National Park, Kenya	4,210	7.88	534.26	553	Leuthold and Leuthold 1973
13. Mkomasi Game Reserve, Tanzania	1,731	9.78	176.99	425	Harris 1972
14. Loliondo Controlled Area, Tanzania	5,423	28.98	187.13	784	Watson, Graham and Parker 1969
15. Serengeti National Park, Tanzania	8,352	134.00	74.00	803	Hendrichs 1970
16. Ruaha National Park, Tanzania	3,909	5.54	705.60	625	Norton-Griffiths 1975
17. Akagera National Park, Rwanda	3,980	32.52	122.39	785	Spinage, Guinness and Eltringham 1972
18. Henderson's Ranch, Rhodesia	2,869	30.17	95.09	406	Dasmann and Mossman 1961
19. Kruger National Park, South Africa	2,383	14.90	159.99	481	Pienaar et al. 1966
20. Willem Pretorius Nature Res., South Africa	3,344	43.43	77.00	520	Borquin 1974
21. Mfolosi Game Reserve, South Africa	4,385	25.06	174.98	650	Mentis 1970

Note: Data after Coe et al. (1976).

assemblages in terms of indicator species of environment type, their size, generation
time, and feeding strategy is likely to prove an invaluable interpretive tool.

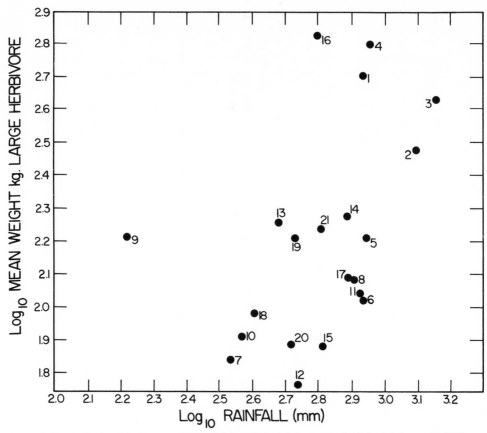

Figure 4.3. Relationship between mean weight (kg) and log rainfall in African wildlife
ecosystems.

Hominids and Paleoenvironments

The extinction of the Pleistocene megafauna has been the subject of considerable
discussion in recent years (Martin 1966, 1973). From the point of view of attempts to
predict paleoenvironments, it must be recognized that, whether it is considered that man
has been responsible for these extinctions in the last 20,000 years or whether climatic
variations have been wholly or partly responsible, we must face the fact that changes in
the structure of mammal communities have been dramatic. During the middle Pleistocene
the mammal fauna of the Narmada Valley of India is said to have comprised no less than
eleven species of *Elephas* (Mukherjee 1974), although this does not necessarily mean that
they were all abundant. In terms of energy flow in an ecosystem it is difficult to see
how a community at this time could have had a higher biomass carrying capacity than we
would expect or predict for the same climate zone today.

It is today extremely difficult to study an ecosystem that has not been influenced
by man either directly or remotely, and it may well be that effects relating to the disturbance of nutrient cycles and allied phenomena often do not indicate a lower carrying
capacity than might be expected under other conditions. At the present time we can ob-

serve arid environments dominated by pastoralists that might be considered as equilibrium systems of which man is merely a part. Yet an examination of the data provided by Coe, Cumming, and Phillipson (1976) for pastoral ecosystems of Northern Kenya indicate that without exception, in all the six areas studied, the domestic livestock biomass, although bearing a significant linear relationship with rainfall (Watson 1972), is all well above the carrying capacity as indicated by wildlife biomass data. Clearly, since man in these systems obtains perhaps 80% of his energy needs from within the ecosystem, the human biomass should also be added to that of the other mammals.

In Southern India ecosystems with sparse vegetation are reminiscent of areas at the same latitude in Africa. Yet when we examine the rainfall data for the Indian location we find that the rainfall is more than twice that which would be received by a similar (at the genera and species level) area in Africa (Coe, personal field observations, 1975). In this case the explanation can largely be accounted for in terms of pressure from domestic animals and more particularly through the continual export of nutrients via the collection of firewood and the utilization of virtually all bovid dung for fuel.

It is often assumed that these dramatic influences are a phenomenon peculiar to modern man, and yet when we look at simple pastoral systems and observe the overexploitation of their surroundings, it is not difficult to imagine that perhaps even the earliest hominids began to have a profound influence on their surroundings as soon as they began to hunt and gather on the plains of Africa. The elephants of Africa (Laws et al. 1975) and the giant tortoises of Aldabra (Bourn and Coe 1976) both have generation times similar to our own and both species show signs of changing (or destroying) their environment. In the past, no doubt, it was the role of the elephant to open up woodland and convert it to grassland before it moved on, yet today there are fewer and fewer places for them to go. To me it seems more than a coincidence that the pastoralists of the Rift Valley have very little oral tradition regarding the people that occupied their country before they did, for perhaps like the elephant man has also come, exploited, overexploited, and then either died or moved on.

Part 3

TAPHONOMY IN RECENT ENVIRONMENTS

There are various observable patterns in fossil and archeological accumulations of bone. Different parts of a skeleton may occur in different abundances, and the bones may be broken or damaged in particular ways. To what extent do these patterns reflect taphonomic processes that are relevant to the reconstruction of paleoecology? In attempting to answer this question a number of workers have examined bones in present environments, where the processes affecting bone accumulation and destruction can be observed in action.

Amboseli National Park in Kenya is ecologically one of the best known areas in sub-Saharan Africa, due primarily to the long-term studies carried out there by D. Western. Behrensmeyer and Dechant Boaz add to this knowledge of the large mammal fauna through their investigation of the skeletal remains that occur in a wide variety of habitats within the park. By comparing what they find in the bone assemblages with what Western has determined for the living ecosystem, they are able to demonstrate how well the potential fossil assemblages represent aspects of the community structure such as species representation and relative abundance. Their study also highlights a number of important taphonomic processes that operate on land surfaces before bones are transported by moving water. The Amboseli ecosystem thus provides an exciting body of data and ideas for comparison with other modern and fossil situations.

Gifford introduces us to the concept and methods of ethnoarcheology, which shares with taphonomy the common goal of reconstructing processes and biases that have affected fossil assemblages. However, ethnoarcheology focuses specifically on humans as agents in the formation of archeological sites, and ways of sorting out their effects from those of all other taphonomic processes. Gifford describes the complexities of a human pastoralist group in Northern Kenya in order to show how the preservable evidence from their culture might be interpreted by a future archeologist. We learn that much of the record may be lost, and that what remains may be strongly biased. Without sufficient understanding of the biasing processes, it is clear that interpretations of early human behavior and ecology can be subject to considerable error. Gifford believes that much can be learned about these processes in modern environments, but she also discusses the importance of caution in using studies of the present to decipher the past. Her work ties in with that of Hill and Klein in presenting guidelines for distinguishing and correctly interpreting evidence for human activity in the fossil record.

Brain gives a clear and instructive example of how careful documentation of modern processes, supplemented by experimental work, can resolve interpretive problems in fossil assemblages. His research and many of its implications are specific to African cave situations, but some of the results and many of the ideas have broader application. In this sense, Brain's paper exemplifies the value of the inductive approach in providing a firm empirical basis for potentially important hypotheses concerning taphonomic processes. Statements such as, "survival of parts will follow an entirely predictable pattern if the destructive forces are known," while based on Brain's analysis of only a few bone assemblages, could be tested in a wide range of cave and other situations. His comments that species lists are not necessarily indicative of the accumulating agent, and that trampling is an important process of bone fragmentation, also point to matters of potentially broad taphonomic significance.

While as yet there are hardly any accepted "rules" in taphonomy, in Hill's paper
we see that there are consistent patterns in the way bones become disarticulated and
damaged during the early stages after death. With more work these patterns may become
recognized in most early post-death situations comparable to the environments described
in the paper. Such information becomes important for interpreting the preburial nature
of a fossil assemblage, and may allow the paleoecologist to distinguish between such things
as attritional and catastrophic death assemblages. Hill also points out that the
evidence for biological activity in the dispersal and destruction of bones represents
another class of information potentially available from the fossil record, namely, information concerning the behavior of organisms that are not themselves preserved. One
of the organisms for which bones may give a considerable amount of paleobehavioral
evidence is man and his ancestors, but perhaps not in the way that previous interpretations of osteodontokeratic cultures in South Africa have suggested.

5. THE RECENT BONES OF AMBOSELI PARK, KENYA, IN RELATION TO

EAST AFRICAN PALEOECOLOGY

Anna K. Behrensmeyer and Dorothy E. Dechant Boaz

Introduction

Many of the problems in paleoecologic interpretation have been defined through
study of fossil bone assemblages. Another approach to the reconstruction of paleocommunities and taphonomic processes lies in the observation of modern ecosystems, where
the kinds of ecological information preserved by organic remains can be precisely defined. This study, carried out in the Amboseli Basin, Kenya, demonstrates the latter
approach. From analysis of a large recent bone assemblage, we have sought to determine
how accurately true species diversity and paleocommunity structure can be reconstructed
from animal remains preserved together in a "potential" fossil assemblage.

Several similar studies have been done on modern marine invertebrate communities,
a notable example being J. E. Warme's "Paleoecological Aspects of a Modern Coastal
Lagoon" (1971), but until recently comparable study of a terrestrial vertebrate bone
assemblage has been deterred by lack of information on the ecologies of areas where
diverse land mammal communities are present and where man's impact on or role in the ecosystem can be determined. Such information has recently been made available by the work
of D. Western (1973) in the Amboseli area, and the results of his study have enabled us
to compare data on living animals with our sample of the modern bone assemblage of the
basin.

Background and Methodology

The aim of the study, initiated in Amboseli Park, Kenya in 1975, was to determine
the effects of natural taphonomic processes on the modern bone sample and to learn which
aspects of the recent living community and environmental interrelationships could be
recognized from the features of this "potential" fossil assemblage. The Amboseli Basin
was chosen for this study for a number of reasons. Movements of species in the larger
vertebrate community have been monitored in a number of different habitats since 1967 as
part of an ecological study by D. Western (1973). The large vertebrate community is
representative of the current plains and woodland ecosystems in East Africa. Many
species of the Amboseli community respond to seasonal fluctuations in resources by migration into and out of the basin, a pattern of behavior which probably reflects long-term
adaptation to East African climates. The basin area is presently a closed drainage
system with alkaline soils conducive to bone preservation. The overall geomorphic and

72

tectonic setting resembles the environment reconstructed for Olduvai Gorge Beds I-II, 1.8-1.5 million years ago (Hay 1976). There are no major rivers in the basin, and the Amboseli bone assemblage serves as a model for "pre- or nontransported" conditions in fossil or recent settings. Finally, there are upper Pleistocene/Holocene fossil deposits (Western and Von Praet 1973) which provide material for comparisons with the recent bone assemblages.

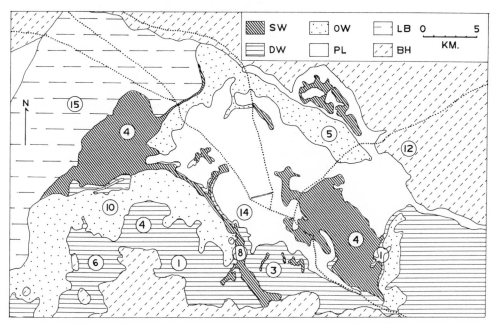

Figure 5.1. Map showing different habitats in the central part of the Amboseli basin, adapted from Western (1973). Circled numbers indicate the number of bone-sampling transects in the habitat areas. Dotted lines represent the major roads. Abbreviations for habitats: SW = swamp, DW = dense woodland (*Acacia tortilis* and *A. xanthophloea*), OW = open woodland, PL = plains (open grassland), LB = lake bed, BH = bush (dense and open).

Six major habitats were selected for study (fig. 5.1, table 5.1). Within each of the habitats, sampling areas were selected, nonrandomly, on the basis of the following criteria: (1) bones relatively common on the surface, (2) ground cover relatively sparse, (3) area accessible to four-wheel drive vehicle. Within each selected area, for all habitats except the swamp, straight-line transects were laid out either east-west or north-south and spaced from .5 to 1.0 km apart. Using compass orientation, the vehicle was driven along the midline of each transect with one observer on top searching for bones with the aid of binoculars. Transect widths were decided according to how well surface bones could be seen on either side of the center line, and this depended on ground-cover density. In open areas transects were 100 m wide (50 m either side of the vehicle) and in more vegetated areas 60 m wide. In most cases transect lengths were determined by the distance required to record twenty individuals, a method which prevented oversampling of any particular area. For the swamp habitat, a 30 m wide strip along the bank was covered mainly on foot. In the open woodland, dense woodland, and bush, much of the area was also covered on foot.

Strict control was kept on the transect width by pacing out bones that lay near the edge of the 50 or 30 m distance limit from the transect center line. To permit later

Table 5.1

Characteristics of the Amboseli Bone Sample

Habitat	Total Area in km^2	Area Sampled (km^2)	Percentage of Habitat Sampled	Total Number of Bones	Estimated Maximum Bones/km^2	Percentage of Bones Buried > 50%
Swamp	60	.98	2	6082	6.21×10^3	6
Dense wood	102	.73	.7	1989	2.72×10^3	4
Open wood	66	1.44	2	2691	1.87×10^3	-
Plains	78	1.60	2	4666	2.92×10^3	5
Lake bed	90	2.66	3	3244	1.22×10^3	5
Bush	204	1.34	.6	1396	1.04×10^3	2
Total	600	8.75	1.46	20068	2.29×10^3 (average)	

Note: The density of bones per square kilometer is the maximum for each habitat since bone sampling areas were chosen initially for their relatively high abundance of surface skeletal remains.

determinations of transect bone densities, only the parts of a carcass which lay inside the boundaries were recorded. Transect lengths were measured by pacing wherever possible, and otherwise from aerial photographs.

Two major categories of data were observed and recorded during the initial field study: (1) the fauna (relative numbers of different species) represented in the surface bone assemblage; (2) the nature of the bone assemblage in terms of skeletal parts represented, damage, rates of weathering and burial, and density of bone debris in the different habitats. All bones seen on the transects were recorded without removing them from the field unless certain identification required comparison with museum collections. "Indeterminate" bones were collected and identified later using the National Museums of Kenya osteology collections in Nairobi. Identification was recorded only when both observers were in agreement.

All information was recorded on standardized data sheets for each bone or carcass encountered on the transects. The following data were taken: (1) taxon (to species if possible); (2) age group (juvenile, subadult, adult); (3) weathering category (the most advanced stage seen in concentrations of bones of one individual); (4) damage (chewing, punctures, destruction by termites, etc.); (5) burial (bones with > 50% of their surfaces covered); (6) whether a single individual was represented by more than one skeletal part; (7) skeletal parts: including teeth present in cranium or mandible, tooth eruption stage; right or left side; proximal, distal, or complete; articulated or separate; and which parts comprised the burial sample.

Fossils encountered on the transects were also noted. General descriptions of soil, ground cover, shade, and live fauna present on each transect were recorded, and measurements of ground cover were taken for open woodland and plain habitats.

Within a transect, minimum numbers of individuals were deduced using several criteria. Differentiating the skeletal parts to taxon served as the primary means of determining minimum numbers. If this failed, or if several individuals of the same taxon were possibly present, conclusions were based on the most common element per individual, i.e., two left femora of the same species were counted as two individuals, but a left and a right, if approximately the same stage of weathering and degree of epiphyseal fusion, were recorded as one individual, even if found at opposite ends of the transect. We attempted not to infer association from proximity, since occasionally the remains of several individuals were found together. Also it seems that scavengers may carry single bones of one carcass to the vicinity of another, where they "exchange," leaving one anomalous bone in a skeletal scatter from a different individual.

We have yet to correct for minimum numbers of individuals over all transects in a habitat, and there is probably some overrepresentation of numbers of individuals in the samples due to the occurrence of bones of the same individual on different transects. With separate transects spaced at least .5 km apart, we assumed that each was independent with respect to minimum numbers of individuals. This assumption is more likely to be valid for small and medium-sized animals than for the largest herbivores. In several instances it was obvious that elephant (*Loxodonta africana*), rhinoceros (*Diceros bicornis*), and giraffe *(Giraffa camelopardalis)* bones from single individuals were spread over several transects. This was corrected for wherever recognized.

In order to determine potential sampling error in recording the bones of smaller animals, short transects were sampled first by the standard method, then resampled by walking along closely spaced traverses. Resampling indicated a definite sampling bias against smaller animals.

Comparison of Faunal Representation in the Bone Assemblage with the Living Community

A faunal list derived from a fossil assemblage gives an impression of the number of different kinds of vertebrates which formed the original community. It is usually assumed that this list is more or less representative of the original community (at least of the large vertebrates), particularly if the fossil sample is large. Using the data from Amboseli, based on a sample of over 20,000 surface and partially buried bones, we can examine how representative a "bone" faunal list is of the number of living species in the environment.

Forty-seven species of wild mammals, weighing at least 1 kg, occur within the central Amboseli Basin (Western, personal communication; Williams 1967)(table 5.2). Five species of domestic animals belonging to the genera *Bos, Capra, Ovis, Equus,* and *Canis,* as well as *Homo sapiens*, also occur in the basin. Table 5.3 compares the number of species represented in the bone assemblage with those represented in the living mammalian community. The total number of taxa (of weight $\geq$ 1 kg) represented in the bone sample is 72%, and in the mandible sample 57%, of the actual number of contemporaneous species. The subsample of species represented by mandibles only has been calculated to simulate what might constitute a typical paleontological sample collected in the classical manner, i.e., one where postcrania were not considered important for taxonomic determination and thus were not kept as part of the collection. A second subsample consisting of bones buried more than 50% gives 43% of the actual number of species.

The number of bones of a species which accumulate in an environment depends initially on the population size and annual death rate of that species. For the same population size, species of small average adult weight produce more offspring and carcasses per year than species of large average adult weight. Consequently, a bone sample should contain a greater number of the bones of small animals than of large animals (if the population sizes are nearly equal), and there is a greater likelihood of finding sekeletal parts representing the total range of small species than of large species in the community.

Table 5.2

Species List for Amboseli National Park

> 15 kg

Loxodonta africana (elephant)
Diceros bicornis (rhinoceros)
Hippopotamus amphibius (hippopotamus)
Giraffa camelopardalis (giraffe)
Taurotragus oryx (eland)
Syncerus caffer (buffalo)
Oryx gazella callotis (fringe-eared oryx)
Equus burchelli (Burchell's zebra)
Connochaetes taurinus albojubatus (white-bearded wildebeest)
Alcelaphus buselaphus cokii (Coke's hartebeest or kongoni)
Kobus ellipsiprymnus (waterbuck)
Tragelaphus imberbis (lesser kudu)
Aepyceros melampus (impala)
Gazella granti (Grant's gazelle)
Tragelephus scriptus (bushbuck)

Redunca redunca (Bohor's reedbuck)
Litocranius walleri (gerenuk)
Gazella thomsoni (Thomson's gazelle)
Phacochoerus aethiopicus (warthog)
Panthera leo (lion)
Crocuta crocuta (spotted hyena)
Hyaena hyaena (striped hyena)
Panthera pardus (leopard)
Acinonyx jubatus (cheetah)
Felis caracal (caracal)
Orycteropus afer (aardvark)
Papio cynocephalus (yellow baboon)
Homo sapiens (man)
Bos taurus (domestic cow)
Equus asinus (donkey)
Ovis aries (domestic sheep)
Capra hircus (domestic goat)

< 15 kg, > 1 kg

Raphicerus campestris (steenbuck)
Rhynchotragus kirki (dik dik)
Felis serval (serval)
Proteles cristata (aardwolf)
Mellivora capensis (ratel)
Canis adustus (common jackal)
Canis mesomelas (black-backed jackal)
Viverra civetta (African civet)
Felis libyca (wild cat)
Otocyon megalotis (bat-eared fox)

Genetta genetta (small-spotted genet)
Atilax paludinosus (marsh mongoose)
Ichneumia albicauda (white-tailed mongoose)
Helogale parvula (dwarf mongoose)
Ictonyx striatus (Zorilla)
Cercopithecus aethiops (vervet)
Galago sensegalensis (bush baby)
Hystrix cristata (porcupine)
Pedetes capensis (spring hare)
Lepus capensis (African hare)
Canis familiaris (domestic dog)

*Skeletal remains found in the bone sample.

Table 5.3

Representation of the Numbers of Living Mammal Species in the Amboseli Basin in the Total Bone Sample

	Number of Living Species	Number of Species in Bone Sample	Number of Species in Mandible Sample	Number of Species in Buried Sample	Number of Species in Hyena Den
Large herbivores Wild ungulates > 15 kg	19	18 (95%)	16 (84%)	13 (68%)	8 (42%)
Small herbivores Wild ungulates (+ lagomorphs, rodents) > 1 kg, < 15 kg	5	3 (60%)	2 (40%)	1 (10%)	1 (10%)
Carnivores > 15 kg	6	6(100%)	5 (83%)	1 (17%)	
Carnivores (including domestic dog) < 15 kg	14	3 (21%)	2 (13%)	1 (6%)	
Primates > 15 kg	2	2(100%)	1 (50%)	2(100%)	
Primates < 15 kg	2	2(100%)		1 (50%)	
Tubulidentates (aardvark, ~ 50 kg)	1				
Domestic ungulates	4	4(100%)	4(100%)	4(100%)	1 (25%)
Totals of above	53	38 (72%)	30 (57%)	23 (43%)	13 (25%)

Note: Mammals < 1 kg (bats, small rodents, and insectivores) are not included. Numbers of living species modified from faunal list of Williams (1967) with the aid of D. Western (personal communication). Weight categories based on average adult weights.

In Amboseli, the bone sample gives a more accurate representation of the true number of large species than of small species living in the basin, contrary to the theory stated above. Large wild herbivore and carnivore species (> 15 kg) have, respectively, a 95% and 100% specific representation in the bone sample, while small herbivores and carnivores have a 60% and 21% representation (table 5.3).

The bias against small animals can be examined in greater detail by plotting relative representation at different body weights. The cumulative graph given in figure 5.2 shows that the number of living species versus the number represented by the bone sample

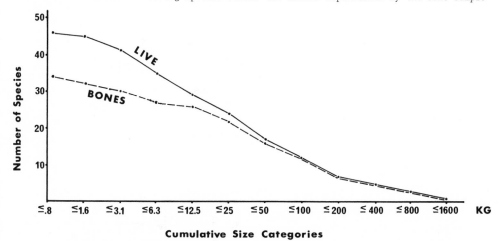

Cumulative Size Categories

Figure 5.2. Cumulative plot of numbers of live species (wild) and those represented by bones in the Amboseli surface assemblage versus body weight categories. The bone assemblage gives accurate representation of extant species numbers for > 100 kg body weight. The bone sample consists of over 20,000 identifiable pieces or approximately 1200 carcasses (minimum numbers of individuals).

converges at body weights between 50 and 100 kg. For animals $\geq$ 100 kg, the Amboseli bone sample accurately represents the numbers of living species, while the percentage representation in smaller size categories fluctuates widely. Similar size biases may hold for species representation in large fossil assemblages which have been altered before burial by processes similar to those in Amboseli. The nature of these processes will be discussed in the next section.

Ungulate Population Structure

The bone assemblage reflects, generally, the relative numbers of the ten most abundant large wild herbivores as shown in figure 5.3 and table 5.4A and 5.4B. The lower histogram in figure 5.3 shows the average live community structure in terms of the relative abundance of the major species. This has been derived from censuses of the living animals over a total of six years (air and ground counts every one to two months; Western 1973 and personal communication). The middle histogram shows the relative numbers of carcasses expected per year from the herbivore populations, based on known annual turnover rates (Western, this volume) and assuming stable populations over the eight-year period. The upper histogram shows the actual relative abundance of the herbivore species in the bone sample based on minimum numbers of individuals.

Herbivore representation can be expressed as the observed carcass frequency (F_S) divided by the expected carcass frequency (F_E), and this can in turn be related to body size as shown in figure 5.4. It is apparent from figures 5.3 and 5.4 that in the weight

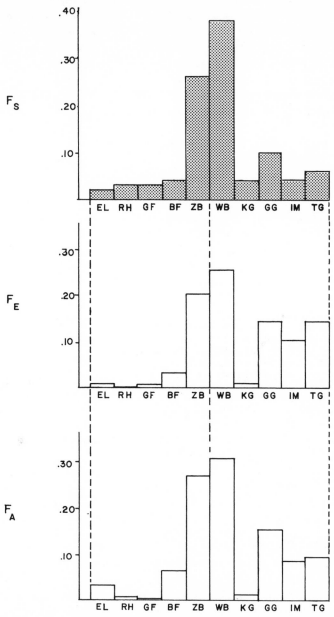

Figure 5.3. Histograms comparing relative frequencies of major herbivore taxa in the living population (F_A) (data from table 5.4A), in the expected carcass assemblage (F_E), and in sampled surface bone assemblages (F_S) for all habitats. F_A and F_E are based on data from Western (personal communication) averaged over 1967-75. Abbreviations for taxa: ZB = Zebra, WB = Wildebeest, KG = kongoni, TG = Thomson's gazelle, GG = Grant's gazelle, IM = Impala, OR = Oryx, ED = Eland, BF = Buffalo, EL = Elephant, GF = Giraffe, RH = Rhino.

Table 5.4A

Comparison of Herbivore Frequencies in the Living Population and in the Carcass Assemblage

	F_A	F_E	F_S	F_M	F_B	Body Weight (kg)
Elephant (EL)	.03	.008	.02	.01	.02	2575
Rhino (RH)	.005	.002	.03	.01	.02	816
Giraffe (FG)	.014	.007	.03	.01	.04	750
Buffalo (BF)	.06	.03	.04	.03	.04	450
Zebra (ZB)	.26	.20	.26	.24	.29	200
Wildebeest (WB)	.30	.25	.38	.49	.39	165
Kongoni (KG)	.01	.01	.04	.03	.04	125
Grant's gazelle (GG)	.15	.19	.10	.08	.04	40
Impala (IM)	.08	.10	.04	.04	.02	40
Thomson's gazelle (TG)	.09	.19	.06	.06	.08	15

Note: F_A = actual frequency of live animals from population censuses (data from Western, personal communication); F_E = expected carcass frequency based on annual turnover rates and assuming stable populations in the basin; F_S = measured frequency of numbers of individuals sampled in the bone assemblage; F_M = measured frequency of population size based on mandibles only; F_B = measured frequency of population size based on buried bone sample (bones $\geq$ 50% buried). See Western (this volume, table 3.2) for original numerical data.

Table 5.4B

Comparisons of Measured versus Expected Carcass Frequencies of the Ten Major Wild Herbivore Taxa

	F_S/F_E	Log $(F_S/F_E$ x 10)	Log Body Weight	F_M/F_E	Log $(F_M/F_E$ x 10)
EL	2.5	1.40	3.41	1.25	. 1.10
RH	15	2.18	2.91	5.0	1.70
GF	4.3	1.63	2.88	1.43	1.15
BF	1.3	1.12	2.65	1.00	1.00
ZB	1.3	1.11	2.30	1.20	1.08
WB	1.5	1.18	2.22	1.96	1.29
KG	4.0	1.60	2.10	3.0	1.48
GG	.53	.72	1.60	.42	.62
IM	.40	.60	1.60	.40	.60
TG	.32	.51	1.18	.32	.51

Note: For abbreviations, see Table 5.4A.

range from 15 kg (*Gazella thomsoni*) to 2575 kg (*Loxodonta africana*), large herbivores (> 200 kg) show a higher than expected carcass representation in the surface bone assemblage and small herbivores (< 100 kg) a lower than expected carcass representation.

The regression of log $(F_S/F_E$ x 10) against log W in figure 5.4 gives a coefficient of determination r = .92 if kongoni (*Alcelaphus buselaphus*) and rhinoceruses are excluded from the calculations. Exclusion of kongoni can be justified because the air counts for the small Amboseli population have a high standard error and may underestimate the actual number in the park (Western 1973 and personal communication). The rhinoceros population

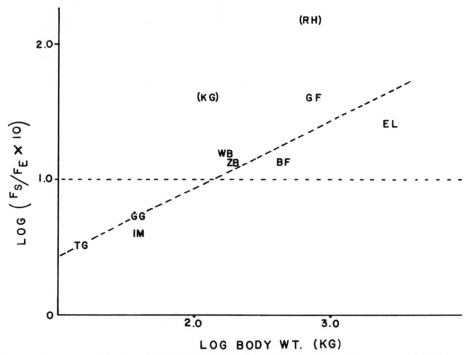

Figure 5.4. Plot of body weight (log) versus the ratio of measured over expected frequency of ten major herbivore taxa in the surface bone assemblage, based on the total carcass sample. Data on expected carcass frequency and live body weight are from Western (personal communication). Taxa with small sample sizes are in parentheses. Regression equation: $y = .48 x - .044$, $r = .92$ (for eight herbivores with adequate sample size).

is in a rapid decline due to human predation, and more carcasses are being accumulated per year than would normally be expected. A decline in the kongoni population could also be partly responsible for its anomalously high F_S/F_E ratio.

Many of the processes which contribute to the body size biases in the Amboseli bone assemblage are clearly evident in Amboseli. Smaller bones are subject to (1) more complete initial destruction by carnivores and scavengers; (2) more rapid rates of surface weathering (Behrensmeyer 1978) and fragmentation due to trampling; (3) more rapid burial by trampling and wind deposition; (4) undersampling due to lower visibility during collection.

Our sampling method has allowed us to correct for bias introduced through the fourth factor, leaving the other three as the major potential causes of differences between expected and actual carcass frequencies. Factors (1) and (2) cause initial size biases against small animal preservation which are likely to influence the composition of the ultimate fossil assemblage. Factor (3) would tend to work in the opposite way, increasing the potential for small bones to become fossils due to their increased likelihood of burial. However, examination of the relative abundance of the major herbivore species in the buried sample (table 5.4A) indicates that "burial potential" has only a minor effect in altering the overall composition of the buried bone assemblage from that which is present on the ground surface. Thus, we can conclude that factors (1) and (2), carnivore destruction and surface weathering and fragmentation, are the most important taphonomic processes causing a size bias in the Amboseli bone assemblage. It should be noted, how-

ever, that burial potential varies enough from habitat to habitat to warrant further
study.

It seems likely that most fossil assemblages in East Africa and elsewhere have
inherited taphonomic size biases both in terms of species representation and relative
population sizes of the different species living in a community. Therefore it is essential to correct for these biases before attempting paleoecological interpretations
regarding species diversity and population structure. As a first step, the relative
abundance of a given species in a fossil assemblage can be compared with the expected
abundance based on its expected annual turnover rate as deduced from its body size.
If relative abundance is lower or higher than expected, evidence must be sought for the
taphonomic processes which have affected the fossil assemblage, in order to determine appropriate correction factors. Clues to these taphonomic processes lie in the characteristics of the bone assemblages themselves: skeletal part representation, nature of
damage (preburial), and weathering stage. Not enough is known at present to provide
specific guidelines for thorough analysis of taphonomic biases in fossil assemblages, but
with more detailed knowledge of size-related taphonomic processes, it may be possible to
estimate some aspects of past community structures.

Representation of Migratory versus Resident Species

Zebra and wildebeest together make up 56% of the living wild herbivore population
(by numbers) and 64% of the carcass assemblage. Both are seasonal migrants into the basic
and generally spend between 50% and 70% of the year there (Western 1973). All species except impala (*Aepyceros melampus*), buffalo (*Syncerus caffer*), and rhino show some movement
into and out of the basin seasonally. These "residents" make up only 14% of the large
herbivore community and 11% of the total bone assemblage. Oryx (*Oryx gazella*) and eland
(*Taurotragus oryx*), which are poorly represented in the bone sample, migrate in during the
wet season, while all other migrants move out of the basin during the wet months (Western
1973). Thus, the bone assemblage is dominated by animals which occupy the basin area
and compete for resources during only part of each annual cycle, mainly the dry season.

The question arises as to whether resident and migratory members of a mammal
community could be recognized in the fossil record. Typically the migratory species have
more concentrated periods of birth and of mortality during droughts. It is possible,
therefore, that species' age structures in fossil samples could help to indicate migratory
species.

Of broader importance to paleoecologic reconstruction is the fact that, while the
species of Amboseli can be regarded as members of the same community, their degree of
actual spatial coexistence and competition varies considerably over a single year. It is
false to assume that animals preserved together in a fossil assemblage were continually
"contemporaneous" or in perpetual competition for resources, since evidence from modern
ecosystems shows that patterns of resource utilization are much more complex than this
assumption implies. In Amboseli, strategies for survival during the dry seasons tend to
concentrate large populations of the different species in the basin area. Because individuals concentrate around water sources, mortality from predation and starvation (as
well as from thirst when water is scarce) is high during the dry season. Consequently, the
bone assemblage strongly reflects dry-season occupation of the basin rather than giving an
average of community structures during both wet and dry seasons.

Carcass Distribution by Habitat

Interpretations of paleoecology depend on the assumptions that animals will usually die where they live and that the skeletal remains will not be thoroughly dispersed and randomized, with respect to habitat, before burial. Data from Amboseli show that relative bone frequencies of some of the major wild herbivore species reflect their habitat preferences. Table 5.5 gives the recorded carcass frequency of ten major herbivore taxa (F_H) per

Table 5.5

Relative Frequencies of Major Herbivore Taxa in the Six Amboseli Habitats

Species	N_H	Swamp F_H	(F_E)	Dense Woodland F_H	(F_E)	Open Woodland F_H	(F_E)	Plains F_H	(F_E)	Lake Bed F_H	(F_E)	Bush F_H	(F_E)
EL	28	.50	(.58)	.21	(.32)	.18	(.10)	.11	(.00)	.00	(.00)	.00	(.00)
RH	31	.23	(.79)	.26	(.03)	.32	(.11)	.13	(.05)	.03	(.00)	.03	(.03)
GF	39	.28	(.44)	.36	(.34)	.10	(.07)	.18	(.01)	.05	(.02)	.03	(.12)
BF	53	.62	(.64)	.15	(.20)	.06	(.05)	.09	(.02)	.04	(.10)	.04	(.00)
ZB	315	.25	(.10)	.08	(.12)	.13	(.17)	.26	(.20)	.18	(.27)	.10	(.13)
WB	452	.33	(.12)	.04	(.16)	.14	(.31)	.27	(.18)	.16	(.17)	.06	(.06)
KG	45	.36	(.34)	.02	(.01)	.16	(.17)	.09	(.06)	.31	(.42)	.07	(.01)
GG	117	.08	(.14)	.13	(.19)	.26	(.19)	.14	(.15)	.06	(.07)	.34	(.25)
IM	44	.05	(.09)	.70	(.79)	.11	(.09)	.09	(.00)	.05	(.00)	.00	(.03)
TG	64	.22	(.19)	.06	(.06)	.06	(.14)	.20	(.26)	.42	(.29)	.03	(.05)

Note: F_H = the number of carcasses of a species recorded in a specific habitat over the total number of carcasses of that species in all habitats (N_H). F_E is the expected relative carcass frequency of each taxon by habitat calculated from 1967-69 and 1973-75 censuses (averaged) by D. Western.

habitat compared with the expected carcass frequency $(F_E$ per habitat based on live distributions from 1967-69 monthly ground counts by Western (1973) and 1973-75 air censuses by D. Western (1973 and personal communication). Histograms comparing measured and expected carcass distribution are given in figure 5.5.

For some taxa, particularly impala and buffalo, there is a close correspondence between the carcass distribution and habitat preferences of the living animals. For others, such as wildebeest and zebra, there is little similarity between distribution patterns of the living animals and their skeletal remains. Broadly speaking, the spatial distributions of those species with the most specific habitat utilization patterns and with little seasonal shift in habitat occupation are most accurately represented by the bone assemblage. For other species, such as zebra and wildebreest, carcass distribution does not appear to reflect habitat preference. A number of factors may be responsible:

1. Predation may tend to concentrate carcasses in habitats which are occupied only briefly during a day, i.e., at waterholes, swamps, or other water sources.

2. Diurnal shifts in habitat utilization occur but are not represented in the daytime live censuses. Since predation occurs more regularly at night, the bone assemblage may be a more accurate record of nighttime patterns of herbivore distribution (censuses record diurnal distribution only).

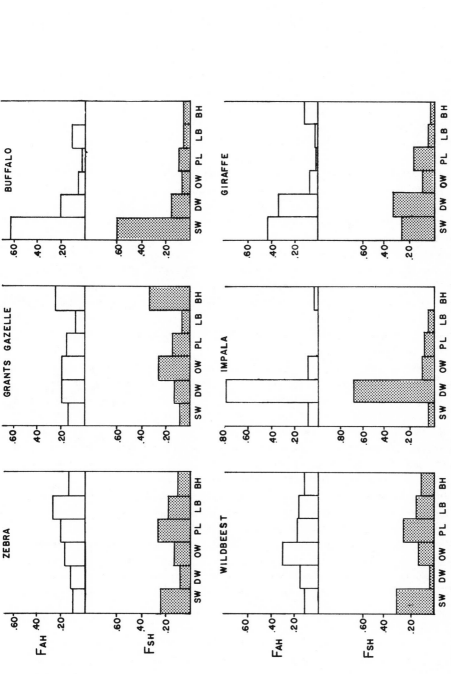

Figure 5.5. Histograms comparing the measured relative carcass frequency (F_{SH}) and the expected relative carcass frequency (F_{EH}) for six herbivore taxa across six different habitats. F_{SH} is based on minimum numbers of individuals recorded in the surface bone sample. F_{EH} is based on 1967-69 and 1973-75 live census data from Western (1973 and personal communication). For explanation of habitat abbreviations see caption to fig. 5.1. Data from table 5.5.

3. Periods of intensified mortality (as during the dry season) may cause concentrations of carcasses in habitats heavily utilized only during these periods.

4. Scavengers and predators may drag dead animals away from the habitats in which
they lived and died.

For zebra and wildebeest, it seems most likely that factors 2 and 3 are important
in causing the observed bone distributions. The low incidence of carcasses in the dense
woodland compared with the plains can be explained by diurnal habitat shifts: woodland is
inhabited during the day and the plains at night, apparently due to the comparative advantages in predator avoidance on the more open ground of the plains habitat. Zebra and
wildebeest also move out of the swamp by night, but high mortality in this habitat during
droughts appears to be the primary reason for high carcass frequencies in the swamp.
Differences between live and carcass frequencies for giraffe and Grant's gazelle (*Gazella
granti*) may also reflect diurnal habitat shifts.

In Amboseli, mortality of the more water-dependent species (chiefly zebra and
wildebeest) is spatially concentrated around water sources and dry-season grazing.
Low-lying areas where plants continue to grow during dry periods also may be more conducive
to bone burial and ultimate fossilization. Zebra and wildebeest form 73% of the individuals in the bone assemblages from the swamp and peripheral plains compared with 64% for
the basin as a whole. Thus, species with seasonally and spatially concentrated mortality
could be overrepresented in many Easter African fossil assemblages relative to their
actual abundances in the original herbivore community.

There is little doubt that short-term fluctuations in habitat boundaries will tend
to blur patterns of bone distribution through time. In Amboseli changes over the past six
years toward more arid conditions in the basin are altering the previous dry-season
distribution patterns of the major herbivores. These changes are clearly recorded in the
bone assemblage (figure 5.5); there are many more carcasses of zebra and wildebeest in
the swamp than predicted only on the basis of the 1967-69 census data. Zebra and wildebeest have moved into the swamp during the past five years (Western, personal communication) and their remains predominate in this habitat, thus altering a bone record that otherwise would indicate their less frequent occurrence in this habitat ten to fifteen years
ago. The actual number of years represented by surface bone assemblages in Amboseli has
yet to be accurately documented, but studies of known-age carcasses indicate that most
skeletal material is no more than fifteen years old (Behrensmeyer 1978).

In East Africa's past and present it is likely that accumulated skeletal remains
create a "time-averaged" picture of the changes which rapid environmental shifts cause
in the habits of members of living communities. Consequently, fossil assemblages will
be limited in how precisely they reflect past environmental changes, habitat preferences,
and community structure. However, evidence for extreme or even moderate environmental
change is usually recorded in the geology of fossil-bearing sediments, and this evidence
can be used to qualitatively assess the degree of time-averaging in any particular assemblage.

Characteristics of the Bone Assemblage

During the Amboseli bone study, individual skeletal part representation, fracturing
and weathering patterns on bones, and species age distributions as reflected in the
skeletal remains were recorded. These data have provided information on the taphonomic
processes affecting surface assemblages prior to burial.

The Surface Assemblage

Overall representation of the various skeletal parts in the Amboseli surface bone assemblage is shown in table 5.6 and figure 5.6. The frequencies of skeletal parts

Table 5.6

Numbers and Frequency (F = Number/Total) of Skeletal Parts in the Amboseli Bone Assemblage

	All Habitats		Buried 50%		Hyena Den		Average (Whole Zebra + Wilde- beest)	
	Number	F	Number	F	Number	F	Number	F
Cranium	(381)	.02	(16)	.02	(4)	.02	(1)	.01
Hemimandible	(504)	.03	(18)	.02	(6)	.02	(2)	.01
Teeth	(5090)	.26	(130)	.16	(53)	.20	(36)	.21
Vertebrae	(4834)	.25	(205)	.26	(26)	.10	(43.5)*	.25
Ribs	(3218)	.16	(97)	.12	(21)	.08	(31)	.18
Forelimb	(1695)	.09	(108)	.14	(50)	.19	(10)	.06
Hindlimb	(1876)	.10	(85)	.11	(56)	.21	(10)	.06
Podial	(1043)	.05	(59)	.07	(36)	.14	(23)	.13
Phalanges	(790)	.04	(59)	.07	(13)	.05	(18)	.10
Indeterminate metapodials	(95)	.005	(11)	.01	--	--	--	--
Total	19,526		788		265		174.5	

Note: Teeth are totaled for those found in crania and mandibles as well as isolated. Forelimb includes scapula, humerus, radius, ulna, metacarpal; hindlimb includes innominate, femur, tibia, fibula, metatarsal.

*Caudals for zebra ~ 16; caudals for wildebeest ~ 14 (Flower 1885).

found have been compared to expected frequencies of skeletal parts based on averaged numbers of parts found in wildebeest and zebra skeletons. The similarity between expected and observed frequencies indicates that postmortem processes (i.e., carnivore behavior, scavenging, and weathering) have not radically changed the representation of skeletal parts. However, results indicate that some elements are consistently scarce while others persist in the assemblage. Ribs, podials, and phalanges are underrepresented in the bone sample over all habitats, and were probably destroyed by carnivores. Crania, mandibles, and teeth show higher representations and did not constitute food items.

In the Amboseli basin, predators and scavengers play a major role in destruction and alteration of articulated carcasses and smaller skeletal units. There is different damage, however, depending on whether the animal was killed or died from starvation, disease, or thirst. During the 1973 drought in East Africa, disease and hunger led to weakening and death of many wild and domestic herbivores in Amboseli. Most of the domestic stock died as a result of the disease and hunger, while the weakened wild herbivores were killed by predators. Preliminary analysis indicates that carcass and bone destruction is consistently greater in the latter case. Vultures, the primary scavengers on starvation-death carcasses in Amboseli, generally leave the bones of large animals intact and articulated. In cases where overall carcass density is high, predators and especially scavengers find an excess of available meat which results in minimized skeletal disarticulation and fragmentation.

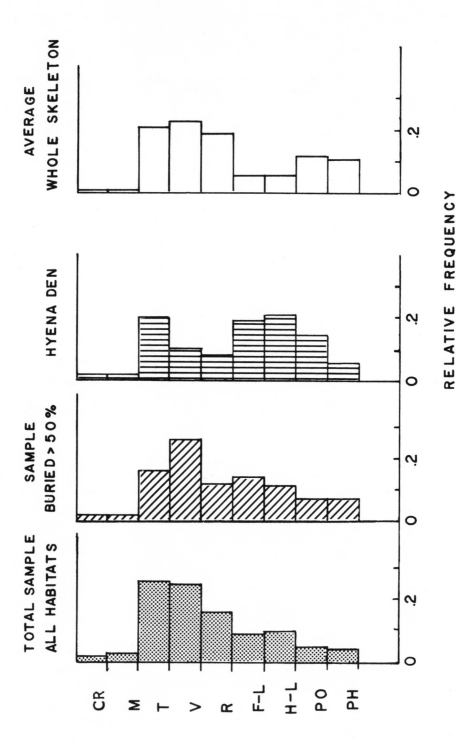

Figure 5.6. Comparisons of the relative frequencies of skeletal parts in the overall surface assemblage and in the buried portion of this as-semblage (those bones whose surface area is more than 50% buried), in a hyena den accumulation (only a portion of the surface den sample), and the relative frequencies which result when the numbers of various skeletal parts from one adult zebra and one adult wildebeest, the most common herbivores, are averaged. Abbreviations: CR = Cranium, M = Mandible, T = Teeth, V = Vertebrae, R = Ribs, F-L = Forelimb, H-L = Hindlimb, PO = Podials, PH = Phalanges, MP = Interminate metapodials. H-L and F-L include metatarsals and metacarpals.

It seems that when predator-prey ratios are low, the predators consistently damage certain parts of the skeleton in the process of eating the prey (see also Hill 1975 and this volume). We found that dorsal borders of innominates, olecranon processes of ulnae, ventral and dorsal rib ends, premaxillae and nasals, and vertebral processes were almost invariably chewed and punctured. Terminal phalanges, caudal vertebrae, and patellae were conspicuously absent from most carcasses. Mandibles were often broken on the left or right side just posterior to the symphysis, and the mandibular angles were usually chewed. These characteristic breakage patterns of the mandible may result when a predator grabs the angle and pulls outward, breaking the corpus in an effort to reach the tongue (the "wishbone" effect).

Because of periodic drought mortality in the Amboseli Basin, there may be a surplus of carcasses at certain times of the year in excess of the needs of the local carnivore population. Under these circumstances, predators and scavengers do not completely destroy carcasses and the types of damage noted above are characteristic. These types of damage, if found in a fossil assemblage, might indicate past conditions of excessive herbivore death (from thirst, starvation, or disease) or predator-scavenger scarcity.

The overall effects of predators and scavengers on bones in Amboseli are notably slight compared with areas such as Ngorongoro Crater (Kruuk 1972) where predator/prey ratios are high and bones are in greater demand as food items. More taphonomic studies are needed in areas where predator pressures are high, but the evidence from Amboseli strongly indicates that the effects of predation and scavenging on bone assemblages will be recognizable in the fossil record.

A fossil assemblage derived from bones accumulated under drought conditions could also give a biased overrepresentation of those animals most affected by periodic drought (the more water-dependent species), as recently noted by Shipman (1975). However, if droughts are infrequent, the contribution made by drought-related bone accumulation to the skeletal assemblage as a whole may not be great enough, as compared to the yearly contribution from normal population turnover rates (e.g., 25% in wildebeest), to significantly bias the assemblage in favor of water-dependent species (Western, this volume).

Future analysis of the Amboseli bone sample will provide information on proximal/distal ratios in limb bones, relative representation of juvenile and adult skeletal parts, differential bone preservation with respect to taxon, and differences between pre- and post-1972 skeletal part representation (largely exemplifying pre- and postdrought conditions, as 1972 marked the most recent onset of drier conditions and periodic droughts in the basin; Western, personal communication).

The Buried Assemblage

A sample of 788 identifiable bones (each buried more than 50%) reveals some interesting trends in differential burial of skeletal parts. In the overall sample (table 5.6) there are more limb elements, podials, and phalanges in the buried sample than in the surface assemblage. Teeth are conspicuously underrepresented.

The primary burial process in the swamp and lake-bed habitats is trampling. This was apparent from our observations of the activities of the living animals, the high degree of bioturbation of the soil, and the vertical or subvertical orientation of the long axes of many partially buried bones. Small, compact bones such as podials are easily forced down into soft soil by heavy feet, but elements such as crania, mandibles, and pelves are either avoided or crushed by the tramplers. Most of our buried sample of crania and mandibles consisted of maxillae, frontlets, and mandible fragments with a few teeth.

Teeth are numerous in the overall habitat sample, but scarce in the buried sample. Under conditions of high surface temperatures and variable degrees of moisture, as in Amboseli, isolated teeth or teeth in alveoli appear to fracture in relatively short periods of time. The only uncracked teeth we observed were either those of recently dead individuals or those unexposed to the air in semiburied skulls or mandibles. Our observations indicate that teeth remain in the mandible and cranium until progressive weathering leads to the enlargement of the alveoli. Once the teeth have fallen out, they may be fragmented and totally buried by trampling (thus becoming invisible to the observer) or degrade entirely. Sieving of areas around fragmented crania and mandibles yielded many more tooth fragments than whole teeth.

There are some differences in characteristics of skeletal burial from habitat to habitat. A higher percentage of teeth were buried in the plains and dense woodland habitats where eolian processes are important in covering bones. There was also a tendency for a greater number of articulated carcasses to be partially buried in windblown silt in these habitats.

Tooth/Vertebrae Ratios

The relative numbers of teeth and vertebrae in fossil assemblages have been used by Behrensmeyer (1975) in assessing whether or not a bone assemblage has been selectively sorted prior to burial. Table 5.7A demonstrates clearly that the teeth/vertebrae (T/V) ratio can vary considerably by habitat within surface bone assemblages. Since none of these samples has been affected by fluvial transport, the range of T/V ratios represents what can be expected for untransported fossil assemblages.

The tooth/vertebrae ratios from Amboseli can be used to help evaluate differences in these ratios in fossil assemblages from the Plio-Pleistocene Koobi Fora Formation of Northern Kenya (table 5.7B)(Behrensmeyer 1975). The T/V ratio in the fossil assemblage

Table 5.7A

Tooth and Vertebra Representation in the Amboseli Bone Assemblage

	SW	DW	OW	PL	LB	BH
Teeth	1485	410	697	1218	951	329
Vertebrae	1661	394	723	1077	660	319
T/V ratio	.89	1.04	.96	1.19	1.44	1.03
Teeth (buried)	45	12	--	60	4	7
Vertebrae (buried)	102	8	--	47	47	1
T_B/V_N ratio	.44	1.50	--	1.29	.09	(7)

Note: T/V ratio = 37/37 = .95 for average whole animal (zebra + wildebeest).

from a deltaic environment falls within the range of ratios for the habitats of Amboseli, indicating that this fossil assemblage was probably subject to taphonomic sorting processes similar to those of the Amboseli Basin (i.e., nonfluvial). For fossil assemblages from floodplain and channel environments, the T/V ratios are well above any in the Amboseli sample. This indicates the dominance of winnowing processes which sorted teeth

Table 5.7B

Tooth and Vertebra Representation in Pleistocene Fossil Assemblages of the Koobi Fora Formation

	Square Frequency, Delta	Square Frequency, Floodplain	Square Frequency, Channel
Teeth	.70	.52	.81
Vertebrae	.58	.15	.26
T/V ratio	1.21	3.47	3.12

Note: Teeth in jaws and skulls not counted, but most in the fossil sample were isolated.

from vertebrae in the fluvial system. The relative importance of all processes which can differentially affect concentrations of teeth and vertebrae is conveniently expressed in the *T/V* ratios. More work is needed, however, in order to understand the nature and significance of these processes in fluvial and nonfluvial environments.

The Hyena Den Bone Assemblage

A currently occupied hyena *(Crocuta crocuta)* den with a large bone concentration was sampled for faunal and body part comparison with the overall surface bone assemblage. The den is in the plains habitat, several kilometers from major swamp and woodland areas. The caliche surface which underlies the plains has collapsed, forming local depressions, and the hyenas have occupied one of these since 1967, with a period of nonoccupancy between about 1969 and 1974. The main underground den opens into a steep-sided depression 40-50 cm below the plains' surface and covers approximately 10 m^2 in area. The depression is elongated N-S and shallows southward. It is littered with bones and has accumulated approximately 20-30 cm of bones and windblown silt since first viewed by D. Western in 1967 (personal communication).

There is no doubt that spotted hyenas are residing in and carrying bones to the den area. Three hyenas (including two juveniles) on one occasion and eight on another were seen in and around the bone-littered depression. Fresh carcass parts were observed on repeated visits to the den.

A mosaic photograph of the depression was made, and all the surface bones in the northern area of the depression were identified. A more detailed study is being done by A. Hill; we were able to identify only a portion of the total bone sample in the time available. The taxa represented in the recorded sample are listed under general headings in table 5.2. Skeletal part representation is given in table 5.6. Fracture and wear features of the bones resulting from hyena activity at the den will be reported elsewhere.

Some species represented in the hyena den bone assemblage were not found in the assemblage recorded from the surrounding plains habitat. These animals include caracal *(Felis caracal)*, oryx, and warthog *(Phacochoerus aethiopicus)*. The calculations in table 5.6 indicate a higher frequency of limb elements and podials and a lower frequency of vertebrae and ribs in the hyena den than in the general sample for all habitats.

In his study on the Topnaar Hottentot people in the central Namib Desert, Brain (1967, 1969) concluded that, as a result of butchery techniques and subsequent carcass consumption by both the Hottentots and their dogs, the remaining bone sample was composed of those body parts which were most dense and most durable. He found that most limb elements and podials have a higher percentage survival rate than vertebrae and ribs, when subjected to destructive forces. Our sample from the Amboseli hyena den probably represents concentration of those bones with the greatest probability of survival, after consumption and gnawing by both adult and young hyenas has destroyed the less durable skeletal elements. These similarities between Hottentot/dog and hyena bone samples suggest that differential body part representation would not be a good character-istic for distinguishing fossil assemblages accumulated by human activities from those ac-cumulated by hyena activities, although natural death accumulations, such as in the general habitat sample for Amboseli (table 5.6)(and perhaps assemblages concentrated by geologic processes), are characterized by body part percentages quite distinct from those of the hyena den.

Microvertebrate Bone Concentrations

Most of the micromammal bones represented in the recorded bone sample were found in the scats of small carnivores (as yet undetermined but probably jackal (*Canus adustus* or *C. mesomelas*) or bat-eared fox *(Otocyon megalotis)*). Twenty fecal samples containing bones were collected, thirteen from the lake bed habitat and seven from the bush habitat. Nine from the lake bed were found lying either on or next to the bones of larger mammals, while the remaining eleven samples were found on the ground or on bushes. The bones of larger mammals may serve as prominent "marking posts" for small carnivores in the lake-bed habitat where trees, bushes, and grass are absent.

One large accumulation of scats found in the bush habitat contained hundreds of cranial and postcranial remains in an area of less than 1 m^2. From a sample of mandibles and maxillae the following taxa have been identified by H. Wesselman: *Tatera robusta, Gerbillus harwoodi* or *pusillus, Steatomys minutus, Mus minutoides, Arvicanthis, Crocidura bicolor, Crocidura hirta*. The rodents mentioned above are primarily seed-eaters adapted to dry savanna habitat; the shrews occupy a variety of habitats ranging from dry bush to wooded grassland, floodplain, and riverine vegetation (Kingdon 1974).

Other less dense concentrations of micromammal, hare (*Lepus capensis*), and bird bones were found under large trees (*Acacia tortilis*) in the open woodland habitat. These remains probably accumulated as a result of the activities of large predatory birds feeding on microvertebrate prey while perched in the branches of trees.

In the Amboseli Basin, where geological processes that could concentrate the bones of small animals (e.g., fluvial winnowing) are lacking, microvertebrate bone ac-cumulations occur primarily as a result of predator activities. In geologically similar situations, such as at Olduvai Gorge during the Pleistocene, pockets of fossil micro-vertebrate bones may have resulted from the activities of such biological agents (see also Mellett 1974).

Summary and Conclusions

The full potential value of bone assemblage data from Amboseli will only be realized when comparable information is obtained from other recent ecosystems, and when fossil assemblages are reexamined in the light of new taphonomic information. At present, the major points relevant to the study of paleoecology are as follows:

1. Species representation in the Amboseli bone assemblage is a function of body size. Predepositional taphonomic processes tend to destroy small bones, reducing the probability of preserving the remains of small species. All live species of 100 kg or greater adult body weight are represented in the bone assemblage, but only some of the

species between 1 kg and 100 kg are represented. The size-related taphonomic processes
in the Amboseli system include destruction by carnivores, trampling, and weathering. (See
also Behrensmeyer et al. 1979.)

2. Relative numbers of carcasses of ungulate species recorded from the bone assemblage of Amboseli do not accurately reflect relative population sizes of ungulate
species recorded from censuses taken in the basin. Comparisons of observed versus expected carcass frequencies for the major ungulate species show a consistent size-
related bias in the persistence or destruction of carcasses. Carcasses of individuals of
greater than 100 kg average adult weight are recorded in greater than expected frequencies
(based on calculated turnover rates for each species), while carcasses of individuals
between 15 and 100 kg average adult weight are recorded in smaller than expected frequencies.

3. Carcasses of migratory, water-dependent species are found in the swamp habitat
(a primary depositional environment) in greater than expected frequencies calculated on
the basis of live census data and species turnover rates. The overabundance of carcasses
in this habitat is due chiefly to seasonal mortality in the wetter areas caused by drought-
related starvation. East African fossil assemblages may also contain higher proportions
of those species most susceptible to drought mortality, depending on the past frequency
of drought relative to normal periods of attritional mortality.

4. Habitat distributions of the major herbivore species are reflected in the
Amboseli bone assemblage. For nonmigratory, habitat-specific ungulate species there is
a close correlation between the habitats preferred by living animals and the habitats of
death. Habitat boundaries shift in response to short-term climatic change, mixing the
bones of individuals that die in one habitat type (i.e., open woodland) with those of
others that later die in the same place after a habitat shift (i.e., to grassland). In
fossil deposits where lithofacies changes can be defined geologically, such as floodplain
versus lake margin, we can expect the most reliable ecological information relating to
the differences between habitats from analysis of the preserved faunal remains within
a specific sedimentary environment.

5. Skeletal part frequencies in the bone assemblage provide information on predepositional taphonomic processes in Amboseli. The effect of predator/scavenger activities
on the overall surface assemblage can be defined in terms of damage and removal of
certain skeletal elements. The relative contribution of predators to the destruction of
carcasses is reflected in the degree of alteration of bone frequencies in our modern
sample, and should be similarly revealed in the remains of fossil communities.

6. About 5% of the surface bones sampled were buried more than 50%, showing that
bone burial is an ongoing process in the Amboseli Basin. The major causes of bone burial
are trampling and eolian sedimentation. Trampling tends to selectively bury compact bones
such as podials. Teeth were less common in the buried sample, probably because of
relatively rapid rates of surface weathering and fragmentation.

7. The hyena den bone assemblage is characterized by skeletal part frequencies
that differ from those of the overall surface assemblage. Destruction by hyenas results
in the concentration of dense limb elements and podials which are resistant to gnawing
and fracturing. Hyena bone accumulations may be recognizable in the fossil record by
characteristic skeletal part frequencies and bone damage which can be attributed to
hyenas.

8. Bones of microvertebrates (< 1 kg adult body weight) are concentrated by the
activities of predators that feed on small prey. Localized patches of owl pellets or

carnivore feces may contain the bones of individuals from a number of species. Species found in such patches may represent prey preferences of the carnivores more accurately than they do the relative population sizes or complete species composition of a localized microvertebrate community.

Because the Amboseli Basin represents a specific geologic setting, climatic regime, and community structure, analysis of the modern bone sample cannot answer all questions relevant to the interpretation of paleoecology for the great variety of existing fossil sites and paleoenvironmental settings. It can serve as a model based on similarity in overall setting, for depositional basins such as Olduvai Gorge (Hay 1976), Peninj (Isaac 1967), and Chesowanja (Bishop et al. 1975; Bishop 1978), and can be compared and contrasted with assemblages from different paleoenvironmental settings. The data provide potentially valuable information on how a bone assemblage reflects species diversity, patterns of seasonal migration and mortality, and overall taphonomic effects of geologic and biologic processes.

In future work, we plan to create artificial fossil samples from the total bone assemblage of Amboseli using random and nonrandom selection procedures and then analyze the relationship of these samples to the known characteristics of the living fauna and community structure. This should help define the limits of resolution of paleoecologic information from relatively small paleontologic samples.

Acknowledgments

We are deeply grateful to David Western, who provided the invaluable ecological background information for Amboseli and who has been generous with help of many kinds throughout the study. The director of parks in Kenya, at the time of this study, Dr. Perez Olindo, generously granted permission for this research, and Mr. Joe Kioko, park warden for Amboseli, provided useful suggestions and encouragement in the field. We would also like to thank Andrew Hill, the National Museums of Kenya, Elizabeth Oswald, and Stuart, Jeanne, Michael, and Rachael Altmann for various forms of help and encouragement, and Hank Wesselman for the time and effort he spent in taxonomic identification of the micromammal sample.

The Amboseli Bone Project was made possible by grants from the National Geographic Society (grant no. 1508). Additional support through National Science Foundation grant no. GS 28607A-1 is also gratefully acknowledged.

6. ETHNOARCHEOLOGICAL CONTRIBUTIONS TO THE TAPHONOMY OF HUMAN SITES

Diane P. Gifford

Introduction

Ethnoarcheology, the study of living human communities by archeologists, has become increasingly common in the past decade. It has developed for many of the same reasons that prompted paleontologists to undertake studies of contemporary biota and geologic processes. Foremost among these is the need to examine in detail the relation of life processes—biological or cultural—to their material and potentially preservable manifestations. Ethnoarcheology and taphonomy overlap considerably in both theory and method due to a common concern with biologic and geologic processes forming deposits of prehistoric materials. The crucial difference between the two fields lies in ethno-archeology's added preoccupation with the role of human behavior in forming archeologic accumulations.

An archeological site in primary context exists and has a certain spatial ar-rangement of materials because of past hominid behavior. The behavioral information intrinsic to this "arrangedness" of archeological sites has few parallels in paleontologic accumulations, with the exception of those deposits preserving organisms in their life contexts.

While it is not the purpose of this paper to detail a history of ethnoarcheology, it may be noted that there are certain parallels between developments in that field and in taphonomy. In general, the initial effect of detailed contemporary studies was to show that the workings of reality are nowhere near as simple as one might assume, and to indicate the practical weaknesses in many elegant explanatory models. However, just as regularities begin to emerge from taphonomic observations, so too in ethnoarcheology the "cautionary note" approach is slowly being replaced by generalizations and testable hypotheses (for review see Schiffer 1978).

This paper will present some findings from my own ethnoarcheologic research among the Dassanetch people on the northeastern shores of Lake Turkana, Kenya. In doing this, I will utilize a conceptual framework which, I believe, provides an excellent means of relating the results of contemporary studies to analysis of archeological sites. This approach, articulated by Binford (1975) and Reid, Schiffer, and Neff (e.g., 1975), views a site as the outcome of a complex, interrelated series of human and natural processes which operated until the time it came under study by the archeologist. A corollary point should be made which is more than the semantic quibble it may seem at

the outset: given this orientation, one cannot accurately speak of an archeological
site's being inhabited. Sites are *created* through certain kinds of human behavior which
produce material consequences—artifacts, structures, hearths, pits, food debris, foot-
prints, and so forth—which occur together at a particular locus in space and time. They
are perceptible in the present due to the interaction of these material relicts with
the biologic, geologic, and chemical processes which tended to preserve rather then
destroy them.

There has been considerable debate in archeology concerning the nature and limits
of ethnographic analogy, and the utility of the concept of uniformitarianism in studies
of prehistoric human behavior (Ascher 1961; Binford 1967; Chang 1967; Gould 1978; Sabloff,
Beale, and Kurland 1973; Schiffer 1975; Tuggle, Townsend, and Riley 1972). Presently
observable cultural systems do not necessarily reflect the total range of prehistoric
systems, and being bound by strict analogy severely restricts the range of possible
interpretations (Binford 1968; Gould 1978; J. N. Hill 1970). Leslie Freeman (1968) has
argued that direct analogies between modern hunter-gatherers and members of earlier hominid
species may be particularly misleading, since in such cases we may be dealing with the
material effects of cultural behavior in adaptive systems vastly different from those of
modern hominids.

However, ethnoarcheology allows the archeologist to go beyond the limits of
strict analogy as an explanatory tool by providing an actualistic arena for framing, testing, and refining general models of human behavior and its material effects. This paper
presents such an approach to certain aspects of human behavior involved in the formation
of archeological sites for which we have reason to believe there is both considerable
antiquity and continuity. Consideration of modern examples of site formation is therefore relevant to explanation of archeological materials of great age. Furthermore,
paradigms of site formation processes developed by archeologists are highly relevant to
taphonomic theory-building.

The first of the following sections will introduce the geography and cultural
context of my study area, sketching the life processes that structure cultural remains
in the region. The succeeding section will discuss human and natural site formation
processes, and their interactions, with particular reference to the potential archeological
record of the Dassanetch people (see also Gifford 1977).

Northeastern Lake Turkana and the Dassanetch

I conducted my research along the northeastern shore of Lake Turkana, Kenya,
between Ileret and Allia Bay, an area about 65 km long (fig. 6.1). This region, well
known for its fossiliferous Plio-Pleistocene sediments, today receives about 300 mm of
rain a year (Butzer 1971). Rainfall occurs in two seasons, the Long Rains falling
roughly between March and May, and the Short Rains between early October and December.
All rivers in the region are ephemeral; runoff after rainfall is swift and massive. The
only year-round local sources of potable water are the mildly alkaline lake and subsurface
water in some major river courses.

In regions not heavily utilized by domestic livestock, the shoreline and inland
zones support a rich mammalian and avian fauna. Along the shore, the most common ungulates
are topi and Burchell's zebra (*Damaliscus lunatus korrigum* and *Equus burchelli böhmi*,
respectively), with some Grant's gazelle (*Gazella granti*) and beisa oryx *(Oryx gazella
beisa)*. Inland, the most commonly encountered ungulates are beisa, oryx, gerenuk

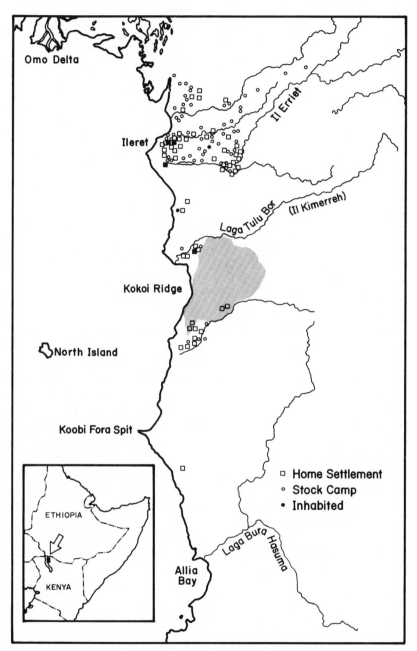

Figure 6.1. The Koobi Fora-Ileret study area, showing the distribution of home settlements and stock camps. Those inhabited during the study period are shown in black.

(*Litocranius walleri*), Kirk's dikdik (*Rhyncotragus kirki minor*), and Grevy's zebra (*Equus grevyi*). Larger carnivores include numerous lions (*Panthera leo*), cheetahs (*Acinonyx jubatus*), hyenas (*H. hyaena* and *Crocuta crocuta*), wild dogs (*Lycaon pictus*), and jackals (*Canis mesomelas*).

The lake itself supports a rich nearshore fauna, including crocodile (*C. niloticus*), soft-shelled turtle (*Trionyx rex*), nilotic terrapin (*Pelusios adansonii*), Nile perch (*Lates niloticus*), two species of catfish (*Clarias lazera* and *Bagrus bayad*), and *Tilapia nilotica*.

The people inhabiting the region are members of the Inkoria territorial section of the Dassanetch tribe. The Dassanetch are a Cushitic-speaking people with a mixed pastoral-agricultural economy. The majority of the tribe live in Ethiopia, in the lower valley and delta of the Omo River. The region south of the border in Kenya contains between 600 and 1,000 people, by my estimate. (For detailed ethnographic information see Carr 1977 and Almagor 1978.)

Various foodstuffs are consumed by the Dassanetch in differing proportions through the seasons. As with most mixed pastoralist-cultivators, the Dassanetch rely upon grain and milk during the time of the Long Rains and immediately thereafter, and upon meat, blood, and some wild food resources during the dry months of late summer and early autumn, when the former resources are exhausted. Family herds, especially sheep and goats, are the major source of meat during the dry season. People without livestock, or with herds too small to cull, must fall back on wild plant and animal resources during the dry season.

Dassanetch without livestock form a separate social and economic class within the tribe. They are called *gal dies*, with alternate meanings of "poor men" and "fishermen." Dassanetch hold that only the very poor would lower themselves to catching and eating fish and lake reptiles, hence the identity of meanings in the word *gal dies*. Lack of pastoral stock, especially cattle, excludes *gal dies* males from most of the important ceremonies and social roles in the society.

While *gal dies* agricultural activities and home settlements are roughly similar to those of other Dassanetch, their patterns of animal protein procurement differ. *Gal dies* men spend considerable amounts of time away from their home settlements, foraging along the lakeshore for fish and reptiles. Animals regularly caught include crocodile, soft-shelled turtle, terrapin, Nile perch, *Tilapia*, and catfish. I saw no evidence of concerted hunting of land mammals during my stay, although sick ungulates and those recently killed by lions were consumed. Informants said that cooperative hunts of hippopotamus, using harpoons and dugout canoes, were conducted in the past. Hippo remains occurred on five of twenty-three *gal dies* camps surveyed between Allia Bay and Koobi Fora.

The differences in economic pattern between pastoralist Dassanetch and the *gal dies* produce several different kinds of sites in the region. The largest sites are family settlements, with dwellings of a number of families related by either kinship or male age-set bonds. In the Ileret area, houses number between fifteen and forty in family settlements. Henceforth, I shall refer to this type of occupation site as "home settlement." Home settlements are located near a reliable water supply, usually a major river course, and close to cultivable land, being at the same time out of the way of possible flash floods. They are occupied for at least several years, and in some cases continually for decades.

Another common type of site created by pastoralists is the stock camp, for either cattle or small stock. These are normally well away from the home settlements, especially in the dry season where forage near to large settlements is exhausted. The camps, staffed by boys and young men, are more transient than other settlements, and usually lack any kind of constructed shelter, having only a hearth, thorn fence pens, and possibly a loosely constructed shade hut for newborn animals. These camps are located within walking distance of either lake or major river course in which wells can be dug, but are situated away from areas liable to flooding.

Gal dies foraging-party campsites are simple, consisting of varying numbers of hearths, food debris—nearly entirely bone—and occasional windbreaks of stone or reed and sapling shade structures. Sites of this type may be created in only a few hours, or parties may camp in one place for over a week, depending upon local fishing conditions and personal preferences. Areas of sites likewise vary widely, from only a few square meters at "lunch stops" to several thousands of square meters at repeatedly occupied locations near good fishing grounds. Such hunting-fishing camps are aligned along the shore, usually within a few meters of the contemporary beachline. Site survey of the Allia Bay-Koobi Fora shore zone located a suite of such campsites on or directly behind beach ridges formed by the 1970 high stand of the lake and another distribution near the 1973-74 beachline.

During my study I examined sites of all these kinds. Some of these data, taken with information gathered on human and site formation processes, will be presented in the following section.

Human Site Formation Processes

In the category of human site formation processes I include all human behaviors contributing to the formation of an archeological accumulation. These include intentional, culturally patterned behaviors such as artifact manufacture and discard and nonintentional formation processes such as loss and trampling. These processes may be divided into three general categories. First, there are those contributing materials to the archeological record. Second, there are processes which determine the spatial patterning of such materials once they have been contributed. Third, there are human activities which interact with natural processes to influence the preservation of specific cultural accumulations.

Only recently have archeologists questioned the notion that artifacts are discarded at their location of use. Lewis Binford, on the basis of his recent work with Nuniamiut Eskimos in Alaska (Binford 1973; Weiss 1975) has concluded that artifact discard locations often may have little correspondence with locations of actual use. This results from what he calls "curate behavior," which involves the retention of all tools for as long as they are useful. The only pieces deliberately discarded are those which are no longer functional. He proposes that the greater the amount of energy invested in an artifact, the less likely it is to be discarded until its use-life is ended. Binford found that among the Nuniamiut the best material index of activities at a site was processing debris, in these cases largely bone. Yellen (1974) found much the same patterns of high tool curation among the San of Botswana.

The Dassanetch are also a high-curation society. All the gourds, pottery and metal vessels, beads, metal implements, and cloth must be acquired through trade, and government restrictions on Dassanetch contact with other tribes has made obtaining such

goods difficult even for the wealthy. These items are used and mended until completely
beyond repair. Observations of abandoned home settlements revealed a low proportion of
artifact discards to food refuse discards. On one abandoned home settlement, artifacts
amounted to only 5% of the total cultural debris. Within the artifact category, most pieces
were nonfunctional fragments; functional items included beads, a few fishhooks, and a tin
crucifix, all of which are of easily lost sizes. This contrasts with the debris of a
pastoral Maasai home settlement near Nairobi, in which over 40% of the debris was artifactual and included such items as food tins, which Dassanetch would never discard because of their rarity and value as containers.

The curate factor must be considered in any estimates of the frequency with
which certain elements, whether lithic, bone, or pottery, can be expected to appear in
the archeological record. Rates of destruction and discard of materials are determined
by cultural practices, which may vary considerably from people to people. For example,
the entry of animal bone into the archeological record depends upon butchery and culinary
habits and upon the rate of retention of bone for other uses. The Dassanetch discard
all bone which survives cooking and consumption, but other peoples, such as the precontact
Eskimo, retained specific bones for use as implements.

Another set of human site formation processes comprises those activities affecting
the spatial distribution of archeological materials. Once artifacts and debris have been
discarded, they can either stay where dropped or be moved from that location. Refuse
which remains close to its point of generation conveys information about the actual
locus where a specific sort of debris-producing activity took place. Refuse subsequently
moved from its point of generation lacks this information. Schiffer (1972) has defined
two useful categories of refuse: debris which is discarded (and remains) at its location
of use is *primary refuse;* refuse discarded away from its location of use is *secondary
refuse*. Both types of refuse reflect human behaviors, but the former directly reflects
secondary refuse disposal activities. Some secondary refuse deposits, such as shell
middens, have been recognized for years. However, it is clear that a number of relatively
recent activity-oriented studies assumed that nearly all clusters of artifacts and debris
on occupation floors are primary refuse (e.g., Binford 1964; J. N. Hill 1968; Longacre 1970).
This may be true in some cases, but ethnoarcheologic observations in widely differing
contexts have shown that secondary refuse is extremely common on human sites.

There appears to be a direct relationship between the duration of time a site is
occupied and the proportion of secondary to primary refuse. Schiffer (1972, p. 162), on
the basis of research in the American Southwest, proposed the following principle:
"With increasing site population (or perhaps site size) and increasing intensity of occupation, there will be a decreasing correspondence between the use and discard locations
for all elements used in activities and discarded at a site." Schiffer's "intensity of
occupation" deserves further analysis. It involves the interplay of several factors, including the rate at which debris is produced and the time over which it is produced.
Given constant debris-producing activities and constant rates of such activities, time is
the major determinant of proportions of primary and secondary refuse on a site.

Dassanetch refuse patterns alter according to duration of occupation. Refuse on
short-term single-occupation fishing camps is nearly all primary, dropped and left at its
location of use. Refuse is predominantly food waste; it lies either at locations of
initial butchering or in and around hearths. This pattern was observed in its development
at two sites I saw during occupation (fig. 6.2), and it is manifest at other similar camps
I did not see occupied. Long-term occupation sites, such as home settlements, display

Figure 6.2. A *gal dies* fishing camp, showing relationship of food refuse to hut and sitting area. The close association of bone refuse and hearths contrasts with the lack of such clear association in long-term occupations (see fig. 6.3).

high proportions of secondary refuse. Women in defensive settlements gather up debris
from meals and other activities within their homes or the house area and throw it over
the fences of the innermost stock pens (fig. 6.3). Houses themselves remain relatively
clear of debris. Excavation of two house floors on a site occupied eight to twelve months
yielded only twenty-four elements on one house floor and eighteen on the other.

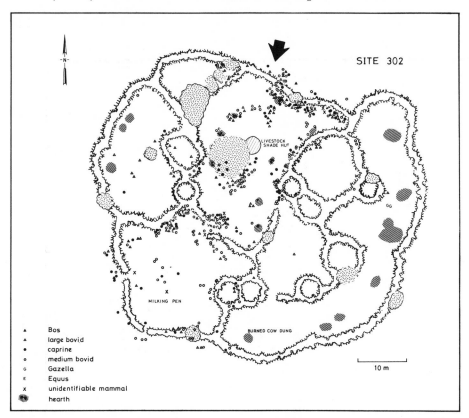

SITE 302

LIVESTOCK
SHADE HUT

GG

MILKING PEN

BURNED COW DUNG

▲	Bos
▵	large bovid
•	caprine
○	medium bovid
G	Gazella
E	Equus
x	unidentifiable mammal
▦	hearth

10 m

Figure 6.3. A Dassanetch home settlement abandoned after about twelve months' occupation in
1969-70. Arrow indicates house area of settlement, from which portable structures were removed on abandonment. Secondary bone refuse accumulates along the thorn fences where it is
thrown after use or cleaning of hearth areas. Such piles of secondary refuse, if excavated
in an archeological site, might be misinterpreted as indicating specific areas of butchering
or cooking.

It is relatively simple to define primary and secondary waste in the ethnographic
present, but in the archeological situation this distinction is not so easy to make.
We first must ask whether prehistoric hominids, especially those of other species, followed our "commonsense" rules of refuse disposal.   Piles of marine mollusk shells, such
as those described from South Africa (Klein 1974), North Africa (McBurney 1967), and
Europe (Clark 1936; Roche 1965) from Upper Pleistocene and Holocene times are usually
taken to have been middens, and thus secondary refuse. But what do we make of smaller
clusters of bone and debitage found outside a shelter? Are these secondary refuse,
or primary refuse of activities undertaken away from the shelter? The stone circle at
DK 1, Olduvai, contains only thirty pieces of bone and worked stone within its perimeter,
with an average density of $1.9/m^2$. Other parts of the excavated floor of similar area
have densities up to $6.5/m^2$ (M. D. Leakey 1971). With sites of this sort can we, in

light of the foregoing discussion, defensibly speak of activity areas, or must we be content with speaking only in generalities about activities inferred to have occurred somewhere in the general area? The obvious point to investigate ethnoarcheologically is whether these two types of refuse have any properties which could be used to differentiate them in archeological situations.

A few researchers have suggested testable hypotheses concerning features that may distinguish secondary refuse. Schiffer (1975) has suggested that clusters of debris containing by-products of two spatially isolated stages in the same activity are secondary refuse. He advances this idea on the basis of his "behavioral chain analysis" of Pueblo Indian corn processing activities, some steps of which are carried out in separate areas of the pueblo. It remains to be seen whether similar behavioral chains can be reconstructed in situations where no direct historical analogies are possible. In other ethnographic situations, however, this approach to identification of secondary refuse may be tested. For example, skin, bones, and ash are often observed together in Dassanetch home settlement refuse heaps. Normally, animals are slaughtered away from the immediate vicinity of the houses, close to the pens, and are skinned there. Cooking, by either boiling or roasting, takes place in the family dwelling, from which bones and excess ash are later removed. The cooccurrence of skin, ash, and bones in the same rubbish heap in this case represents the bringing together of debris from at least two locales through refuse disposal behavior.

The proximity of food processing debris, especially bone, to hearths may provide a reliable index of refuse type. In contemporary Dassanetch sites, hearths of short-term sites are surrounded by primary refuse of food consumption; those of long-term sites are not. This spatial relationship depends on the habitual use of fire in processing food. There are numerous archeological sites for which this relation probably holds, especially if bones display evidence of scorching or burning.

My own research has led me to believe that some primary refuse can be buried in the substrate of a site during the actual period of occupation. Such burial, the result of human trampling, may protect smaller elements from further disturbance or removal by human and other agencies. I became interested in the depositional potential of trampling during excavation of site 20, a single-occupation *gal dies* camp I saw created in November 1973 and completely buried by sands and silts during the spring rains of 1974 (Gifford and Behrensmeyer 1977). I plotted and identified all bone refuse visible on the surface on the day of its abandonment, along with all features. When excavating the buried scatter the following summer, I recovered almost ten times as many elements as had been visible on the surface in the previous autumn (1954 compared to 200 pieces). Although a proportion of these more numerous elements may have been fragments of originally mapped skeletal units, such as catfish crania or turtle shells, many were of sizes and types not observed at all during initial mapping.

Two lines of evidence convince me that these additional bones are not derived from some preexisting subsurface bone assemblage, but are actually smaller elements or fragments of the carcasses collected during the documented occupation of the site. First, the density distribution of the additional bones, including burned elements, so closely parallels that of the originally plotted sample that the essential unity of the two assemblages seems virtually certain. Second, these bones are all attributable to the species, and only to the species, noted in the original surface assemblage on the occupation site, though various elements are present in the two samples in differing proportions. For example, cranial elements of *Lates* and *Tilapia* were observed in the surface sample, but no rib and fin-ray elements were noted. The excavated sample, on the other hand, contained eighty-two ribs and fin rays. In the case of crocodile remains, all excavated elements could be matched to one of four different-sized individuals whose crania were noted in the original surface observations.

Many pieces were recovered from a sandy matrix, which is what remained of the occupation substrate, and were about 3 cm in maximum dimension, with some exceedingly narrow pieces reaching 5 cm. The camp had been occupied only four days, by eight persons, and this apparently was sufficient time for activities of the occupants to cause subsurface migration of most smaller elements. Yellen (1975) noticed similar trampling phenomena on sites. My students and I are currently conducting experiments to more closely define the role and results of trampling as a site formation process.

If this phenomenon is common on human sites with loose substrates, there are several implications for archeological interpretation. Relatively swift, size-dependent sorting of bone into surface and subsurface zones creates a situation where natural agents such as scavengers and weathering may operate to bias only surviving surface bone samples, as described by Yellen (1975). Furthermore, there may be a significant relationship between trampling and preservation of certain size classes of primary refuse. On the basis of students' observations of contemporary American discard behavior, Schiffer (1978) proposed that size is a major factor determining whether an item becomes primary or secondary refuse. Elements less than 2.5 cm (about 1 inch) in maximum dimension appear highly likely to become primary refuse, even when most larger items are discarded away from their location of use. This is close to the average maximum dimension of subsurface pieces recovered from site 20. If small size turns out to be a consistent feature of both primary refuse and elements migrating into an unconsolidated substrate, regardless of whether the elements are stone, bone, or any other artifactual material, a consistent method for defining primary refuse areas will have been found.

Site location is another situation in which human behavior and natural processes interact to influence the preservation of cultural materials. My research at Lake Turkana indicates that people may consciously and consistently favor some depositional environments and avoid others. The Dassanetch are well aware of the relationship between geomorphic setting and geologic process. They accordingly locate their semipermanent settlements in areas of minimal potential for disturbance. Home settlements and stock camps are situated along major watercourses (fig. 6.1) but are deliberately placed well away from areas liable to even minor flash-flood effects. Low-lying areas that collect rainwater are avoided, as are zones liable to inundation by minor lake transgressions. This pattern of site location amounts to avoidance of all environments with high potential for water-transported sedimentation. I observed about twenty-one semipermanent sites and only three were located in areas liable to fluvial inundation. These were all short-term dry season camps, located on a large island in the main course of the Il-Erriet River, one a large camp of about thirty-nine families and the others two small stock camps. These were abandoned prior to the autumn rains. All three sites were almost completely destroyed by floods in the spring of 1974, when the Il-Erriet rose to crests of about 4 m and inundated the entire island.

Study of a fifteen-year series of abandoned pastoralist sites in the Il-Erriet valley, dated through interviews with former inhabitants, indicates that bone of medium to large ungulates lasts at least fifteen years on the surface in well-drained areas. Surface bones from the oldest site, inhabited between 1957 and 1960, had not reached the stage of total disintegration (Behrensmeyer's Weathering Stage 5; Behrensmeyer 1978), but

many elements were flaky and fragile (Stage 4). Subsurface bones were not subject to
weathering of the same intensity (figure 6.4), but my impression was that eolian processes would ultimately expose these to weathering too.

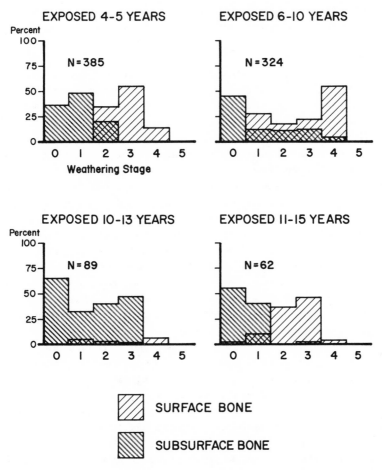

Figure 6.4. Ileret settlement bone weathering data contrasting surface and subsurface
components of sites of varying ages. Subsurface bones are less weathered than those which
remain exposed on the surface. Weathering stages go from 0 (fresh) to 5 (heavily weathered
and falling apart).

Projecting from this sample, I estimate that within about twenty years surface
bone on such sites would have completely disappeared. The ultimate outcome of continued
exposure of pastoralist sites to wind deflation and weathering processes is the massive
reduction of already scanty traces of human behavior. Resulting site inventories would
consist largely of hearthstones, scattered beads, glass shards, occasional hammerstones,
and, perhaps, corroded metal fragments—a much impoverished suite of cultural remains
providing no direct clue to economics.

Short-term camps in the region, on the other hand, did not appear to be so

preferentially located with regard to geomorphic setting. *Gal dies* sites normally lie
along the shoreline, but their precise locations vary considerably in their potential for
inundation. Some sites are on rocky promontories which would not be inundated by a
rise of several meters in lake level. Others lie on the very edges of extremely low
gradient beachlines, and can be inundated by only a few centimeters rise in lake level.
One such site, a hippo kill and butchery camp, was partially covered by the lake within
a week of the kill in September 1972; hippo bones closest to the lake remained covered
until exposed by falling lake level in January 1974. Several other sites lay in the
deltas of ephemeral streams, and one such site was completely covered in silt and sand,
with minimal disturbance of the original scatter, within four months of its creation
(Gifford and Behrensmeyer 1977).

The important point is that short-term camps, being less preferentially located
with regard to geomorphic setting than long-term settlements, sometimes lie in environments favorable to speedy and gentle deposition of the sort which preserves sites well.
Were the lake in a predominantly transgressive phase instead of its present regressive
one, more lakeshore sites would be gently inundated and covered by lacustrine sediments.
Of the twenty-one sites along 16 km of shore south of Koobi Fora, eighteen would be
covered by a 3 m rise in lake level. The Olduvai Bed I sites may represent a very
similar situation to that observed in modern *gal dies* sites near Koobi Fora, with
deposition beginning soon after sites were created, as a relatively minor transgression
of the ancient lake covered the scatter of debris with fine-grained sediments (Hay 1973).

From the standpoint of potential archeological preservation, it appears that
short-term sites in the Koobi Fora-Ileret region are much more likely to be buried
swiftly and with minimal disturbance of material than are long-term sites. These
sites reflect the protein procurement activities of a small group of people whose economy
is actually atypical of that of the Dassanetch as a whole. No population figures exist
for the *gal dies*, but they probably are no more than 5% of the total Dassanetch population. The fishing camps represent only one part of *gal dies* economic and settlement
patterns, and taken by themselves represent a biased picture of even that group's economy.
Were the current cultural and depositional situation to continue for an appreciable amount
of time, preserved sites would reflect a very biased sample of all sites once extant.
Gal dies camps, with their inventory of fish and lake reptiles and their low numbers of
hearths, would give no indication of the pastoral-agricultural adaptation of the
Dassanetch people, nor of their aggregation in settlements of considerable size.

Generalizing to archeological situations, it is possible that site location processes can produce considerable bias in preserved sites. As long as people consciously
avoid staying for prolonged periods in areas likely to be inundated, and prefer to make
their long-term camps in places of minimal aqueous or eolian depositional activity,
these sites are less likely to be "well preserved" than short-term camps situated in
more active depositional environments. In constructing regional models of prehistoric
hominid adaptations, this potential bias must be considered. The pastoralist-fishermen
site dichotomy of the Dassanetch situation proves a clear illustration of the point. Even
societies lacking either pronounced socioeconomic differences or strongly seasonal
variations in subsistence base may be represented in a biased way by differential preservation of short-term sites. It is possible that some "small band" models of prehistoric
social organization have been derived from archeological occurrences biased in this
manner. Yellen (1974) notes that the shorter the period of time people stay in a place,
the less likely it is that the full range of debris-producing activities of the group

will occur there. This produces in the short-term site only a small sample of overall activities.

Archeologists concerned with the possibility of this type of bias in their regional samples can profit from reconstruction of regional paleogeography along the lines of Hay's work at Olduvai Gorge (Hay 1973, 1976). With the aid of such reconstructions they may be able to assess the potential bias in their site sample. If bias in favor of smaller—perhaps short-term—sites appears possible, it may profit researchers to look for other, more poorly preserved sites in order to compare these with "well-preserved" occurrences in terms of their constituents, geologic context, and, possibly, behavioral significance.

Conclusion

Concern with the potential sources of bias in the archeological record, and with realistic interpretation thereof, certainly antedates the beginnings of ethnoarcheology as a field of study. However, ethnoarcheology has an advantage in addressing these problems because of its potential for systematic study of actual relations of humans to materials.

The key to elucidation of the past by studies of the present lies in assuming a comprehensive approach to the study of process and effect. By closely defining site formation processes, one can frame and test hypotheses concerning areas of knowledge which at present remain hazy. Clearly, several of the statements regarding trampling and site location behaviors made in this paper are easily rephrased as hypotheses to be submitted to cross-cultural tests.

Additionally, contemporary studies can be expected to expose weaknesses in assumptions underlying the entire endeavor of interpreting the past. The role of ethnoarcheology in delineating how curate and discard behavior may affect the formation of archeological deposits is a prime example of such a contribution. In this way, ethnoarcheology can assist the further refinement and clarification of archeological theory.

The similarities, both substantive and historic, between ethnoarcheology and taphonomy are many. Both seek, through the study of contemporary processes, to derive knowledge of ancient life processes from the preserved sample of prehistoric evidence. Many current taphonomic studies deal with the paleontological version of site formation processes (e.g., Dodson 1971; Behrensmeyer and Dechant Boaz, this volume; Bishop, Hanson, and Hill, this volume). These studies are all maximally effective when put into a conceptual framework based upon the preserved sample—in this case, the fossil deposit.

Acknowledgments

I thank the Office of the President of the Republic of Kenya for permission to conduct research, and the Trustees of the National Museums of Kenya for their sponsorship and aid. Field work was supported by National Science Foundation Dissertation Research Grant GS-40206, and living expenses were paid by a National Defense Education Act (Title IV) Graduate Fellowship. The Wenner-Gren Foundation for Anthropological Research financed my air fare to Kenya, and part of the basic data processing for this paper was supported by a University of California (Berkeley) Dean's Fund Fellowship, a Regent's Fund Award, and computer time allocations from the U.C. Computer Center.

My thanks are given to Dr. Glynn Ll. Isaac, University of California, Berkeley, and to Mr. Richard E. F. Leakey, Administrative Director of the National Museums, for their invaluable advice and material aid. I thank Dr. A. Kay Behrensmeyer for sharing her bone weathering code with me prior to its publication. My field assistants, Mr. Andrea Kilonzo and Mr. Jack Kilonzo, deserve thanks for their patience and hard work.   I also thank Robert M. Poor, University of Chicago, for his critical reading of preliminary drafts of this paper. Dr. Michael B. Schiffer provided me with copies of his and his associates' articles. Dr. Roger Wood (Stockton State College) kindly provided identifications of terrapins collected at Lake Turkana.

My greatest debt of gratitude is to the Dassanetch people of Ileret.

7. SOME CRITERIA FOR THE RECOGNITION OF BONE-COLLECTING
AGENCIES IN AFRICAN CAVES

C. K. Brain

Introduction

From the paleontological point of view, caves serve as important receptacles
for the preservation of bone over long periods of time. Bones preserved in cave sediments have contributed significantly to an understanding of the course of human evolution
and to a knowledge of the fauna associated with early man.

Any paleoecological interpretation based on remains found in caves relies heavily
on an understanding of the way in which the bones entered the deposit. It is therefore
desirable to know what the main bone-accumulating agencies in African caves are and,
where possible, to define criteria whereby the various agencies can be recognized.

The process of bone accumulation in a cave will depend very much on the form of
the cave itself, but one fact must be faced: any cave which has been open for thousands
of years is likely to have had bones brought to it in a variety of different ways. Some
remains may have come from the natural deaths of animals in the cave itself or within
the catchment area of the entrance; others may have been brought there by porcupines
which used the cave as a lair, still others may represent the occupational debris of
human inhabitants, while another group could well result from the activities of predators
and scavengers. The bone accumulation in any one cave could easily reflect the action of
all these agencies. However, the presence of bones in particular types of caves is
likely to result from certain agencies to the partial or complete exclusion of others. In
interpreting a bone assemblage, therefore, the aim is to isolate the recurrent and
dominant agents of accumulation which may have been active at the time in question.

The potential significance of studies on bone accumulations in African caves has
not been appreciated for very long, and relevant literature is meager. This paper will
present a few personal observations made in caves in various parts of Southern Africa.
Such observations made on bone accumulations by various investigators, including myself,
have led to the belief that the following are among the more important bone-accumulating
agencies in Southern African caves: (a) porcupines; (b) hominid hunter-gatherers; (c)
hyenas and other scavenger-predators; (d) leopards.

Natural deaths may be of special significance in certain death-trap type caves,
but since entire skeletons should be preserved as a result, this agency is likely to be
easily recognizable.

Some observations on collecting agencies will now be made and a few criteria suggested for the recognition of such agencies.

Porcupines as Bone Collectors

It is rather surprising that a vegetarian rodent should prove to be an important collector of bones, yet I suspect it may be true that porcupines carry more bones to African caves than does any other species of animal. Although it was remarked upon forty years ago that porcupine lairs could be detected by the presence of numerous gnawed bones near their entrances (Shortridge 1934), attention was first focused on this strange behavior of porcupines during the 1950s when Prof. R. A. Dart and Mr. A. R. Hughes were trying to make sense of the remarkable bone accumulation in the gray breccia at Makapansgat (Hughes 1954; Dart 1957). About that time Hughes started his observations on contemporary animal lairs, including those of porcupines. In September 1954 he visited the Kalahari Gemsbok National Park in the Northern Cape and examined several lairs in the banks of the dry Auob and Nossob rivers there. Two of these were clearly inhabited by porcupines, and from them Hughes (1958) took a total of 147 bones and other objects which the porcupines had apparently hoarded. The two sites were both in the east bank of the Auob River, respectively 18 and 26 kilometers upstream from the main camp at Twee Rivieren.

In May 1956 Hughes returned to the Kalahari Park and located a third porcupine lair in the east bank of the Nossob River, eight miles upstream from Twee Rivieren. On 21 May 1956 he removed from it 1420 bones and other objects which appeared to have been collected by the porcupines.

Some direct information on porcupine hoarding behavior came to light at about the same time. In 1956 a young male porcupine by the name of Aristotle was very much part of the staff of the zoology department at Rhodes University in Grahamstown. Aristotle made his lair in the photographic darkroom but had free run of the laboratory and surroundings. Before long he started to stock his lair with objects collected during his nightly wanderings—bones, a tortoise carapace, "a wicker basket, an enamel dish, a duster, a piece of flex, a drain pipe, a buck horn and a piece of softboard" (Alexander 1956). Of particular interest, however, was Dr. Anne Alexander's observation that the porcupine showed no interest in fresh bones with meat on them—in fact, the purpose of the gnawing was soon apparent: it had to do with wearing down of the incisor teeth rather than with nutrition. The African porcupine (*Hystrix africae-australis*), like other rodents, has open-rooted incisors which grow throughout the life of the animal; they require regular attrition to keep them at usable length. And so it seems that porcupines have developed a behavior pattern which requires them to collect dry bones and other hard objects and hoard them in their lairs. While resting during the day in their lairs, the porcupines select some of their favored objects and gnaw them. The collecting behavior appears to have become a compulsion—they will bring back far more objects than they can possibly use, and do not get round to gnawing anything like all the treasures they collect.

An interesting comparison between a porcupine collection and a bone assemblage from a human occupation site was made more recently by Hendey and Singer (1965). Two adjacent caves in the rocky wall of the Gamtoos Valley of the Eastern Cape were examined before a dam-building operation destroyed them. The one, designated AK 1, appears to have been a human occupation site, containing, in addition to bone, abundant evidence of culture and fire. The other site, AK 2, which took the form of a low crack in the rock, appeared to have been a porcupine lair. The "human site" yielded 11,056 bone pieces, of which 8887 (or 80.4%) were too fragmentary to be identified; the "porcupine lair" produced 1105 pieces, of which only 465 (or 42.1%) were unidentifiable. There was a striking difference, therefore, in the degree of fragmentation of the bones—this is reflected also in the

weight of each sample. The 11,045 bone pieces from AK 1 weighed 20.9 kg, giving an average weight per piece of 1.9 g; the 1105 pieces from AK 2 weighed 85.3 kg, indicating an
average weight per piece of 77.2 g. Clearly, the bones in the porcupine lair were
are more complete than those in the human site. Likewise, only 0.3% of the identified
bones from AK 1 showed signs of porcupine gnawing, compared with 60% of those from AK 2.

In this study, Hendey and Singer pointed out for the first time the differences
likely to be encountered between bone assemblages from human and porcupine habitation
sites. These indications have been borne out by subsequent study.

Through the generosity of Mr. A. R. Hughes, I have been able to examine his collections from porcupine lairs in the Kalahari Park and have also made personal investigations at the lairs themselves. Some results will now be discussed.

The Nossob Porcupine Lair and Its Evidence

This site was first visited by Hughes in May 1956 when the following notes were
made by him (personal communication):

The lair was situated in one of a number of intercommunicating solution cavities in the
calcrete at the top of the east bank of the Nossob River 8 miles from the base camp at
Twee Rivieren in the south of the Kalahari Gemsbok National Park. In the front of the
entrance was a 72 sq. ft. area of loose ground 6 ins. - 1 ft. in depth which contained 411
porcupine quills and many very small hair-like quills, gnawed bones, skulls and horns of
large and small animals, 9 tins, pieces of iron, bottles, 105 pieces of wood up to 3 ft.
in length and other articles collected by porcupines. The inside of the lair was about
425 sq. ft. in extent and tapered away from the entrance for 36 ft. to a small opening
in the calcrete surface superadjacent to the river bank. On the inside of the lair,
especially the sides near the entrance, were scattered the bones, skulls and horns of
animals which had died during the last 25 years as was indicated by the presence of
the horns of domestic rams and cows; domestic cattle had not lived in the neighbourhood
since 1932 when the park was proclaimed.

Inside the lair was a circular raised piece of clean ground well consolidated by the
porcupines that had lain there sleeping . . .

I have been able to visit the site on two occasions—in March 1968 when all
bones and other objects were again removed from the lair, and in June 1969 when the sketch
plan shown in figure 7.1 was drawn. The site itself and abundance of bones inside its
entrance are shown in figure 7.2 and 7.3. Apart from the main entrance at A on the plan,
the lair has three openings to the surface. The roof is low, varying from 18 to 24 in.
(46 to 61 cm), a situation which makes retrieval of bones difficult. Two porcupines were
visible in the further recesses of the lair at the time of my visits; they shook their
quills in a threatening manner but were not otherwise troublesome. More discomfort was
occasioned by the voracious tampan ticks which emerged in large numbers from the dusty
substrate of the lair while the investigators were creeping around inside it.

The 1956 collection, made by Hughes and kindly put at my disposal by him, consisted of 1328 pieces as listed in table 7.1. By 1968, a further 384 objects were available for collection, making a total of 1712. These were all bone, horn, or tortoise
shell, with the exception of seventy-one pieces of wood and seventeen metal objects (iron
bars, rusty tines, enamel mugs, and the like).

As reflected in table 7.2, the remains in the whole collection came from a minimum
of 106 individuals, bovids such as springbok, gemsbok, hartebeest, and wildebeest being
particularly well represented on the basis of their horns.

The study of the bone assemblage from the Nossob lair has provided information on
a number of questions:

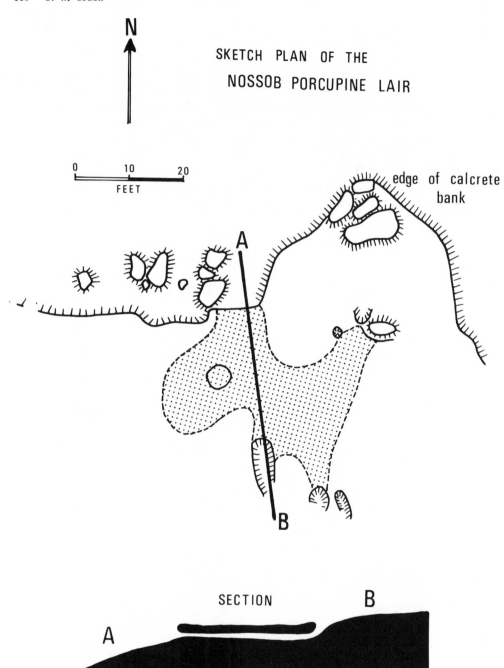

Figure 7.1.   Sketch plan of the Nossob porcupine lair.

Figure 7.2. The main opening to the Nossob porcupine lair.

Figure 7.3. Bones and horns inside the entrance to the Nossob porcupine lair.

1. What is the rate of accumulation of objects in a porcupine lair? The Nossob
lair was cleared by Hughes in May 1956 and again by myself in March 1968, giving a time
lapse of almost 12 years. During this period 380 objects were taken to the lair, a
rate of about 32 per year. I was able to observe two porcupines in the lair on two occasions, but there may well have been others in the deeper recesses which were out of
sight. The rate of collection of bones would be dependent on their availability; I suspect that the Nossob locality would generally provide an abundance of bones for porcupines
to collect.

2. Do the bones in the porcupine collection reflect the natural abundance of
animals from which they come? It is a fortunate fact that the Nossob lair is situated in
the Kalahari National Park, an area for which census data on the contemporary bovid
populations are available. These unpublished data have been kindly provided by Dr. G.
De Graaff (De Graaff, Bothma, and Moolman, unpublished) of the National Parks Board. The
census figures for springbok, gemsbok, hartebeest, wildebeest, eland, steenbok, and

Table 7.1
Faunal Remains Collected from the Nossob Porcupine Lair

	Collection 1956	Collection 1968	Totals
Gemsbok *(Oryx gazella)*			
Cranial pieces	7	2	9
Horns	21	7	28
Springbok *(Antidorcas marsupialis)*			
Cranial pieces	12	5	17
Horns	46	26	72
Wildebeest *(Connochaetes taurinus)*			
Horns	5	3	8
Hartebeest *(Alcelaphus buselaphus)*			
Cranial pieces	2	—	2
Horns	11	6	17
Steenbok *(Raphicerus campestris)*			
Horns	2	4	6
Duiker *(Sylvicapra grimmia)*			
Cranial pieces	2	—	2
Goat *(Capra hircus)*			
Horns	2	1	3
Cattle *(Bos taurus)*			
Horns	5	—	5
Indet. bovid postcranial pieces	405	243	648
Equid			
Cranial pieces	1	—	1
Postcranial pieces	6	4	10
Lion *(Panthera leo)*			
Cranial pieces	1	—	1
Hunting dog *(Lycaon pictus)*			
Cranial pieces	3	3	6
Mongoose *(cf. Cynictis sp.)*			
Cranial pieces	2	—	2
Skunk *(Ictonyx striatus)*			
Cranial pieces	—	1	1
Indet. carnivore postcranial pieces	2	5	7
Porcupine *(Hystrix africae-australis)*			
Cranial pieces	1	—	1
Tortoise *(Testudo occulifera)*			
Carapace pieces and scales	56	3	59
Ostrich *(Struthio camelus)*			
Postcranial pieces	7	4	11
Eggshell pieces	7	—	7
Large bird (gen. and sp. indet.)			
Postcranial piece	—	1	1
Miscellaneous bone fragments	364	31	395
Bone flakes	277	28	305
Pieces of wood	68	3	71
Metal objects	13	4	17
Total	1328	384	1712

duiker are given in table 7.3 and may be compared with the minimum numbers of individuals reflected in the porcupine collection. The relevant numbers for these species are as follows: springbok, 40; gemsbok, 15; hartebeest, 9; wildebeest, 5; steenbok, 5; and

Table 7.2

Minimum Numbers of Individuals of Taxa Identified in the Nossob Porcupine Lair Bone Collection

	Collection		Totals
	1956	1968	
Springbok	25	15	40
Gemsbok	11	4	15
Hartebeest	6	3	9
Wildebeest	3	2	5
Steenbok	2	3	5
Duiker	2	0	2
Cattle	3	0	3
Goat	1	1	2
Horse	2	2	4
Hunting dog	1	3	4
Mongoose	2	0	2
Skunk	0	1	1
Lion	1	0	1
Porcupine	1	0	1
Tortoise	6	3	9
Ostrich	1	1	2
Bird	0	1	1
Total	67	39	106

Table 7.3

Census Figures and Percentages for Herbivores in the Kalahari National Park

Species	August	November	February	April	Annual Average	Percent of Total
Springbok	17560	18228	24041	12894	18181	40.2
Gemsbok	11183	7384	16073	15915	12639	27.9
Hartebeest	14178	1408	5200	5011	6449	14.2
Wildebeest	6378	2230	4067	2522	3799	8.4
Eland	1433	611	6569	1235	2462	5.4
Steenbok	1402	1073	1051	1645	1293	2.9
Duiker	262	381	710	497	463	1.0
Total	52396	31315	57711	39719	45286	100.0

Note: From Kalahari Park area census 1973-74 (De Graaff et al.).

duiker, 2. Eland has not been identified on cranial remains, but at least three individuals of an eland-size bovid are represented in the postcranial parts.

Percentage abundance of the various species is plotted in figure 7.4, from which it is evident that the minimum numbers of individual animals represented by the porcupine-collected remains do indeed mirror the actual abundance of the antelope species.

An interesting point here is that, although more springbok are represented by their

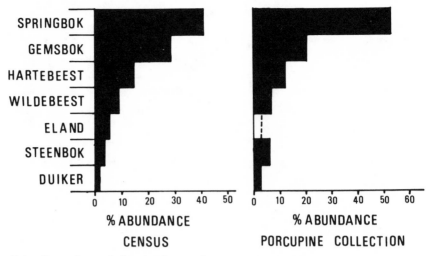

Figure 7.4. Comparison of the abundance of mammal taxa in censuses of the live populations in Kalahari National Park and in the Nossob porcupine collection.

horns than are larger antelope, the postcranial remains of the larger bovids are more numerous than are their smaller counterparts. For convenience, southern African bovids have been divided into four size classes (Brain 1974) according to weight: Class I, up to about 50 lbs (23 kg); II, 50-200 lbs (23-91 kg); III, 200-650 lbs (91-295 kg); and IV, above 650 lbs (295 kg). Springbok fall into class II, gemsbok, hartebeest, and wilde-beest in class III.

Table 7.4 lists the bovid skeletal parts in the 1968 collection from the Nossob lair (unfortunately I do not have data on bovid size classes for the larger 1956 collec-tion). On the basis of horns, the 1968 collection contains parts of 15 class II antelope (springbok) and 9 class III ones (gemsbok, hartebeest, and wildebeest), yet, as is ap-parent in table 7.4, class II antelope are represented by 132 pieces and class III by 183. The reason for this discrepancy is probably that the horns of almost every antelope killed survive and may be collected by porcupines, but that much less is left of the skele-ton of a smaller antelope after predator and scavengers are finished than is the case with larger prey.

3. What parts of bovid skeletons are represented and do these represent the most resistant elements in the skeleton? Table 7.5 lists the parts of bovid skeletons repre-sented in the two Nossob collections. As will be seen, there is wide representation of almost all skeletal parts. On the basis of horns there appears to have been a minimum of eighty-one individual antelope which contributed to the sample. Working with this total it is possible to calculate how many individual parts of the skeletons there should have been if none was lost and all were collected. These figures are given in the third column of table 7.6. Surprisingly enough, following pelvic pieces, atlas and axis verte-brae have the highest survival figures, while caudal vertebrae have the lowest. Following through the vertebral column, the sequence of percentage survival from highest to lowest is as follows: atlas (30.9%), axis (28.4%), other cervicals (18.8%), lumbars (15.6%), sacral (8.6%), thoracic (5.0%), caudal (0.1%).

These survival figures very probably reflect the robusticity of the different vertebral types but may also have been influenced by preferential selectivity by the

Table 7.4

Bovid Skeletal Parts from the 1968 Nossob Porcupine Lair Bone Collection

Part	Antelope Size Class			Totals
	I	II	III	
Horn pieces	0	28	17	45
Other cranial pieces	0	7	10	17
Vertebrae: atlas	0	2	6	8
axis	0	5	9	14
cervicals 3-7	0	17	19	36
thoracic	3	7	6	16
lumbar	2	20	18	40
sacral	0	2	2	4
caudal	0	0	0	0
Rib pieces	0	5	0	5
Scapula pieces	0	3	16	19
Pelvic pieces	3	7	20	30
Humerus: complete	0	1	1	2
proximal	0	1	2	3
distal	0	2	2	4
shaft	0	1	0	1
Radius and ulna: complete	0	2	3	5
proximal	0	2	3	5
distal	0	0	0	0
shaft	1	0	0	1
Carpal bones	0	0	1	1
Metacarpal pieces	0	1	2	3
Femur: complete	0	0	0	0
proximal	0	1	10	11
distal	0	0	8	8
shaft	0	1	2	3
Tibia: complete	0	1	4	5
proximal	1	2	4	7
distal	0	0	4	4
shaft	1	0	0	1
Calcaneus	0	1	1	2
Astragalus	1	1	2	4
Metatarsal pieces	0	3	1	4
Metapodial pieces	2	1	7	10
Phalanges	1	8	3	12
Total	15	132	183	330

Table 7.5

Representation of Bovid Skeletal Parts in the Nossob Porcupine Lair Collection

Part	Collection		Total
	1956	1968	
Cranium: complete skulls	0	0	0
maxillary pieces	11	4	15
mandibular pieces	19	5	24
isolated teeth	27	0	27
other pieces	16	9	25
horn pieces	112	45	157
Vertebrae: atlas	17	8	25
axis	9	14	23
cervical	40	36	76
thoracic	40	13	53
lumbar	36	40	76
sacral	3	4	7
caudal	1	0	1
Ribs	38	5	43
Scapula pieces	24	18	42
Pelvis pieces	36	30	66

(continued)

Table 7.5—Continued

Part	Collection 1956	1968	Total
Humerus: complete	2	2	4
proximal	2	3	5
distal	12	4	16
shaft	3	1	4
Radius and ulna: complete	0	5	5
proximal	13	5	18
distal	9	0	9
shaft	1	1	2
Carpal bones	12	5	17
Metacarpal pieces	7	3	10
Femur: complete	1	0	1
proximal	5	11	16
distal	6	8	14
shaft	7	3	10
Tibia: complete	3	5	8
proximal	6	7	13
distal	9	4	13
shaft	3	1	4
Calcaneus	16	2	18
Astragalus	15	4	19
Other tarsal bones	9	0	9
Metatarsal pieces	10	4	14
Metapodial pieces	14	10	24
Phalanges	48	12	60
Sesamoids	1	0	1
Total	643	331	974

Table 7.6

Percentage Survival Figures for Bovid Skeletal Parts in the Nossob Collection Compared with a Hottentot Goat Bone Sample

Part	Nossob Sample:81 Bovid Individuals			Hottentot Goat Bone Sample Percent Survival
	Number Found	Original Number	Percent Survival	
Horn pieces	157	162	96.9	94.7
Pelvic pieces	66	162	40.7	9.0
Atlas vertebrae	25	81	30.9	6.3
Axis vertebrae	23	81	28.4	7.4
Scapula pieces	42	162	25.9	9.2
Cervical vertebra, 3-7	76	405	18.8	1.2
Maxillae	15	81	18.5	26.3
Metatarsal, proximal	26	162	16.0	10.3
Metatarsal, distal	26	162	16.0	5.3
Lumbar vertebrae	76	486	15.6	2.7
Half mandibles	24	162	14.8	30.7
Radius and ulna, proximal	23	162	14.2	17.1
Metacarpal, proximal	22	162	13.6	8.4
Metacarpal, distal	22	162	13.6	6.0
Tibia, proximal	21	162	13.0	3.4
Tibia, distal	21	162	13.0	19.0
Humerus, distal	20	162	12.3	21.5
Astragalus	19	162	11.7	4.2
Calcaneus	18	162	11.1	3.7
Femur, proximal	17	162	10.4	4.7
Femur, distal	15	162	9.3	2.4
Sacral vertebrae	7	81	8.6	0.5
Radius and ulna, distal	14	162	8.6	5.8
Phalanges	60	972	6.2	0.9
Humerus, proximal	9	162	5.5	0
Thoracic vertebrae	53	1053	5.0	0.9
Ribs	43	2106	2.0	3.4
Caudal vertebrae	1	810	0.1	0

porcupines. Perhaps porcupines simply prefer a substantial-looking, chunky cervical vertebra to a scrawny-looking thoracic one with its long neural spine.

The individual parts of a bovid skeleton vary considerably in strength characteristics and resistance to destructive treatment. Under a given destructive regime, the
individual parts will survive in proportion to their robusticity. Study of the survival
of goat bones, subjected to Hottentot and dog feeding action (Brain 1967, 1969, 1976) ,
has shown which parts of the skeleton are able to withstand destructive treatment better
than others. Table 7.7 lists the various parts of the goat skeletons and then provides

Table 7.7

Numbers of Different Skeletal Parts and Percentage Survival in the Hottentot Goat Bone
Sample (See Text and Brain 1967, 1968, 1976)

Part	Number Found	Horns Excluded		Horns Included	
		Original Number	Percent Survival	Original Number	Percent Survival
Horns	360	--	--	380	94.7
Half mandibles	117	128	91.4	380	30.7
Maxillae	50	64	78.1	190	26.3
Humerus, distal	82	128	64.0	380	21.5
Tibia, distal	72	128	56.3	380	19.0
Radius and ulna, proximal	65	128	50.8	380	17.1
Metatarsal, proximal	39	128	30.4	380	10.3
Scapula	35	128	27.4	380	9.2
Pelvis half	34	128	26.6	380	9.0
Metacarpal, proximal	32	128	25.0	380	8.4
Axis	14	64	21.9	190	7.4
Atlas	12	64	18.8	190	6.3
Metacarpal, distal	23	128	18.0	380	6.0
Radius and ulna, distal	22	128	17.2	380	5.8
Metatarsal, distal	20	128	15.6	380	5.3
Femur, proximal	18	128	14.1	380	4.7
Astragalus	16	128	12.5	380	4.2
Calcaneus	14	128	10.9	380	3.7
Ribs	170	1664	10.2	4940	3.4
Tibia, proximal	13	128	10.1	380	3.4
Lumbar vertebrae	31	384	8.1	1140	2.7
Femur, distal	9	128	7.0	380	2.4
Cervical vertebrae, 3-7	12	320	3.8	950	1.2
Phalanges	21	768	2.7	2280	0.9
Thoracic vertebrae	21	832	2.5	2470	0.9
Sacrum	1	64	1.6	190	0.5
Caudal vertebrae	0	1224	0	1900	0
Humerus, proximal	0	128	0	380	0

Note: Minimum number of individual goats: 64 (horns excluded), 190 (horns included).

figures for their percentage survival. Horns had the highest survival followed by mandibles,
maxillae, and distal humeri. Proximal humeri and caudal vertebrae were found to have a
nil survival value. Survival may be correlated with the compactness of the bone, expressed
as specific gravity, and with fusion times of the epiphyses of the long bones. In fact,
the survival of parts will follow an entirely predictable pattern if the destructive
influences are known.

In figure 7.5 the percentage survival of bovid skeletal parts in the Nossob sample
are plotted and contrasted with that for the same parts in the Hottentot goat bone sample.
The Nossob survival sequence does not follow the Hottentot pattern; to my mind it does not
follow the pattern which could be predicted for skeletons subjected to a good deal of

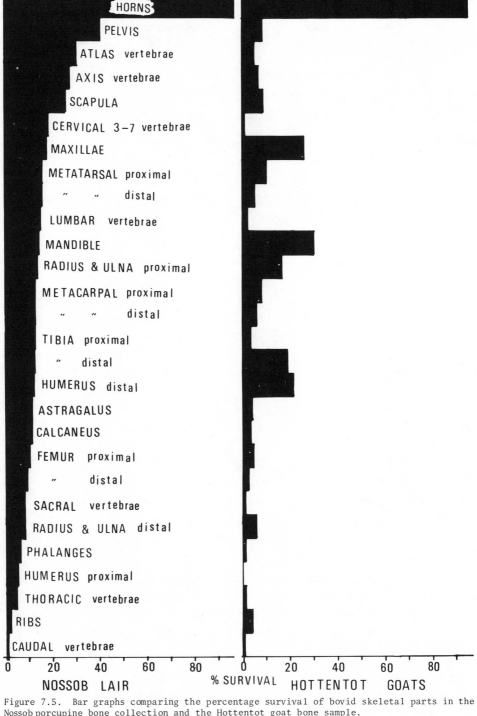

Figure 7.5. Bar graphs comparing the percentage survival of bovid skeletal parts in the Nossob porcupine bone collection and the Hottentot goat bone sample.

destructive treatment. For instance, the ratio of proximal/distal humeri in the Hottentot sample is 0/82; in the Nossob sample it is 9/20. It is perfectly clear that the bovid skeletons in this part of the Kalahari Park have not been subjected to extreme scavenging pressure; complete bones frequently survive as do delicate elements. The residue left for porcupines to collect therefore does not necessarily consist of only the most resistant elements.

4. Do porcupines preferentially collect objects of a particular size and weight? In order to answer this question, each individual object in the Nossob collection was weighed and measured, the greatest length per piece being recorded in inches. The results given in tables 7.8 and 7.9 are plotted in histogram form in figure 7.6. It will be seen

Table 7.8

Numbers of Gnawed and Ungnawed Bones, Listed by Weight Category, in the Nossob Porcupine Lair Collections

Weight (g)	1956 Collection			1968 Collection			Overall Totals
	n Ungnawed	*n* Gnawed	Total	*n* Ungnawed	*n* Gnawed	Total	
0- 10	388	283	671	31	30	61	732
10- 20	45	100	145	19	34	53	198
20- 30	21 486	70 524	91 1010	11 68	19 116	30 184	121
30- 40	16	42	58	4	18	22	80
40- 50	16	29	45	3	15	18	63
50- 60	6	20	26	6	15	21	47
60- 70	6	35	41	5	20	25	66
70- 80	5 25	15 99	20 124	4 21	10 69	14 90	34
80- 90	4	16	20	4	12	16	36
90-100	4	13	17	2	12	14	31
100-150	14	59	73	9	41	50	123
150-200	13	31	44	4	20	24	68
200-250	2	16	18	2	9	11	29
250-300	5	13	18	3	8	11	29
300-350	3	5	8	0	4	4	12
350-400	5	3	8	0	2	2	10
400-450	1	6	7	0	2	2	9
450-500	1	2	3	0	0	0	3
500-550	1	3	4	0	0	0	4
550-600	1	3	4	0	1	1	5
600-650	0	3	3	0	1	1	4
650-700	1	1	2	0	0	0	2
700-750	0	2	2	0	0	0	2
Total	558	770	1328	107	273	380	1708

that, as far as weight is concerned, the great bulk of the sample falls in the 0-50 g category, though some objects up to 750 g in weight are included in the sample. Concerning length, the bulk of objects have maximum dimensions of between 1 and 6 inches (2.5 to 15.2 cm), though some up to 36 inches (91.4 cm) (mostly gemsbok horns and pieces of wood) do occur.

5. Do porcupines preferentially gnaw objects of a particular size? Tables 7.10 and 7.11 provide figures for the percentages of gnawed bones in each weight and length class. It is immediately apparent that the percentages of gnawed bones in the smallest weight and length classes are *lower* than in the larger categories. The conclusion therefore is that, although porcupines collect large quantities of small bone pieces, they prefer to gnaw the larger ones. It is probably more difficult for a porcupine to

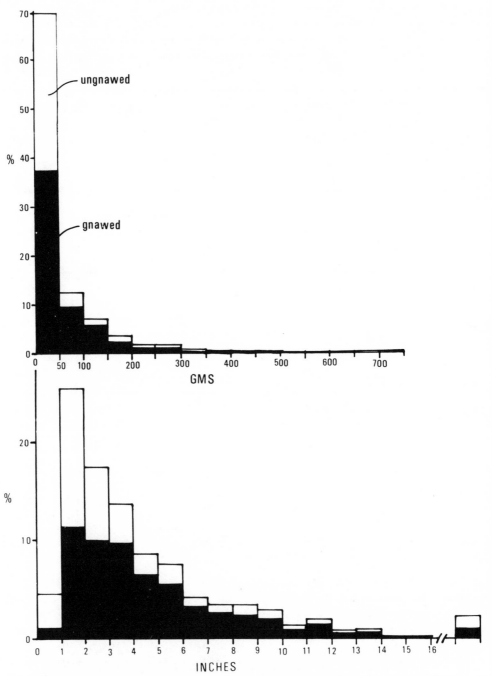

Figure 7.6. Histograms comparing percentages of gnawed and ungnawed bones in different weight (grams) and size (inches) classes from the Nossob porcupine lair.

hold a small object between its forepaws while gnawing than is the case with a larger one.

6. Are the collected bones generally fresh, or weathered? Any curator of an

Table 7.9

Numbers of Gnawed and Ungnawed Bones, Listed by Size Category, in the Nossob Porcupine Lair Collection

Inches	1956 Collection			1968 Collection			Overall Totals
	n Ungnawed	*n* Gnawed	Total	*n* Ungnawed	*n* Gnawed	Total	
0- 1	59	14	73	2	3	5	78
1- 2	216	185	401	19	16	35	436
2- 3	113	134	247	15	37	52	299
3- 4	55	121	176	13	47	60	236
4- 5	26	71	97	9	41	50	147
5- 6	21	65	86	15	34	49	135
6- 7	11	34	45	6	20	26	71
7- 8	9	30	39	6	14	20	59
8- 9	13	30	43	3	13	16	59
9-10	7	19	26	8	17	25	51
10-11	3	10	13	4	7	11	24
11-12	6	21	27	2	6	8	35
12-13	4	4	8	0	8	8	16
13-14	1	8	9	1	7	8	17
14-15	1	2	3	0	1	1	4
15-16	1	3	4	0	0	0	4
16-17	1	4	5	0	0	0	5
17-18	1	1	2	0	0	0	2
18-19	2	0	2	0	0	0	2
19-20	0	2	2	1	1	2	4
20-21	0	0	0	0	0	0	0
21-22	0	1	1	0	0	0	1
22-23	0	1	1	0	0	0	1
23-24	2	2	4	2	0	2	6
24-25	1	0	1	1	0	1	2
25-26	0	1	1	0	0	0	1
26-27	0	1	1	0	1	1	2
27-28	0	0	0	0	0	0	0
28-29	0	0	0	0	0	0	0
29-30	0	0	0	0	0	0	0
30-31	1	0	1	0	0	0	1
31-32	0	0	0	0	0	0	0
32-33	4	0	4	0	0	0	4
33-34	0	1	1	0	0	0	1
34-35	0	0	0	0	0	0	0
35-36	0	1	1	0	0	0	1
36-37	0	1	1	0	0	0	1
37-38	0	2	2	0	0	0	2
38-39	0	1	1	0	0	0	1
39-40	0	0	0	0	0	0	0
Total	558	770	1328	107	273	380	1708

osteological collection will have had the experience that, unless bones are defatted before being stored, they will continue to exude grease indefinitely. An interesting point about the Nossob procupine collection is that only a very small number of bones (no more than 15 out of a total of 1620) showed appreciable traces of fattiness at all. I have little doubt that if the bones had been brought to the lair in a fatty condition, this fattiness would have remained indefinitely in the shaded conditions of the lair. I have this conviction as a result of a bone weathering experiment (Brain, in preparation) which I set up more than ten years ago at the Namib Desert Research Station in South-West Africa. Two series of fresh bones were placed in a wooden frame; one series was fully exposed, protected only by wire netting. The other series was protected from direct sun by surrounding screens of 1/8 inch (.3 cm) pegboard with holes at 1 inch (2.5 cm) centers. The frame was set up in a fully exposed situation on 1 December 1965; effects were observed

Table 7.10

Totals and Percentages of Gnawed Bones in Each Bone-Weight Class for the Nossob Porcupine Lair

Weight Class (g)	Number of Bones etc.	Percentage of Total	Number of Gnawed Bones etc.	Percentage of Total	Percentage of Gnawed Bones etc. this Weight Class
0- 50	1194	69.9	640	37.5	53.6
50-100	214	12.5	168	9.8	78.5
100-150	123	7.2	100	6.4	81.3
150-200	68	4.0	51	3.0	75.0
200-250	29	1.7	25	1.4	82.8
250-300	29	1.7	21	1.3	72.4
300-350	12	0.7	9	0.5	75.0
350-400	10	0.6	5	0.3	50.0
400-450	9	0.5	8	0.5	88.9
450-500	3	0.2	2	0.1	66.7
500-550	4	0.2	3	0.2	75.0
550-600	5	0.3	4	0.2	80.0
600-650	4	0.2	4	0.2	100.0
650-700	2	0.1	1	0.1	50.0
700-750	2	0.1	2	0.1	100.0
Total	1708	99.9	1043	55.2	

Note: Overall percentage gnawed objects = 61.1.

Table 7.11

Totals and Percentages of Gnawed Bones in Each Size Class for the Nossob Porcupine Lair

Length Class (inches)	Number of Bones etc.	Percentage of Total	Number of Gnawed	Percentage of Total	Percentage Gnawed Bones etc. in this Size Class
0- 1	78	4.6	17	1.0	21.8
1- 2	436	25.5	201	11.8	46.1
2- 3	299	17.5	171	10.0	57.2
3- 4	236	13.8	168	9.8	71.2
4- 5	147	8.6	112	6.6	76.2
5- 6	135	7.9	99	5.8	73.3
6- 7	71	4.1	54	3.2	76.1
7- 8	59	3.5	44	2.6	74.6
8- 9	59	3.5	43	2.5	72.9
9-10	51	3.0	36	2.1	70.6
10-11	24	1.4	17	1.0	70.8
11-12	35	2.1	27	1.6	77.1
12-13	16	0.9	12	0.7	75.0
13-14	17	1.0	15	0.9	88.2
14-15	4	0.2	3	0.2	75.0
15-16	4	0.2	3	0.2	75.0
16+	37	2.2	21	1.2	56.8
Total	1708	100.0	1043		

at yearly intervals and some specimens were removed on 21 March 1973 for photography, after a time lapse of several years. Two halves of a pig mandible showed striking differences. One half, which had been exposed to full sun, was bleached and cracked while the other, which had been shaded from the sun in a well-ventilated situation, retained enough fat after seven years storage for dust to be adhering to it. The defatting process of the fully exposed bone was complete within one year, and I would suspect within three months. Defatting in a shaded situation may take decades.

On the basis of these observations, I have no doubt that the bones generally brought to the Nossob lair by porcupines are ones which have been naturally defatted through surface exposure. In fact, the porcupines show a very marked preference for bleached, defatted bones.

Study of the objects in the Nossob lair has made it possible to isolate a number of criteria whereby a porcupine-collected assemblage might be recognized. It will be interesting to see how well these suggested criteria stand in the light of future studies.

Food Remains of Hominid Hunter-Gatherers in Caves

Many studies have been made of bones resulting from hunter-gatherer occupation of caves. They have served to emphasize the fact that people are opportunists, feeding on whatever animal resources the particular environment has to offer and which the people have access to. Depending on circumstances, people will feed on a variety of animals ranging from elephants to invertebrates. The species list for a bone assemblage may therefore be of little value in identifying the bone-collecting agency. Similarly, fragmentation of the bones, though likely to be extreme in Stone Age human food remains, is very variable; it is dependent not only on deliberate bone breakage for the extraction of marrow, but also on the effect of people walking round on bones which happen to be lying on the cave floor.

Other criteria are fortunately diagnostic. These are as follows: (a) <u>Association with human culture material</u> in a cave which has been occupied by Stone Age people for a long time, the abundance of cultural material in any given layer is likely to be mirrored by an associated abundance of bone fragments. An example of this situation is shown in figure 7.7 where data from the Pomongwe Cave in Rhodesia, excavated by Mr. C. K. Cooke (Cooke 1963), are plotted. Over an estimated occupation period of 70,000 years, the abundance of artifacts per level is broadly mirrored by an abundance of bone fragments. (b) <u>Effect of fire on the bones</u>. Preservation of bone fragments in the ash of people's camp fires in caves is common and represents good evidence for human involvement. Charring of bones in caves is more likely to have resulted from human agency than from natural causes.

Food Remains of Hyenas and Other Scavenger-Predators

Although hyenas may have been responsible for accumulating large quantities of bones in European caves, their role in an African context has been vigorously challenged by Dart (1957) and Hughes (1954). Nevertheless, evidence is slowly accumulating that several species of African hyena may be important as collectors of bones in lairs. Sutcliffe (1970) has drawn attention to bone collecting by spotted hyenas, *Crocuta crocuta*, in East Africa, while studies on *Crocuta* in the Eastern Transvaal by Bearder (personal communication) have indicated that bones may accumulate around the breeding lairs of these animals (see also Behrensmeyer and Dechant Boaz, this volume).

Observations in the Kahalari Gemsbok National Park by Mills (unpublished) have shown that brown hyenas *(Hyaena brunnea)* can be important collectors of bones around their breeding lairs. Although such lairs are often in ant bear (aardvark) holes, they might be in caves as well. Mills listed the prey items brought back by brown hyenas to five dens in the Kalahari Park. Parts of 91 mammals were retrieved of which no less than 28 were other carnivores; carnivores therefore accounted for 30.8% of all the hyenas' prey items in this study. The carnivore species involved were these: black-

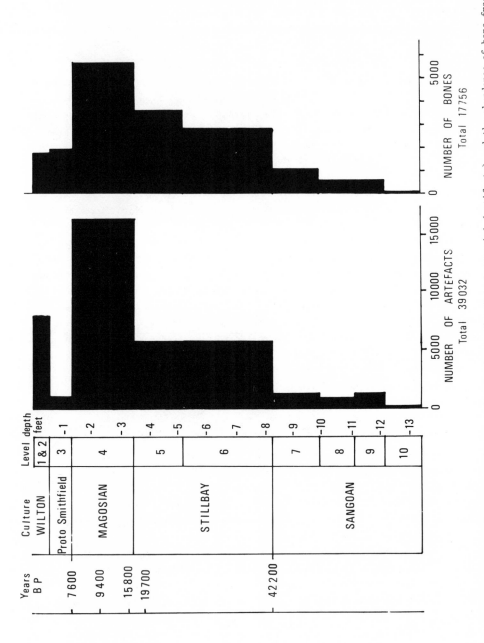

Figure 7.7. Bar graphs showing the correspondence of the abundance of cultural material (artifacts) and the abundance of bone fragments in the successive levels of the Pomongwe Cave in Rhodesia (Cooke 1963).

backed jackal (7 individuals), bat-eared fox (6), aardwolf (4), honeybadger (2), caracal (1), and unidentified fox or jackal (8).

Mills's observations suggest that brown hyenas are rather specialized predators of other carnivores; they certainly do take food (including bones) to their lairs, and a criterion for the recognition of such collections might well be the high proportion of carnivores represented in them.

Quite recently I was able to visit a brown hyena breeding lair about 30 miles northeast of Pretoria in the company of Prof. J. D. Skinner, who is currently studying these hyenas in the Transvaal. The site will be fully described elsewhere. Suffice it to say that the two hyena prey items which I have so far observed there were both other carnivores: one was a large domestic dog, the second a caracal. Both had been partly consumed.

It is clear that further field observations on bone collecting by hyenas would be worthwhile and rewarding. The late Dr. L. S. B. Leakey once told me of bone accumulations made by striped hyenas (*Hyaena hyaena*) in East Africa which he had seen. I have personally visited a number of spotted hyena lairs in various parts of Southern Africa in the hope of finding bone collections. Bones were generally present, to be sure, but so also were porcupines. Their own hoarding activities obscured the possible results of hyena bone collecting.

Leopards as Bone Collectors in Caves

The nature and completeness of leopard food remains are likely to be strongly influenced by the availability of prey and the competitive pressure exerted on the leopards by hyenas. It seems that spotted hyenas have little difficulty in relieving leopards of their prey unless this is placed in an inaccessible spot; for this reason leopards tend to store their prey in trees while feeding. Where trees overhang caves, as has been suggested for the Swartkrans situation (Brain 1970), leopard food remains may fortuitously enter and be preserved in such caves.

With the help of Mr. A. F. Port, I have been able to study several contemporary caves in South-West Africa currently used by leopards as lairs and retreats. These will be described in full elsewhere, but one will be mentioned here as an example of the sort of cave used by leopards and the nature of remains found in it.

The Quartzberg leopard lair is situated on the farm Verloren in the rugged, Hakos Mountain country of the pro-Namib desert. Here a rocky gorge in mica schist cuts through a spectacular seam of pure white quartz, containing large inclusions of calcite, some of which have been dissolved out to form a large cave, a sketch plan of which is given in figure 7.8. The entrance, about 12 ft. (3.7 m) high, leads past some large fallen blocks of rock into a substantial chamber with a high roof (fig. 7.9). At the back of this, a low entrance takes one into a second chamber, low-roofed and dark, tapering back to a tunnel on the right-hand side.

On the occasion of our visit in March 1968, we disturbed a leopard on the great blocks of rock in the cave's entrance. It withdrew into the lair and we were able to follow its tracks through the cave as far as the dark tunnel, where we decided on a policy of peaceful coexistence rather than confrontation.

The lair was divided up into three zones, the light, the twilight, and the dark

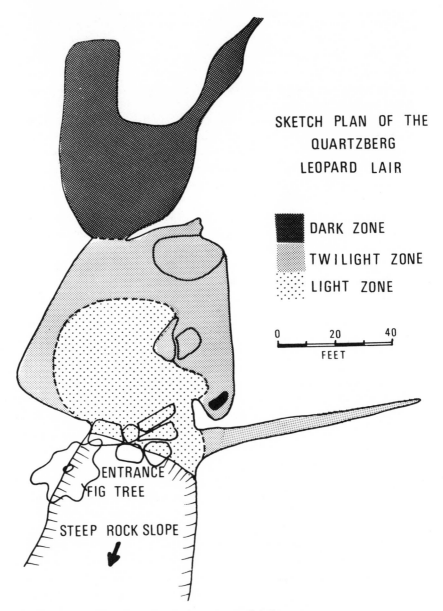

Figure 7.8. Sketch plan of the Quartzberg leopard lair.

as shown on the plan in figure 7.8. Bones were collected from each of the zones and marked separately. From the entrance blocks, where the leopard had been resting and feeding, we collected reasonably fresh remains of two klipspringers and older remains of a female baboon (fig. 7.10).

These and all the other remains from the lair fell naturally into two groups: (a) unweathered remains with adhering tissue, derived from animals which were killed or died in the cave, and (b) clean defatted bones, clearly derived from outside the lair,

Figure 7.9. View from inside the Quartzberg leopard lair. The entrance is to the right
and the figure is standing on the large blocks used by the leopard as a feeding site. The
transition from the light to the twilight zone is shown.

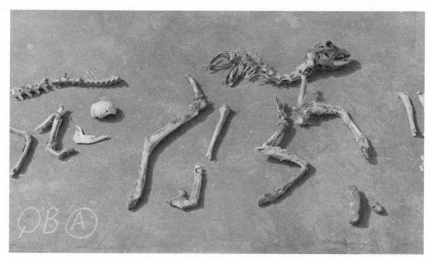

Figure 7.10. Remains of two klipspringer and a baboon from the entrance blocks of the
Quartzberg lair.

showing abundant signs of gnawing, and almost certainly carried in by porcupines, which
use the lair as well.

Table 7.12 lists the bones found in the Quartzberg cave. Of the 211 pieces, 147
are regarded as the probable porcupine-collected component and 64 as the probable
leopard-collected one. As shown in table 7.13, it is estimated that the porcupine
component came from a minimum of six individuals while the minimum number of individuals
in the leopard-collected component is fifteen.

In the presumed porcupine component, 80.2% of the bones were gnawed by porcupines,
while 23.4% in the presumed leopard component showed such marks, suggesting that the

Table 7.12

Quartzberg Leopard Lair: Table of Bones Found (Overall Total = 211)

A. Probable Porcupine-Collected Component

Gemsbok *(Oryx gazella)*	
Cranial pieces	3
Zebra *(Equus zebra)*	
Cranial pieces	2
Postcranial pieces	3
Ox *(Bos sp.)*	
Cranial piece	1
Baboon *(Papio ursinus)*	
Cranial piece (mandible)	1
Indet. bovid postcranial pieces	106
Indet. bone pieces	20
Bone flakes	11
Total	147

B. Probable Leopard-Collected Component

Klipspringer *(Oreotragus oreotragus)*	
Cranial pieces	6
Postcranial pieces (some articu-lated)	32
Steenbok *(Raphicerus campestris)*	
Cranial pieces	1
Postcranial pieces	2
Gemsbok *(Oryx gazella)* calf	
Postcranial pieces (articulated)	3
Domestic calf *(Bos. sp.)*	
Postcranial piece	1
Zebra *(Equus zebra)* colt	
Postcranial piece	1
Baboon *(Papio ursinus)*	
Cranial piece	1
Postcranial pieces	2
Dassie *(Procavia capensis)*	
Cranial pieces	12
Postcranial pieces	3
Total	64

Table 7.13

Minimum Numbers of Individuals of Taxa Identified in the Quartzberg Leopard Lair Bone Collection

A. Probable Porcupine-Collected Component

Gemsbok: adult	1	
calf	1	
Cattle: calf	1	
Mountain zebra: adult	1	
colt	1	
Baboon	1	/6

B. Probable Leopard-Collected Component

Klipspringer: male	2	
female	2	
Steenbok: male	1	
Gemsbok: calf	1	
Cattle: calf	1	
Mountain zebra: colt	1	
Baboon: female	1	
Dassie	6	/15

Incidence of Porcupine Gnawing on Bones	No. Gnawed	No. Ungnawed	Total	Percentage Gnawed Bones
Probable porcupine-collected component	118	29	147	80.2
Probable leopard-collected component	15	49	64	23.4

porcupines may have picked up some of the dry remains from leopard kills at the cave
entrance and gnawed them in their part of the lair.

Table 7.14 gives some figures on the distribution of bones in the Quartzberg cave

Table 7.14

Distribution of Bones Collected by Different Agents in the Quartzberg Leopard Lair

	Presumed Porcupine-Collected Component		Presumed Leopard-Collected Component	
	Number	Percentage	Number	Percentage
Light zone	5	3.4	23	35.9
Twilight zone	57	38.8	15	23.4
Dark zone	85	57.8	26·	40.6
Total	147	100.0	64	99.9

relative to the three lighting zones. Only 3% of the bones in the presumed porcupine-
collected component were found in the light zone, while 40% of the leopard component
came from this zone. These data suggest that, in a cave, porcupines prefer the twilight
and dark zones, avoiding the well-lit area, while leopards appear equally at home in the
light, twilight, and dark zones. They probably do most of their undisturbed feeding at
the entrance to the lair, but have no hesitation in retreating into the darkest interior
when necessary.

In a leopard-collected bone assemblage, it is the specific damage done to certain
bones that is most diagnostic. Such damage, first observed in the leopard lairs, has
since been verified as leopard-typical in a series of controlled feeding experiments. It
is unfortunately beyond the scope of this paper to discuss such aspects now, and results
will be published in full elsewhere.

Conclusion

The successful interpretation of a bone accumulation in a cave implies that the
taphonomist should understand how the bones found their way to the fossilization site.
Generally, several bone-accumulating agencies have been at work. Can these be recognized
by the nature of the bone assemblage itself?

Field studies of the sort described here will allow us to build up an inventory
of criteria whereby the various bone-collecting agencies in caves can be recognized. Such
criteria will be useful in the interpretation of bone accumulations in African caves and
may indicate guidelines for distinguishing taphonomic agents in similar situations on other
continents.

Acknowledgments

This work was undertaken with the financial support of the Wenner-Gren Foundation
for Anthropological Research, to which grateful acknowledgment is made.

I am particularly grateful to Mr. A. R. Hughes for allowing me to study his bone
collection from the Nossob porcupine lair and for helpful discussions; also to the
National Parks Board for permission to work in the Kalahari National Park, and to Dr.

G. De Graaff and Mr. M. G. L. Mills for allowing me access to unpublished material. Mr. A. F. Port's help with the study on South-West African leopard lairs was invaluable, and I was grateful to Mr. O. P. M. Prozesky and Dr. A. C. Kemp for companionship in the field.

My wife and daughters helped me a good deal with the tedious handling of the bone collections; my thanks are also due to Mrs. E. A. Voigt, Dr. E. S. Vrba, and Miss S. Kuluva. Mrs. S. M. Beukes kindly typed the manuscript.

8.   EARLY POSTMORTEM DAMAGE TO THE REMAINS OF SOME

CONTEMPORARY EAST AFRICAN MAMMALS

Andrew P. Hill

Introduction

The condition of vertebrate animals found in fossil sites is substantially different from their condition in life, and these differences constitute the subject of taphonomy.   Fossil accumulations are but samples of the living community from which they were derived, and to understand the paleocommunities it is necessary to understand the sampling process and the biases it may have introduced.   Often essential matters regarding the history of formation of an assemblage may be inferred from the condition of its constituents.   A correct evaluation of the biases involved may make it possible to correct for them and to reconstruct the past ecology.

Also, the modifications themselves may be of more direct paleoecological interest. Changes to the animal remains have been produced by environmental factors operative after death.   Consequently, the nature of the modifications may provide information about the action of these factors, including some that may otherwise be unrepresented in the local fossil record.   When, for example, the agent of modification has been man, then details of behavior such as feeding and tool use may be detectable from features of the assemblage, such as relative representation of skeletal parts and taxa, and damage.

Thus a prerequisite to the successful interpretation of fossil assemblages is knowledge of the distinctive characteristics of accumulations of bone formed in different contexts by different processes—by human feeding, catastrophic deaths, natural traps, moving water, and others.   Many major changes occur relatively shortly after death.   This paper describes some of the postmortem changes suffered by various modern species of mammal in East African open environments, with particular emphasis upon damage to parts of the skeleton.   This information, when considered with other similar observations of different modern situations, may provide a useful base for the evaluation of fossil assemblages.

Concurrent with damage to bone are other changes, involving loss of soft parts, disarticulation, and the dispersal of skeletal elements.   Mention is briefly made of these features before a description is given of the kinds of damage that take place.   The relevance of these observations to the interpretation of fossil assemblages in general, and alleged osteodontokeratic cultures in particular, is then discussed.   But firstly a brief account is given of the nature of the environments and species of animal studied, and the agents to which they have been subjected after death.

Environments and Animals Studied

If the results are to be useful in paleoecological interpretation it is important in work of this kind to specify fairly closely the environment of death of the animals being described, and the postmortem influences to which they have been subjected. Even if specific modifications cannot be directly related to particular environmental agents, they can then at least be related to a general context. All the areas studied are ones where the influence of man in the recent past has been minimal. The postmortem factors affecting the animals described here have been carnivores, both predators and scavengers, birds such as vultures, and climatic weathering.

Information has been collected from three main areas: East Lake Turkana, Kenya; Kabalega National Park, Uganda; Rwenzori National Park, Uganda. The presence of an asterisk by a species name in the lists below indicates that details of damage to its skeleton are to be found in tables 8.1-8.7.

East Lake Turkana

Lake Turkana (formerly Lake Rudolf) lies between 2°25' N and 4°35' N, extending for about 250 km along the rift valley from the northern border of Kenya. Butzer (1971b) provides a useful summary of the geography of the lake area at the present and in the recent past. The land to the east is semidesert in character; rainfall has a mean annual value of 308 mm, concentrated to two maxima in April and November. The highest mean monthly temperature occurs in January. Butzer quotes a mean of 33.4° C. The lowest mean monthly temperature is in August, at 21.2° C, but the daily range in temperature exceeds this annual variation. Only a sparse vegetation is supported by the stony soils. It consists of grass with *Acacia* or *Commiphora* bush for the most part, and riverine woodland along the mostly dry river channels. The coastal zone is a narrow strip of grassland which is able to support a greater density of large animals than the more arid regions inland.

Apart from the effects of climate, after death the remains of animals on land may be subjected to the action of various birds, including all six species of vulture found in Kenya, of which *Gyps rueppelli* is the most common. There are also a few marabou storks *(Leptoptilos crumeniferus)*. Carnivores and scavengers include lion *(Panthera leo)*, leopard *(Panthera pardus)*, cheetah *(Acinonyx jubatus)*, spotted hyena *(Crocuta crocuta)*, striped hyena *(Hyaena hyaena)*, jackal *(Canis spp.)*, and wild dog *(Lycaon pictus)*. The animals whose remains were studied were lion, cheetah, zebra *(Equus burchelli)**, hippopotamus *(H. amphibius)**, oryx *(O. beisa)*, topi *(Damaliscus korrigum)**, and gazelle *(Gazella granti)*.

Kabalega National Park

Kabalega National Park is situated in the western rift, in northwest Uganda. Climatically it is subhumid to semiarid, with rainfall maxima in April (144 mm) and October (182 mm), and a mean annual value of about 1160 mm. The mean annual maximum temperature is about 30° C. The soils generally have a low humic content (< 1% C) and a pH of 5-7. The vegetation is predominantly grassland or bushed grassland.

The carnivores that kill or subsequently damage other animals include lion, leopard, and spotted hyena. The vultures are mainly *Gyps bengalensis*, and marabous are common. Animal remains studied in the Park were those of warthog *(Phacochoerus aethiopicus)**, Uganda kob *(Kobus kob)*, hartebeest *(Alcelaphus buselaphus)*, and oribi *(Ourebia ourebi)**.

Rwenzori National Park

Rwenzori National Park is also in the western rift valley, in the southwest of
Uganda. The climate is dry subhumid. As in the Kabalega Park, rainfall is concentrated
into April (80.26 mm) and October (89.66 mm) maxima, with a mean annual value of about 669
mm. The mean annual temperature is about 29° C. The soils are humic (1.5-3% C) with
a pH of 5-5.6. They support a dominantly grassland vegetation.

Predators include lion and spotted hyena, and vultures and marabou storks are
also important in the breakup of carcasses after death. The animals studied were elephant
(*Loxodonta africana*)*, hippopotamus*, and buffalo (*Syncerus caffer*)*.

Observations

The following account is based almost exclusively upon observations of series of
individual carcasses of animals. Fairly extensive quantitative work was also done in
these areas upon the nature of whole assemblages of bone recovered from different land
surface environments. That work provides details of taxonomic diversity, disarticulation of skeletons, proportions of skeletal parts occurring, and relative completeness and
proportions of proximal and distal ends of long bones; and discusses this in the light of
similar data from selected Cenozoic fossil sites. These results are presented elsewhere
(Hill 1975; Hill, in preparation) and hence, while relevant to some of the considerations of this paper, they are not discussed here.

For some individuals examined the date of death was known, and some it was possible to examine a number of times at intervals to record successive changes. The
dates of death of buffalo in the Rwenzori National Park ranged from a few days before the
initial examination, to up to 53 months previously. Similarly, for topi at East Lake
Turkana, examinations were made from the time of death to 328 days afterward for a series
of individuals. This permits the estimation of the approximate rates at which various
postmortem changes take place. Data for some of the other species are less good, and for
some such as elephant and warthog nothing at all concerning dates of death is known.

Loss of Soft Parts, Disarticulation, and Dispersal

As a carcass loses its soft parts, the bones of the skeleton become disarticulated
from each other and scattered. Under the conditions being considered here, this occurs
fairly rapidly after death. In the case of a medium-sized bovid carcass lying on a land
surface, within a day or so most of the flesh usually has disappeared, and the forelimbs
have become separated from the rest of the carcass. This is due mainly to any carnivore
that might have caused the death, and to the subsequent depredations of vultures and
other scavengers. After a few weeks the remains are largely disarticulated, and some
elements, especially parts of the forelimbs, have been scattered so widely as to remain un-
discovered despite extensive searching around the site of death. Skin and ligaments remain attached to some parts of the skeleton of animals that are a year dead, and some
elements are still articulated at this stage. In one buffalo two cervical vertebrae were
still attached after fifty-three months. The rate and manner of disarticulation seem to
be affected by size, anatomy, and the local situation. Small bovids were disarticulated
and scattered rapidly. The sequence in which skeletons break up seems to be largely
controlled by differences in the anatomy of the various bones and joints. Within a single
species, therefore, this sequence tends to be more or less constant. Similarly, variations

found between species in the manner and rate of disarticulation can be attributed to inter-
specific anatomical differences. The carnivore skeletons examined appeared to be in a
greater state of articulation than bovids would have been at the same date after death.
The vertebral column of a lion, for example, was fully articulated, and this may be correlated with the greater amount of ligament strengthening the axial skeleton of active
predators.

In addition to these underlying anatomical factors there may be other determinants
of the disarticulation sequence that make a detailed knowledge of it potentially useful to
paleoecology. The disarticulation pattern may be diagnostic of different conditions of
environment and behavior. For example, it may illustrate those features that are unique
to various human butchery patterns. It may also explain aspects of differential representation of skeletal parts in fossil accumulations.

Observations of individual skeletons can provide some information regarding the
disarticulation pattern of a particular species, but there are difficulties with this
method. Recording the relative frequency with which different skeletal joints occur in
large assemblages of modern bones lying on land surfaces permits the construction of a
statistical model of disarticulation. This has been done at East Lake Turkana for the
topi remains on the delta flats of the Laga Tulu Bor, and the results are presented as a
flow diagram in figure 8.1. This agrees well with observations elsewhere, including those
made of individual carcasses. Starting with a complete skeleton on the left of the
diagram, and moving towards the right, successive numbers indicate the order in which
different parts of the skeleton become detached. For example, at (1) the whole front
limbs become separated from the body; the next stage (2) is the loss of the caudal vertebrae, and so on. Further details regarding the method and the sequence of disarticulation
is given in Hill (1975, 1979), where comparisons are also made with the order of disarticulation resulting from an instance of butchery of bison by man.

These remarks apply to disarticulation on land, where the principal agents of
dispersal are carnivores. In the case of hippopotamuses floating dead in Lake Turkana,
disarticulation was very rapid and the dispersal of the bones by water was considerable
as they fell from the carcass.

Damage

After death the bones of animals in the circumstances described here become quite
extensively damaged. Most of the impetus to considerations of damage to fossil bone
has come from its interest to anthropology, insofar as it might represent food debris or
bone tools, or reveal paleopathology. Apart from setting up criteria for distinguishing
various hominid accumulations from others, a knowledge of types of damage is important
to paleoecology in general. By providing a possible means of identifying the agent or
agents concerned, such information may supply clues to explain related features such as
accumulation and differential representation of parts of the skeleton. Damage may lead
to the complete destruction of some bones, or even all the bones, of some species. It
may also provide information about environmental conditions and agents otherwise unrepresented in the local fossil record. The recognition of damage, particularly distortion,
is vital if measurements are taken of critical specimens to support taxonomic or functional notions (see Walker, this volume).

Notes on the type of damage to be found in a number of East African mammal species
are given in tables 8.1–8.7. The species concerned are elephant, warthog, hippopotamus,

Table 8.1. Damage to Bones of Loxodonta africana

	Cranium	Mandible	V. Cerv.	V. Thor.	Ribs	Scapula	Humerus	Innominate
Missing or broken off	zygoma		ventral centra/ centra epips.	ventral or all of centra/ centra epips.		dorsal+ anterior margin of glenoid	proximal end	
Chewing	margins of tusk alveoli	asc. ramus + condyle	vent. centra/n. sps./dors. + vent. transverse processes	n. sps.	proximal + distal ends	blade + spine margins/tip acromion/ crenln post margin blade	supinator ridge/dist ant.artic. /crenln prox.shaft	suprailiac border (crenln)/ symph. part of pub. + ischium
Depressed fracture				n. sps.		posterior margin of blade	prox. shaft	suprailiac border
Teeth marks						glenoid/ spine		
Cracking	parietals /frontals			n. sps.		anterior fossa		
Perforation	occiput/ occ. crest/ above nasal aperture							blade of ilium
Fracture							shaft, horizontal (crenln)	
Erosion								

and a large, medium, and small bovid—buffalo, topi, and oribi. However, the discussion following is based on observations of all thirteen species listed under "Environments and Animals studied" above. The tables make no pretense to being complete; indeed, for the elephant not all the bones of the skeleton were available. They give an idea of the possibilities of kinds of damage found in the situations under discussion. Consistency was shown in the type of damage suffered by individuals of the same species, even those from different areas. For example, damage to hippopotamuses from East Lake Turkana and Rwenzori Park was identical. This suggests that the descriptions may be regarded as typical. Further information appears in Hill (1975).

Most existing classifications of bone damage are unsatisfactory in some way. The main determinant of the way a bone is damaged seems to be the anatomy of the bone itself, and it is difficult to identify and record all the additional features that are potentially paleoecologically significant, especially when working in the field. More definitive classifications should be descriptive at two discrete levels at least. One scheme should employ mutually exclusive categories that are purely descriptive of morphology, and quite divorced from any knowledge or assumptions regarding the agents or processes that have

Table 8.2A

Damage to Bones of *Equus burchelli*

	Cranium	Mandible	V. Cerv.	V. Thor.	V. Lumb.	Sacrum	Ribs	Scapula	Rad/Uln	Metacarpal
Missing or broken off	occ. crest /nasals	coronoid process/ condyle	anterior margin of neural spine	neural spines/ lateral processes		post. part /tips n. spines/ lateral processes				
Chewing		coronoid process/ condyle	neural spine	n. sps./ lateral processes /ventral centra	neural spines/ lateral processes	margins of missing parts	proximal and dis- tal ends			
Depressed fracture				neural spines	base of centra					
Teeth marks	dorsal surface							distal margin		
Cracking	parietals/ alveolar margin/ teeth	ascend- ing ramus/ tooth enamel	lateral process				longi- tudinal	distal+ pre- scapular margins	longi- tudinal	longi- tudinal
Perfora- tion	ext. face of maxilla /parietal	ascend- ing ramus	lateral process							
Fracture										
Erosion	ext. face of maxilla /zyg./dors. orbit/ posterior parietals	condyle	spine margins/ post. face of centrum	edges	edges					

Note: Axis: one intact after one year, ligaments remaining.

Table 8.2B

Damage to Bones of Equus burchelli

	Innominate	Femur	Tibia	Podials	Metatarsal	Phalanges
Missing or broken off	part of ilium	greater + third trochanter				
Chewing	margin of ilium and ischium	greater + third trochanter /distal artic.	anterior + exter- nal lateral ridges	tubero- sity of calcan- eum		distal end
Depressed fracture	ilium	greater + third trochanter /internal condyle	prox. end of anterior ridge			distal end
Teeth marks						
Cracking	ilium	surface of distal artic.			longit- udinal	
Perforation						
Fracture		diagonal, midshaft	diagonal, midshaft			
Erosion						

produced such morphologies. The scheme should be comprehensive, in that it could be used adequately to describe bone damage found in all situations, fossil and modern. A secondary level of description should consist of inferences about the conditions or the relevant agent implied by particular morphologies. As more modern bone assemblages are investigated, which have been produced in different circumstances, then it will be possible to distinguish those differences indicative of particular environmental processes from the background of damage characteristics due to the nature of the bones themselves. Then it will be possible further to refine classificatory schemes.

The scheme that is used here does not conform to these ideals. The categories used are not entirely mutually exclusive, nor are they all strictly morphological descriptions, as one or two genetic expressions have been used where little doubt exists. However, I feel justified in using the scheme in a broad treatment such as this, as it serves to illustrate the general nature of damage, and to reveal recurrent patterns that have fairly unequivocal implications for paleoecological questions.

Missing or broken off describes which parts of a skeletal element have been broken off or lost, by whatever cause—chewing, trampling by other animals, and so on (figs. 8.2 and 8.3).

Chewing is applied to the broken margins that seem unequivocally to be due to the gnawing of animals. Characteristically such edges are uneven and jagged, revealing internal cancellous bone. Often the margins are crenulated (figs. 8.2 and 8.3).

Table 8.3A

Damage to Bones of Phacochoerus aethiopicus

	Cranium	Mandible	V. Cerv.	V. Thor.	V. Lumb.	Sacrum	Ribs	Scapula	Humerus
Missing or broken off	styloid procs. /parts of canine alveolus + zygoma/ant. nasals	anterior edge of ascending ramus/ condyles	n. sps. + lat. procs./ centra epips.	n. sps./lat. procs./ant. + some post. zygs./ centra epips.	centra epips.		capitula/ tubercles	distal part/ edge of spine	prox. end/ greater tuberosity /part of head
Chewing									
Depressed fracture								coracoid process	
Teeth marks	dorsal orbit							distal part/ glenoid margin	greater tuberosity/ head
Cracking	canine alveolus/ ext. face of M3 alveolus/ nasals/ zyg.	alveoli/ asc. ramus /vent. horiz. ramus/C split sagittally	neural spines				longitudinal	blades	longitudinal
Perforation	tympanic bulla/ nuchal crest/ premax.	horiz. ramus below M3						distal part	
Fracture			edges	rib facets	neural spine edges/ zygs.	edges	capitula/ tubercles /shaft		prox. epip. /midshaft curved diagonal
Erosion	nuchal crest/ zygoma	condyles					capitula/ tubercles	glenoid margin	condylar margins/ artic. surfaces

Table 8.3B

Damage to Bones of _Phacochoerus aethiopicus_

	Radius	Ulna	Metacarpal	Innominate	Femur	Tibia	Metatarsal
Missing or broken off	distal end	distal end /olecranon process		ends of ilium + ischium	proximal end/head + greater trochanter		
Chewing				ends of ilium + ischium			
Depressed fracture		olecranon process/ shaft		ilium + ischium margins	proximal end/dis- tal condyles	edge of proximal facets	
Teeth marks		olecranon process	shaft	ilium + ischium	distal artic.	proximal + distal ends	
Cracking		longitu- dinal; forming large openings		margin of ischium	longi- tudinal	longi- tudinal	longitu- dinal/ mosaic on artic. sur- faces
Perforation							
Fracture	midshaft horizontal jagged	midshaft		jagged	midshaft, jagged, horiz. diag. or stepped	below proximal epip., diagonal	
Erosion	edges of proximal artic. surfaces	edges of olecranon process/ artic. surfaces	edges		edges of artic.	surface of prox. artic.	edges

Notes: Cranium: Juvenile separated into calvaria and individual facial bones. Innominate: None intact. Acetabulum with parts of ischium and ilium common; and broken fragments. Podials and Phalanges: Intact.

Depressed fractures are often found associated with chewed margins. They are punctate, usually round depressions or perforations in the surface of the bone, where plates of surrounding bone have collapsed into the cavity so produced. They are usually small (5-10 mm) and are no doubt produced by the teeth of carnivores. Where a broken edge of bone crosses one of these, only a semi-circular collapsed area is left.

Teeth marks is a term reserved for parallel edged grooves formed on the surface of bone.

Cracking is applied to relatively deep cracks that develop in the cortex of bone, presumably as a result of climatic weathering, or sometimes in associ-ation with more specific damage (fig. 8.2). Longitudinal cracking develops parallel to the long axis of a bone. It seems ultimately to lead to openings forming in long bones, and along the horizontal ramus of a mandible it may result in the loss of the ventral edge. Horizontal cracking is at right

Table 8.4A

Damage to Bones of *Hippopotamus amphibius*

	Cranium	Mandible	V. Cerv.	V. Thor.	V. Lumb.	Sacrum	Ribs	Scapula	Humerus
Missing or broken off	premax./I+C /tymp. bulla /jugal + paroccipital processes	coronoid process/ condyle	lat. sp. tips/centra epips.	tips+ant. edges of lat. sps. /centra epips.	tips+ant. edges of lat. sps./ centra epips.	anterior n. sps./ anterior epip.		suprascap. + ant. edges /proximal part of spine	parts or all head
Chewing	ant. maxilla /alv. margin of C + PM + M/ dors. + post. orbit edges	vent. + dors. + post. asc- ramus/alve- olar edge of C					tubercles		edge or all head/ greater tubero- sity
Depressed fractures				centra					
Teeth marks									margin of head
Cracking	general/ M crowns		thin areas /mosaic on articular surfaces		spines		longit- udinal	blade	
Perforation	typmanic bulla		thin areas						
Fracture							shaft, jagged		
Erosion	zygomatic suture		edges of articular surfaces	edges of epips. + zygs. + rib facets	ventral part of centra/ zygs.	edges of lateral spines	edges of tubercles		edges + surface of head

Table 8.4B

Damage to Bones of Hippopotamus amphibius

	Rad/Uln	Innominate	Femur	Tibia	Podials
Missing or broken off	dorsal olecranon/ part of distal artic.	edge of ilium + ischium	anterior margin of internal condyle		
Chewing	olecranon /distal artic.	margins of ilium + ischium + symph. pub. (crenlns.)	anterior margin of internal condyle		
Depressed fractures					
Teeth marks		margins of ilium + ischium			
Cracking	longitudinal	general	longitudinal/ mosaic on condyle	longitudinal	
Perforation					
Fracture					
Erosion			head	surface + edges of prox. facets	articular surfaces

Note: Cranium: chipping of tooth enamel. Metapodials and Patella: intact.

angles to the long axis. Mosaic cracking occurs on the articular surfaces, where a series of irregular shallow cracks break up the surface into a fine tesselated pattern.

Perforations are formed in areas of thin bone by cracking and general weathering, though sometimes canine teeth may be responsible (figs. 8.2 and 8.4).

Fracture is a term used here to describe breaks in the main body of a bone. The nature of cracking in a bone often appears to control the character of midshaft breaks. Smooth diagonal fractures are found in bones that in life are subjected to torsional stress, such as humerus and tibia, where the surface of the break is smooth and at a relatively low angle to the long axis of the bone. The surface in which the broken edge of the bone lies need not always be flat. Often it is curved or twisted, presumably corresponding to what have elsewhere been referred to as "spiral fractures." Horizontal fractures occur across the bone at right angles to its long axis. This occurs particularly in such bones as metapodials in which stresses are mainly compressional, and which also show horizontal cracking. Interference of this and longitudinal cracking may produce step fractures, in which the edge is broken into a sequence of more or less right angle steps. Steps may also occur on what are otherwise smooth

Table 8.5A
Damage to Bones of *Syncerus caffer*

	Cranium	Mandible	V. Cerv.	V. Thor.	V. Lumb.	Sacrum	Ribs	Scapula	Humerus
Missing or broken off	tip premax. + nasals/post. dors. occip./ ant. maxilla /post. zygoma	coronoid proc. /vent. horiz. ramus/ant. edge asc. ramus	lat. sp./ tip + post edge n. sp. /ventral processes	n. sp./vent. proc./epip. centra	lat. proc. centra/rib facets/lat. n. sp./ part of centra/ zygs.	posterior element/ edges of lat. sps. /n. sps.	distal ends/ post. margin	distal blade/ edge of spine + acromion	prox. end /median epicondyle /gt. tub.
Chewing							facets		
Depressed fractures		condyle	neural arch/base of lateral processes	neural spines/ centra	anterior part of centra	edges of lateral spines	shaft		
Teeth marks		posterior edge of ascending ramus							
Cracking	teeth enamel (chipping)	ant. edge asc. ramus /teeth (chipping)	general	general/ mosaic on rib facets			longit- udinal	longit- blade	longit- udinal/ mosaic on artic. surfaces
Perforation	tympanic bull/sides of pterygoid /ventral alisphenoid							post- scapula fossa (teeth?)	
Fracture							proximal shaft, sharp diagonal		shaft, jagged
Erosion	edges of foramen magnum/ horns (slight)	condylar surface	edges/ zyg. surfaces	zyg.	edge of centra/rib facets		surfaces of facets		articular surfaces

Table 8.5B

Damage to Bones of Syncerus caffer

	Rad/Uln	Metacarpal	Innominate	Femur	Tibia	Podials	Metatarsal
Missing or broken off	distal end/tip of olecranon	distal end	suprailiac edge/post. ischium/ sacral edge of ilium	greater troch./ dorsal int. condyle	proximal end	tuberosity of calcaneum	
Chewing			(as for missing)				
Depressed fracture			posterior ischium				
Teeth marks		shaft					
Cracking	longit- udinal	longit- udinal		longit- udinal/ mosaic, head surface		mosaic, articular surfaces	
Perforation							
Fracture	shaft, diagonal step	shaft, diagonal		shaft	shaft, diagonal		shaft, diagonal
Erosion	articular surfaces		dorsal acetabulum /area of symphysis	edges of epips.			

Note: Phalanges intact.

diagonal fractures. Jagged fractures have more irregular edges than the other types recognized. They are possibly produced by chewing.

Erosion is applied to wear along sharp edges that exposes the cancellous bone. It is probably due more to weathering than to the activities of animals. In some bones it can result in perforations.

Surficial cracking and flaking is one of the categories of damage not listed on the tables as it ultimately affects all bones. It is the peeling off of very thin flakes of surface bone, and is due to climatic weathering.

On the basis of these observations some general comments can be made. Extensive damage occurs to bones in natural conditions shortly after death. It often conforms to repeated patterns and is primarily caused by carnivore activity, by climatic weathering, and by the chemical effects of soil and plants. Different patterns of damage exist, and the variations seem not so much controlled by the agent of damage as by the nature of the bone itself. This is correlated with the age of the individual, the size of the species concerned, and its anatomy. The last two factors determine the similar way in which individuals of the same species become damaged.

The age of the animal is important, as is shown by the case of the warthog, be-cause the unfused bones or epiphyses of young individuals will fall apart, exposing them

Table 8.6A

Damage to Bones of Damaliscus korrigum

	Cranium	Mandible	V. Cerv.	V. Thor.	V. Lumb.	Sacrum	V. Caud.	Ribs	Scapula	Humerus
Missing or broken off	premax./ant. nasal /tips of horn-core	ventral horiz. ramus /coronoid proc./part of angle	n. sp.	n. sp.	n. sp./ lateral process	n. sp./ posterior part		distal end	dist. end/ V-shaped frag. from prescap. fossa	head
Chewing			lateral process							
Depressed fractures				centra					proximal end	
Teeth marks										
Cracking		rami/ incisors							general	longit- udinal
Perforation									postscap. fossa	
Fracture		diagonal from P2/ asc. ramus/ vent. horiz. ramus					midcentra	shaft, jagged		horiz./ diagonal/ part sagit- tal
Erosion	ear region		lateral process	edges						

Table 8.6B

Damage to Bones of Damaliscus korrigum

	Rad/Uln	Innominate	Femur	Tibia	Metatarsal
Missing or broken off	olecranon	suprailiac margin/ posterior ischium	head/gt. troch. /dorsal internal condyle		
Chewing			head/gt. troch. /patellar region head/lat.		
Depressed fractures			epicondyle internal condyle		
Teeth marks					
Cracking	longitudinal		longitudinal	longitudinal	longitudinal
Perforation					
Fracture	shaft, curved diagonal		shaft, prox. near epip. curved diagonal	shaft, curved diagonal stepped	
Erosion			edges	edges	

Note: Cranium: One intact after 314 days; another with no sign of cracking after 328 days. Scapula: One intact after 328 days; another with cracking beginning after 314 days. Sacrum and Metapodials: Some intact after 314 days. Podials and Phalanges: Generally intact.

to further damage. Also, some old animals have a tendency to resorb bone, perhaps for dietary needs, and this may influence susceptibility of bones to breakage and other damage.

Different species show varying patterns of damage mainly due to differences in overall size and anatomy. Within the bovids, for instance, the gazelle and oribi are broken up far more, and far more rapidly, than is buffalo. It should not be assumed however, that the larger an animal is the more readily it will resist damage. The elephant carcasses provide examples of very large animals that have been subjected to a great amount of carnivore damage. It seems more probable that destruction pressure operates to eliminate more carcasses from the large than from medium-sized animals. Small carnivores such as jackals have no difficulty in consuming small creatures such as gazelles, but they do not possess the apparatus for damaging much larger beasts. Large carnivores such as hyenas are more capable of dealing with the bones of larger animals (fig. 8.3). As the most nutrient part of the bone consists of the fatty cancellous tissue and marrow, then the larger the bone the greater will be the volume of this in relation to the relatively hard cortex that has to be penetrated. Thus, to such a carnivore, dealing with large bones is a more efficient activity than attacking smaller elements that have a higher proportion of cortical to cancellous bone. If this impression is true, medium-sized animals are likely to suffer less damage than either small or large ones.

The ways in which anatomy is correlated with skeletal damage are for the most part obvious. Breakage is often correlated with soft anatomy, as bones become damaged in the course of removing flesh from a carcass. Damage to the sacrum and innominate is mainly of this sort. Possibly the common breakage in bovids of at least one mandible

Table 8.7A

Damage to Bones of *Ourebia ourebi*

	Cranium	Mandible	V. Cerv.	V. Thor.	V. Lumb.	Sacrum	Ribs	Scapula	Humerus	Rad/Uln
Missing or broken off	most of facials/ styl. proc. /edge of pterygoid	coracoid	lateral process	n. sps./ tips of lateral process /zygs.	n. sps./ lateral process	posterior elements/ n. sps.		all but spinal base + glenoid	head/ gt. tub.	
Chewing										
Depressed fracture					ventral centra	posterior ventral surface			proximal	
Teeth marks		dorsal edge of diastema								
Cracking		along horiz. ramus/ teeth								longit-udinal
Perforation		horiz. ramus below M's								
Fracture	zygoma	diastema/ between M1 + M2/ asc. ramus vertically					shaft, jagged		diagonal curved + part sagittal	diagonal shaft, diagonal sharp
Erosion	horn cores			edges of centra	edges of zygs.	edges of lateral process	edges of capitula/ tubercles		edges of gt. tub. + head	anterior edge of olecranon

Table 8.7B

Damage to Bones of Ourebia ourebi

	Metacarpal	Innominate	Femur	Tibia	Metatarsal
Missing or broken off		suprailiac crest/ ischial tuberosity			
Chewing					
Depressed fracture		around aceta- bulum	external face of greater trochanter		
Teeth marks					
Cracking			longi- tudinal	longi- tudinal	longi- tudinal
Perforation					
Fracture	stepped, into long thin splints	through obturator foramen	diagonal stepped	diagonal	stepped, into long thin splints /saggital
Erosion			edge of prox. epip. + patellar area	epip. fusion lines	

Note: Cranium: Some entirely fragmented, such as horn cores + frontlet; fragments of maxilla; ear region; condyles + occipital around foramen magnum. Podials and Phalanges: Intact.

in the region of the diastema is caused by the removal of the tongue (fig. 8.5). Brain (1967, 1969) has shown the importance of the relative time of epiphysis fusion and specific gravity of ends of limb bones for their survival. Such features as perforations on the crania of pigs and elephants (fig. 8.2) are due to the widely separated internal and external surfaces of some of the cranial bones.

Bone that has escaped or survives the depredations of carnivores lasts a relatively long time. Its preservation depends largely upon the local situation. Bones of a buffalo that had been buried in a mud wallow were in almost perfect condition after fifty-two months. Unburied bones that are shielded by a favorable microclimate, provided perhaps by vegetation, are more likely to survive than are those exposed to harsher conditions. Some bones from animals at least one year dead still show no signs of cracking, and surficial cracking and flaking had not greatly affected a buffalo skeleton forty-eight months after death. Remains of buffalo almost five years after death were still in a quite robust condition.

The importance of considerations such as these to aspects of paleoecology is clear. Differential damage to the remains of different species can result in their reduced representation, or even nonrepresentation, in an assemblage of fossils. Quite apart from other factors, this suggests that the relationship of the minimum number of a particular species, based upon some skeletal element, to its past abundance in the paleocommunity cannot be known with great precision. Neither can fossil assemblages be

MODEL OF <u>Damaliscus korrigum</u> DISARTICULATION

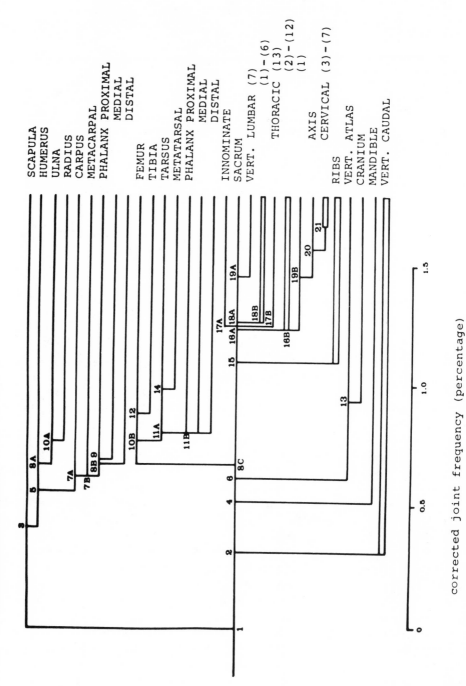

Figure 8.1. Model of topi (*Damaliscus korrigum*) disarticulation, East Lake Turkana delta flats.

Figure 8.2. *Loxodonta africana* cranium. Rwenzori National Park. Date of death: eleven months previously. Zygoma missing, chewing around margins of incisor alveoli, cracking, and development of perforations in occipital region and above nasal aperture.

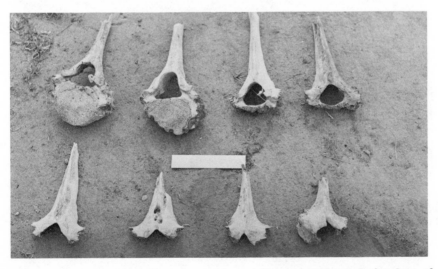

Figure 8.3. *Loxodonta africana* thoracic vertebrae. Rwenzori National Park. Date of death: unknown. Damaged neural spines, and chewed or completely missing centra.

assumed to represent instants in time if animal remains coexisting on land surfaces vary from freshly dead to five years, and probably at least twice this in age.

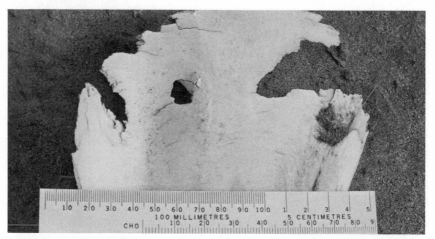

Figure 8.4. *Oryx beisa* scapula. East Lake Turkana. Date of death: unknown (same individual as Figure 8.5). Perforation on blade, characteristic damage to distal end, with
triangular areas cracked off.

Figure 8.5. *Oryx beisa* mandible. East Lake Turkana. Date of death: unknown. Diagonal
break of right mandible in the region of the diastema, cracking along length of left
horizontal ramus.

Osteodontokeratic Cultures

The patterns of damage that are produced in the natural state seem to bear a
ʳclose similarity to those that elsewhere have been ascribed to osteodontokeratic cultures
of fossil hominids, particularly that from Makapansgat, South Africa. Some comments have
already been published concerning this (Hill 1975).

Dart (1957 and elsewhere) describes and figures the damage suffered by various
bovid bones in the assemblage from this locality. He attributes the damage to the use of
certain bones as tools by australopithecines. However, nearly all the features shown by

these bones can be found in the modern bovid remains considered here, which all come from situations in which man has not been involved. Crania from Makapansgat seem to be more damaged than all but the small ones discussed here; however, most of the modern examples record only short-term changes on modern land surfaces.

Dart believes damage to horn cores has been caused by hominids, but similar modern examples have been found in the topi. Chipping of tooth enamel on maxillary fragments is suggested to be due to their use as scrapers. He notes that mandibles have lost the angle and the ascending ramus (allegedly because of their use as "blades") and the ventral side of the horizontal ramus, and they are often broken across the region of the diastema (for use as "jabbing and penetrating implements"). The incisors have been "removed." All these features are present in the modern remains (fig. 8.5). Dart suggests that the loss of neural spines and the generally damaged condition of most vertebrae implies their use by australopithecines for purposes of a "pounding and levering character," and explains that centra devoid of their processes could have been used as projectiles. From the illustrations that he provides, the damage shown appears consistent with that to the modern examples, with breakage and damage occurring to similar parts (fig. 8.5). Caudal vertebrae are soon lost from the modern bovid carcasses examined, having been totally eaten by carnivores. Dart relates their absence in the Makapansgat collection to the use of tails by australopithecines as "signals and whips outside the cave."

Makapansgat calcanei show damage analogous to that of modern ones. Dart believes they were damaged in the course of the disarticulation of the hindlimb at this point by the hominids, so that the foot could be used as a double-ended club. Similarly, damage to the ends of the femur, tibia, and humerus is thought to be due to their use as "flails and clubs." Breakage to the olecranon process of the ulna is supposedly caused by its use as a club, and the sharp diagonal breaks of the ulna shaft are deliberate modifications to create daggers.

The splitting of metapodials into splints is also thought to be so frequent as to be deliberate, and spiral fractures of the humerus shaft are alleged to be the results of the purposive removal of small and variously useful flakes of bone.

Reference to the accompanying tables will reveal patterns of damage very similar to those that Dart describes. These patterns are consistent and repeated. This is not to say that no bones in the Makapansgat assemblage were modified or at least used by australopithecines. Osteodontokeratic elements in cultures exist, but if the osteodontokeratic culture of *Australopithecus* is to be sustained, then reliable morphological criteria need to be established for discriminating such implemental modifications from the effects of hominid feeding, and both of these from the patterns of bone destruction outlined above, formed in situations in which hominids have had no part.

M. D. Leakey (1971) considers certain modifications to bone at Olduvai to be due to hominids. They include abrasion and polishing of broken ends of zebra ribs, and flaking of broken ends of their limb bones. There are similar features in hippopotamus, where limb bones have been flaked to a point, and shaft fragments have been found with evidence of chipping flaking along the edges. Bones comparable to these were not found in the modern carcasses examined.

Summary

An examination of animals' skeletons after death, and of the changes they undergo, is clearly of importance to paleoecology. It can give basic information regarding the effects of different animals and various environmental conditions on disarticulation and damage, features of which may be diagnostic when applied to fossil and archaeological situations. One major area of its application is in providing details of the sorts of changes produced by factors other than man. These will permit the more accurate assessment of characteristics produced by hominids and of theories concerning human behavior to which they give rise. In its early stages this branch of taphonomy will be more valuable in a critical sense, in falsifying or imposing constraints upon paleontological and archeological theories. But we may hope that further investigations will result in the ideas of taphonomy being more generally applicable, suggesting fresh directions for paleoecology, rather than predominantly defining the resolution of its existing hypotheses.

Acknowledgments

Most of this work was carried out while I was employed as Research Assistant to the late Professor W. W. Bishop at Bedford College, University of London. It was financed by a research grant to me from the Wenner-Gren Foundation for Anthropological Research, New York, and by grants to Professor Bishop from the Wenner-Gren Foundation and the Natural Environments Research Council. The Koobi Fora Research Project, directed by Richard Leakey, National Museums of Kenya, provided field maintenance at Lake Turkana and gave help with travel expenses. I thank members of the Research Project for their encouragement, and particularly Kelly Stewart and Libby Nesbitt-Evans who helped a lot with the field work. In the Rwenzori National Park I was permitted to use the resources of the Nuffield Unit of Tropical Animal Ecology. I thank the then director, Dr. K. Eltringham, and other members of the unit for their interest and help. I thank the governments of Kenya and of Uganda for granting research permission.

Part 4

TAPHONOMY IN THE LABORATORY

The study of bones in modern environments can focus on specific processes that may have affected fossil assemblages. However, attempts to discover general principles or to test hypotheses may be complicated by the many possible explanations for observed phenomena. Experimental taphonomy can further limit the number of variables in test situations to help isolate those that have been important in the formation of fossil assemblages. Moreover, through experimentation, preexisting theory from other disciplines can be used to formulate models to test the effects of these variables in natural situations.

Many fossil vertebrate accumulations are found in river deposits, and there can be little doubt that the bones have been affected in some way by moving water. Hanson shows that the behavior of bones in artificial and natural currents is to some extent predictable. Some bones move more easily than others, and the ratios of different elements indicate the transport history of the whole assemblage at any particular locality. The reasoning in Hanson's paper is based upon general principles of hydraulics, and the strength of its systematic theoretical treatment of bones as sedimentary particles is clearly demonstrated. For the first time, detailed equations are formulated to characterize and predict bone transport, and Hanson presents some empirical evidence that indicates his equations can be applied to natural stream situations. The paper provides a foundation for further experimental work which should enable us to define more precisely the many variables affecting bone transport. It also presents a generally applicable model for organizing all taphonomic processes and effects into a manageable conceptual framework.

Using hominid and other primate fossils as examples, Walker shows how a paleontologist may cause his own biases, depending on how he observes and analyzes his fossils. He illustrates a number of techniques available to help the functional anatomist clarify taphonomic processes which may affect his morphological interpretations of bones. In addition to providing words of caution concerning the pitfalls of using fossils as anatomical specimens, Walker presents some new and potentially exciting information about how dietary information can be derived from fossil teeth. Using the scanning electron microscope to examine details of tooth enamel and scratches on tooth wear surfaces, he has been able to distinguish between animals with different diets. Such direct information on food habits is of great potential importance to paleocommunity reconstruction.

A second method of extracting dietary information from teeth, involving geochemical analysis of tooth enamel, is described by Parker and Toots. The chemistry of vertebrate fossils has been relatively neglected as a source of paleoecological information. This paper shows that much can be learned from trace elements concerning diet and evolutionary history, providing these elements have not been altered during diagenesis beyond the interpretive powers of the geochemist. If the trace elements have undergone change, then this may provide another sort of information on the environment of burial and processes of fossilization. In either case, Parker and Toots demonstrate how quantitative information on chemical components of fossils may provide a direct pathway to intriguing taphonomic or paleoecological conclusions.

Hare gives further evidence for the usefulness of geochemical work in taphonomy. Whereas Parker and Toots attempt to study stable trace elements incorporated into teeth and bones during the lifetime of an animal, Hare examines the changes that take place in the organic components of bones during fossilization. He discusses laboratory simulation of the processes of weathering and fossilization, showing how the effects of temperature and moisture cause changes in the composition of amino acids in bone collagen. Collagen is often preserved in fossil bone, hence the nature of weathering prior to mineralization can theoretically be read from fossils. Hare also emphasizes the complexity of the biochemical processes involved, and shows why the racemization of amino acids in bones may be less useful as a means of age-dating than as an indicator of original temperature and moisture conditions in environments where bones accumulated and became fossilized.

9. FLUVIAL TAPHONOMIC PROCESSES: MODELS AND EXPERIMENTS

C. Bruce Hanson

Introduction

The taphonomic method of paleoecological reconstruction entails working back from a fossil assemblage to the ecologically significant aspects of the parent population (Olson, this volume). The accuracy of the reconstruction is limited by the extent of knowledge about the natural system of taphonomic agents which selectively collected and preserved the fossil sample. The required knowledge includes both a specific and a general component. Features of the assemblage itself, its associated sediments, and the broader paleogeomorphologic context hold clues to the specific taphonomic agents responsible for the existence and biases of the particular assemblage. Unambiguous interpretation of the clues, and the higher-order inferences leading to the eventual reconstruction, require a general knowledge about the individual and collective effects of a broad range of taphonomic agents.

Experimental studies, as well as direct observation of natural processes and their clearly attributable effects, can provide basic information about individual taphonomic agents, but if the information is to be generally applicable, it should be tied to a testable theoretical framework. An understanding of the collective effects requires a higher-level framework, a general model which reflects the possible controls, sequences, interactions, and interdependence of the individual agents and groups of agents comprising taphonomic systems—a conceptual map to guide the specific reconstruction.

A particularly common problem encountered in the study of terrestrial fossil assemblages is recognizing and accounting for the taphonomic effects of various physical processes associated with fluvial environments. Existing data and concepts in the fields of sedimentology, hydraulics, and geomorphology bear on the solution of this problem. Quantification of the behavior of vertebrate remains in terms of basic parameters important to these fields of study could provide keys to a rapid increase in the understanding of taphonomic processes.

This paper is an attempt to draw together some theoretical and experimental investigations of the fluvial taphonomic processes of entrainment, transport, burial, and destruction of vertebrate remains within a framework which shows the relationships of these processes to each other, to other processes in taphonomic systems, and to some of the basic variables of fluvial hydraulics. The direction of approach is from the general systems to the specific processes, presented as though they were accessible for direct

examination. Reconstruction of past systems, using evidence associated with fossil assemblages, is the eventual goal, but is beyond the scope of this paper.

Models of Taphonomic Systems

A General Mode

Taphonomic processes cause four basic kinds of continuous change in organisms and
their remains. A limiting value of the measure of each change may be specified so as
to define an *effectively* discontinuous or discrete change. The kinds of continuous
change[1] and their potential "discontinuous" consequences are as follows:

1. Alteration of behavior in res-
    ponse to potentially lethal
    agents   Death

2. Disorganization   Destruction

3. Movement   Import, export

4. Burial, unburial   Permanent burial

The occurrence of a "discontinuous" change may be regarded as a change of state for the
affected organism or part. Except for death, the choice of the limit of change defining
each "discrete" state is arbitrary. The choice will be influenced by the temporal and
spatial scale of the particular investigation or discussion and by the level of organization of remains and degrees of resolution to be considered.

Boundaries designated to define import-export and permanent burial will also define a segment of the earth's surface and near-surface. Geographic boundaries could be
chosen to reflect the areal distribution of taphonomic agents as well as the desired
level of resolution. The boundary separating temporary from permanent burial corresponds
to the boundary between the unstable and stable parts of the lithosphere, the unstable
part being that in which burial is possible, but the probability of physical disturbance
(as by erosion, bioturbation, or mass movement) within a specified time period is greater
than some arbitrarily small value.

The "discontinuous" changes add remains to or subtract them from the defined surface segment in accordance with the scheme depicted in figure 9.1. This scheme leads
to a basic accounting equation for the segment:

$$(\text{Death} + \text{Import}) - (\text{Destruction} + \text{Export} + \text{Permanent burial}) = \begin{array}{l}\text{Net change of quantity}\\ \text{in storage}.\end{array}$$

Accounting could be done in terms of biomass, absolute numbers of individuals, or numbers
of elements of particular kinds or taxa. It will generally be most convenient, however,
to view each term of the equation as a time-rate of change, that is, number of "discontinuous" changes of that kind per unit time.

A taphonomic system consists of two parts corresponding to Müller's subdivisions of
taphonomy (Müller 1951, 1963; see also Lawrence 1968, 1971) (1) a biostratinomic part
comprised of a mosaic of one or more surface segments directly or indirectly linked by
import routes to a particular segment of interest, together with the taphonomic agents
operating within those segments; and (2) a diagenetic part; the stable lithosphere containing organic remains, subjacent to the mosaic, and the agents operating within it. Subsequent discussion in this paper concerns only the biostratinomic part.

---

1. The importance and interrelated nature of these four processes were recognized
by Richter (1928*b*). He supplied a name for the study of each process and regarded their
collective study as a subfield of his *Aktuopaläontologie*.

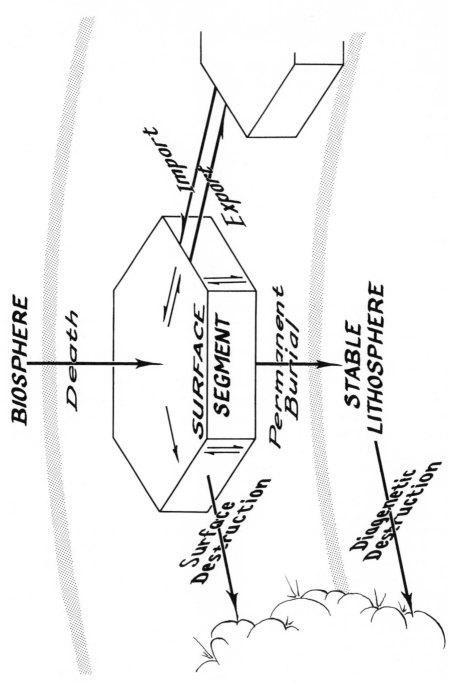

Figure 9.1. Flow of carcasses and parts with respect to a geographically defined element of a taphonomic system. Within the biostratinomic part of the element (between the shaded bands), four basic kinds of change may cause flow to and from the surface segment: the net rate of flow to the segment approaches zero as the system approaches dynamic equilibrium. Movement, degradation and changes in degree of burial may occur within the surface segment, as suggested by the small arrows.

Each of the basic kinds of change might be effected by any of several agents. For example, either import or export may be accomplished by animals, wind, water, downslope movement (gravity), or some combination of these. Conversely, an imposed environmental condition or set of related conditions may lead to more than one kind of change, simultaneously, alternately, or differentially: within a short time and distance, a flooding stream may drown an animal, bury a tooth, carry a limb bone into the reach, excavate and carry a vertebra out, and crush a skull. These multiple effects of related causes, and multiple causes of similar effects, so common in taphonomic systems, might be be sorted out and investigated with the aid of appropriately detailed models of the systems and subsystems. Depending on the nature of the investigation, subsystems may be delimited by the subarea or the process, agent, or change of interest. This approach will permit closer examination of particular phenomena while keeping them in the broader context of the system.

A Model Emphasizing Fluvial Processes

Flowing water has clearly been an important but complex part of the taphonomic systems which produced many of our assemblages of terrestrial fossil vertebrates. A better understanding of its effects on these assemblages hinges on an integrated understanding of the multiple, interdependent roles it plays in the transport, destruction, and burial of vertebrate remains.

The relationships of these processes to each other and to certain other taphonomic agents and processes are shown schematically in figure 9.2. The diagram is intended to represent a segment of any size or position with respect to sample or input points: an infinitesimal length or area increment, a convenient unit size, or a drainage basin. This model is functionally identical to that shown in figure 9.1, but the surface segment itself is subdivided into three states, the input is regrouped into fluvial and nonfluvial components, and diagenetic destruction is not considered. The primary storage state includes carcasses and parts which have not undergone significant fluvial transport within the modeled segment; secondary storage is occupied by those that have been transported by moving water. They are separated here because the fact of transport may be associated with a different response to subsequent events. Elements in either storage may be temporarily buried. The biosphere, other surface segments, destruction, and permanent burial are also separate states. Some of the states may be combined or further subdivided for specific applications; the configuration shown here is an arbitrary compromise between convenience and completeness.

A transition is any instantaneous "discontinuous" change which removes an element from one state and adds it to another. Except for import and export, the rates of these transitions are proportional to the number of elements in the state preceding the transition, but not on the number in the succeeding state, unless the concentration of elements is very high. Physical or chemical degradation of magnitude less than that which defines destruction may proceed within any of the states.

Those elements in fluvial transport have, by definition, some transport velocity, v_t, in the downstream direction. The number of elements in transport per unit of stream length is the linear transport density, δ_t. The product $v_t \delta_t$ equals the transport rate, T, which is the number of elements passing a given point per unit time. The transition rates for fluvial import and export are therefore equal to the transport rates at the upstream and downstream ends of the segment, respectively. As with the other transition rates, they may also be expressed as a proportion of the total number of elements in the

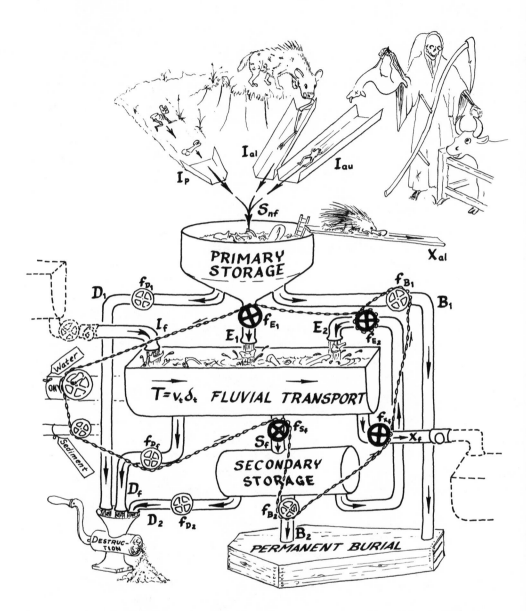

Figure 9.2. Flow of carcasses and parts through a fluvial surface segment. Compare with
figure 9.1: The containers labeled Primary Storage, Fluvial Transport, and Secondary Storage are states within the surface segment. Capital letters with subscripts symbolize rates
of flow between states: B, burial; D, destruction; E, entrainment; I, import; S, storage;
X, export. T is rate of fluvial transport within segment. Subscripts to capital letters
refer to agent or previous state: 1, primary storage; 2, secondary storage; al, allopod;
au, autopod (both terms defined by Vrba, this volume); f, fluvial; nf, nonfluvial; p,
nonfluvial physical. Functions which control transition rates are indicated by f with sub-
script. Black valves mark transitions at which biases with respect to fluvial transportability may be induced. The chain linkage to the master valve on the water pipe symbolizes
the dependence of certain of the transition rates on the discharge (volume of water per
unit time) passing through the segment.

immediately preceding state, but unless the system is in equilibrium or the segment very short, the proportionality coefficient may vary with the linear density distribution as well as with extrinsic factors.

The valves, labeled f with subscripts in the lower half of figure 9.2, represent the regulation of the transition rates by extrinsic factors: they are functions which yield the probability that an element will make the transition within a unit time interval, given the properties of the element and the parameters of the imposed conditions. Together with the number of elements in the preceding state, this probability value (the "setting of the valve") determines the rate of flow to the next state. The black valves designate transitions which are, in part, functions of the fluvial transportability of elements. Many of the regulators are linked to (but not exclusively controlled by) hydraulic factors such as depth, width, stream velocity, shear stress, and sediment load, which in turn are related to discharge (Leopold and Miller 1956; Leopold et al. 1964), as suggested by the chain linkage in figure 9.2. The experiments and conclusions of Voorhies (1969), Dodson (1973), Behrensmeyer (1975), Boaz and Behrensmeyer (1976), and in other parts of this paper provide an embryonic understanding of the response of bones to some of these hydraulic parameters. Other extrinsic physical factors in the fluvial realm, such as channel geometry and lateral migration rate, aggradation or degradation of the channel and floodplain, bed forms, and clast size undoubtedly influence rates of transport, burial, and destruction. But these and the hydraulic factors are all interrelated and, at least for our modeling purposes, might eventually be reducible to a small number of independent geomorphologic and hydrologic parameters.

In passing from one state to the next, a heterogeneous assemblage of elements becomes sorted or biased according to the responses of individual elements to the particular control conditions. Assemblage composition in each of the states is affected by one or more of the regulators in the system, each one sorting according to its own criteria. The degree of sorting may be primarily time-dependent (e.g., destruction of elements in primary or secondary storage) or distance-dependent (gains and losses accrued during fluvial transport) under a given set of extrinsic controls. The relative importance of these two categories of sorting to the composition of an assemblage is thus dependent on the long-term transport velocity of the assemblage and its components.

Other papers in this volume shed valuable light on rates and controls of transitions in both the biological and physical realms. Many of the included concepts and data could be integrated into the type of model discussed here to build predictive models analogous to real systems of moderate complexity. Such models would have potential utility in testing hypotheses (initially including those embodied in the models themselves), pinpointing problems for further research, and, perhaps deriving more powerful models and more general concepts.

An Assemblage Classification Based on Fluvial Transport History

The model suggests that biases related to transportability will be induced at any of the transitions between the storage and fluvial transport states (the black valves in figure 9.2). A biased fraction of the assemblage in any of these states will make the transition to the next state, and the complementary fraction, an assemblage with the opposite bias, will remain in the initial state. Four types of assemblages can result, as shown in figure 9.3, depending on whether or not a bias for or against transportable elements has potentially been induced (two criteria, four combinations).

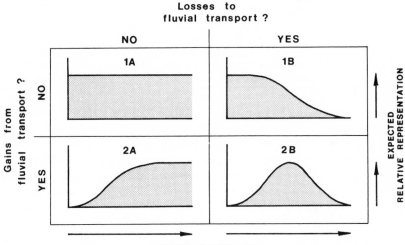

Figure 9.3. Genetic classification scheme based on fluvial transport history of assemblage. Curve associated with each category shows the general form of bias to be expected in an assemblage whose history is described by the given combination of events. Relative representation of elements (number present divided by number expected) should plot on upper boundary curve if hydraulic transport is the only biasing agent. Other kinds of bias (e.g., selective destruction) will tend to scatter data points within the shaded area below each curve. An assemblage may include a mixture of categories (and have a more complex bias curve) if different parts of the assemblage have different histories.

Type 1:   Not transported by moving water; least transportable elements completely or pre-
          ferentially retained.   (Assemblages in primary storage or buried directly from
          primary storage.)

   Type 1A: Never exposed to moving water; easily transported elements also retained.
              Examples: Assemblages accumulated in pond or lake bed, natural trap, or
              dry cave, or assemblages buried by airfall volcanic or eolian sediments.

   Type 1B: Exposed to moving water; transportable elements potentially removed or
              depleted. Examples: Residual accumulations in flooded human occupation
              sites, flooded carnivore, scavenger, or collector dens, interdistributary
              delta surfaces, or flood plains.

Type 2:   Transported by moving water; least transportable elements potentially left up-
          stream. Assemblages in fluvial transport or secondary storage states, or
          permanently buried assemblages derived from secondary storage.

   Type 2A: Transported into very low-energy, "dead-end" environment, or including
              instantaneous downstream limit of dispersal; easily transported elements
              preferentially represented. Examples: Assemblages in crevasse splays,
              delta foresets and bottomsets, basins with internal drainage, or deep
              depressions encountered along a waterway.

   Type 2B: Exposed to moving water after fluvial transport from initial site; trans-
              portable elements potentially removed or depleted. Examples: Assemblages
              on channel bed or point bar.

      The above outline is subject to three qualifications: (1) The predictions concern-
ing biases with respect to transportability of individual elements will hold only if the
represented skeletons were completely disarticulated prior to the initial transport event.
Otherwise, the term "elements" should be replaced by "units," a unit being a complete

carcass or any physically connected portion of it.   (2) The predicted biases may not be apparent in the observed assemblage if the parent assemblage included only a narrow transportability spectrum of elements.   (3) Nonfluvial factors may produce biases which are at least qualitatively similar to fluvial biases.   For instance, the goat bone assemblage in Hottentot food remains (Brain 1976 and this volume) is biased against low-density, small bones, approximately the same elements which would tend to be underrepresented in a fluvial type 1B assemblage.   For these reasons, the form of apparent bias in a fossil assemblage is not, in itself, reliable evidence of the transport history unless these potential problems can be independently discounted.

This classification, as presented, is a genetic and not a descriptive one, so classification of a site according to this scheme constitutes an intermediate step in taphonomic interpretation.   Reconstruction of the actual history (necessary for classification) should depend on evidence from sedimentology, stratigraphy, and perhaps element damage if clearly attributable to fluvial transport.   From this, the general form of bias due to hydraulic sorting can be anticipated and taken into account if relative abundances of elements or taxa are to be used in subsequent taphonomic or paleoecological interpretation.   The kinds of higher-order paleoecological objectives that will be attainable, and the additional steps needed to attain them, will partly depend on the transport category into which the assemblage is found to fit.

Shotwell's assumption that the degree of element bias exhibited by a particular taxon reflects transport distance (Shotwell 1955) must be reconsidered in the light of transportability-dependent biases.   While the assumption may be approximately valid for animals within a restricted size range, larger or smaller animals may exhibit greater or lesser bias arising only from prevailing hydraulic conditions during transport and burial. Wolff (1973) noted this effect on an assemblage from Pleistocene delta distributary channels.

Fluvial Transport Criteria

Derivation and Predictions

Central to an understanding of fluvial taphonomic systems and their vertebrate fossils is an adequate understanding of the complementary processes of entrainment, movement, and deposition of vertebrate remains by flowing water.   Direct observation of these processes by Voorhies (1969), Dodson (1973), and Boaz and Behrensmeyer (1976) has demonstrated the varied but approximately ordered responses of different skeletal elements to particular conditions of flow and substrate.   Behrensmeyer (1975) attempted to establish a more general, quantitative relationship based on the assumption that a bone will respond to flow in a manner similar to a sediment clast having the same settling velocity.   This approach has appeal because of its potential for direct tie-ins with sediment movement.   However, our subsequent flume experiments (discussed below) have revealed disparities between theory and observation, in part due to differences in stable orientations of the bones relative to current direction in the flume as opposed to the settling experiments.

There remains considerable room for improvement in our understanding of these processes, both in terms of generality and degree of resolution.   Continued experimentation would improve the data base, but cannot cover all the natural possibilities.   General, predictive patterns of behavior may continue to elude us unless the basic independent variables are identified and related to behavior through a rational (rather than empirical) mathematical statement.

The needed expression should relate the probable state of motion to measurable features and geometrical relationships of the bone, the fluid flow, and the bed. The closely related problems of sediment entrainment and transport are the subject of a large body of literature (see Raudviki 1967 for a fairly recent synthesis and bibliography), and the importance of sediment transport to fluvial taphonomy makes it desirable that the bone transport expressions be compatible with established sediment transport relationships. But the differences between bones and stones with respect to physical attributes, concentrations, and observed behavior preclude the direct application of sediment transport equations to our problem.

The approach outlined below begins with basic equations of fluid flow and resistance forces and leads to a relationship which partially fulfills the stated requirements. It is predictive over a specified range of conditions.

The force per unit of cross-sectional area, F, exerted on an object by a fluid moving at a velocity U, can be expressed by the Drag Law:

$$F = \frac{C_D}{2} \rho \ U^2 ,$$ (1)

where C_D is the drag coefficient and ρ is the fluid density. The velocity of water near the bed of a stream is not uniform, however, but increases upward. In the absence of a laminar boundary layer, the mean velocity, $\bar{U}$, at a given height, z, above the zero velocity height, z_0, is given by

$$\bar{U} = \frac{U_*}{K} \ \ln \frac{z}{z_0} ,$$ (2)

where U_* is the shear velocity and K is the von Kármán coefficient of turbulent exchange.[2] The force per unit area corresponding to the mean velocity *at height z* above the reference level can then be calculated by substituting equation (2) into equation (1); thus

$$F = \frac{C_D \rho}{2} \left(\frac{U_*}{K} \ \ln \ \frac{z}{z_0} \right)^2$$ (3)

Assume that an object resting on the stream bed has a rectangular crosssection of width b, height c, and area $A = bc$ as shown in figure 9.4. The total fluid force, F_f, on the object may be calculated by integrating equation (3) over this area, from the height, z_b, of the bed surface to $z_c = c + z_b$:

$$F_f = \frac{C_D \rho}{2} \left(\frac{U_*}{K} \right)^2 \ b \int_{z_b}^{z_c} (\ln \frac{z}{z_0})^2 \ dz.$$ (4)

Using the center of the roughness elements (the top layer of particles of diameter k_s in a fixed, uniform bed) as the reference height, $z_b = k_s/2$ and $z_0 \doteq k_s/30$. By performing the integration and substituting these values for z_b and x_0 , we obtain

---

2. This relationship is usually expressed in a form such as

$$\bar{U}/U_* = 8.5 + (2.3/K) \ \log_{10} \ (z/k_s)$$

(Briggs and Middleton 1965). The constant 8.5 (relative velocity at the tops of the roughness elements) can be replaced by $(2.3/K) \log_{10}(k_s/z_0)$, allowing the terms to be combined (see Rouse 1950, p. 101-3). Converting to natural logarithms and multiplying throughout by U_* we get equation (2), an expression which is more amenable to the subsequent integration.

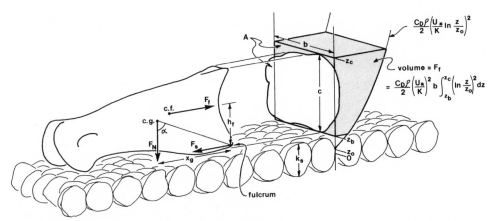

Figure 9.4. Fluid force, F_f, whose quantity is approximated by the volume of the shaded figure, must exceed maximum frictional resistance force, F_S, to initiate sliding. For rolling of the bovid phalanx to begin, the moment (force times length) of fluid force about the fulcrum must exceed the moment of normal force, F_N, assumed equal to submerged weight. The length of the moment arm for fluid force is the height, h_f, of the center of fluid force, c.f., through which this force can be considered to act. The moment arm for normal force has a length, x_g, equal to the horizontal distance from the fulcrum to the center of gravity, c.g. The irregular cross-sectional area, A, of the bone is equal to the rectangular area of height c (the bone height) and width b. Distances along the vertical z axis are measured from the center of the top layer of bed particles. Other symbols and formulae are explained in the text.

$$F_f = \frac{C_D \rho}{2} \left(\frac{U_*}{K}\right)^2 A \ f(c,k_s) \ , \tag{5}$$

where the function

$$f(c,k_s) = \left(1 + \frac{k_s}{2c}\right)\left\{\left[\ln\left(\frac{2c}{k_s} + 1\right) +1.71\right]^2 + 1\right\} - \frac{2k_s}{c} . \tag{6}$$

If the height, c, of the object is between about one and thirty times the size of the roughness elements, this function is closely approximated by

$$f(c,k_s) \doteq 11.1 \ (c/k_s)^{1/3} \ . \tag{7}$$

The shear velocity is related to fluid density and shear stress τ (the tangential fluid stress on a unit area of the bed) by $U_* = \sqrt{\tau/\rho}$, hence $\tau = U_*^2 \rho$. Substituting this into equation (5) yields

$$F_f = \frac{C_D \tau}{2K^2} \ A \ f(c,k_s) \ . \tag{8}$$

If a bone resting on a stream bed is subjected to a gradually increasing fluid force (e.g., increasing shear stress), it will eventually move in one of two modes—either sliding or rolling. If the first movement is by sliding, this will occur when the fluid force just exceeds the maximum static force of resistance due to friction between the bone and the bed. This resistance force F_s, can be expressed in proportion to the normal force, F_N , between the bone and the bed: $F_s = \mu_s F_N$, where μ_s is the sliding coefficient. If action of the moving water on the bone does not contribute to the normal force, and if the bed is horizontal, the normal force is the submerged weight of the bone, W_s; hence

$$F_s = \mu_s W_s \ . \tag{9}$$

At the point of incipient sliding, the forces are equal and the ratio R of the available fluid force to maximum static resistance force equals 1:

$$R = F_f/F_s = F_f/W_s\mu_s = 1. \tag{10}$$

If $R > 1$, the bone will move, and the converse also holds.

     If the initial movement is by rolling rather than sliding, the moments of force about the fulcrum will just balance at the time of incipient motion; that is, the fluid force times the length of its effective lever arm (h_f) will equal the submerged weight times the length of its effective lever arm (x_g), and

$$R = F_f h_f/W_s x_g = F_f/W_s (x_g/h_f) = 1 \ . \tag{11}$$

The quantity x_g/h_f is seen to be analogous to the sliding coefficient and may be defined as the rolling coefficient $\mu_r = x_g/h_f$. By defining a general resistance coefficient μ to equal either the sliding coefficient or the rolling coefficient, equations (10) and (11) can be combined into a single expression: $R = F_f/W_s\mu$. Substituting the equations for fluid force (eqs. [8] and [6] or [7]), we can write general expressions which theoretically relate the state of movement (implied by the force ratio R) to pertinent parameters of the bone, bed, and flow:

$$R = \frac{C_D a}{2W_s\mu}\frac{\tau}{K^2}\left\{\left\{(1 + \frac{k_s}{2c})\left\{[\ln(\frac{2c}{k_s} + 1)+1.71]^2 + 1\right\} - \frac{2K_s}{c}\right\}\right\}, \tag{12}$$

and its approximation

$$R \doteq 11.1\frac{C_D A}{2W_s\mu}\frac{\tau}{K^2}\left(\frac{c}{k_s}\right)^{1/3} = 11.1\left(\frac{C_D Ac^{1/3}}{2W_s\mu}\right)\left(\frac{\tau}{K^2 k_s^{1/3}}\right). \tag{13}$$

In the last expression, the parameters in the first set of parentheses pertain to the bone and those in the second set pertain to the flow and bed conditions. These parameters will be further discussed below.

     These equations carry with them a number of assumptions and approximations which were explicitly or implicitly drawn into the derivation. It is clear that the assumptions are not strictly valid over much (perhaps most) of the range of natural conditions. Nevertheless, the derived relationships will hopefully serve two purposes: (1) to yield approximations of bone behavior at a better level of resolution than that presently attainable; (2) to provide a rational baseline which can be empirically tested and refined. The areas in which refinement probably is most needed are those outside the realm of the assumed conditions. The assumptions are:

     1. Vertical distribution of fluid velocity is adequately described by equation (2): Velocity gradient is logarithmic near the bed and flow is turbulent and fully rough (laminar sublayer disrupted).

     2. Water depth is much greater than element height.

     3. There is an insignificant amount of sediment in transport.

     4. Bones rest on top of roughness elements: They are not partially or completely buried and are not sheltered or impeded by bed features (including their own scour features) or anomalous obstructions.

5. Normal force between bone and bed is greater than zero and equal to submerged weight of bone. Hydrodynamic lift, net reactive forces due to deflection of flow, and dispersive grain pressure (Bagnold 1954) do not significantly affect normal force. Bed is horizontal.

Conditions which meet all of these assumptions will hereafter be referred to as "ideal conditions."

Natural conditions which do not meet these assumptions will not render the transport relations invalid, but will introduce systematic inaccuracies of predictions. If conditions deviate from those of assumptions 1, 2, or 3, elements may be on the verge of movement at values of R greater or less than 1, but the amount of error should be related almost exclusively to height, and should be apparent if observed behavior is expressed on an R vs. c plot (see figure 9.5). Deviations from assumptions 4 and 5, however, may cause errors related also to element shape. Although these are more difficult to eliminate, it should still be possible to correct empirically any resulting inaccuracies for particular "problem" elements.

The statement that bones will move under ideal conditions only if $R > 1$ must be slightly modified because of the assumed and almost inevitable condition of turbulent flow: Equation (2) predicts the *time-averaged* velocity of flow, but because the flow is turbulent, instantaneous velocities exceeding the mean will cause intermittent movement of elements whose calculated force ratios are slightly less than 1.

Approximations inherent in the force ratio relationships may also lead to slight disparities between prediction and observation.

Approximations:

1. Zero velocity height, z_0, equals $k_s/30$. Other values and expressions for this empirical relationship are discussed by Meland and Norrman (1966). The alternative values would produce slightly different constants in equation (6) and subsequent derivations, but with minor net effect.

2. Fluid force on the bone equals that on an object with the same drag coefficient, cross-sectional area, and height, but with a rectangular section. The resulting error is probably quite small for circular or elliptical sections, but may be as high as 15% for triangular sections.

Most of the parameters included in equations (12) and (13) require additional explanation:

The submerged weight, W_s, is the only included bone parameter which is independent of the orientation of the bone relative to the bed and the current direction. This quantity can be expressed in terms of more basic attributes: $W_s = V(\rho_b - \rho)g$, where V equals bone volume, ρ_b is the bulk density of the bone, ρ is the density of water, and g is gravitational acceleration. Behrensmeyer (1975) discusses these factors and provides data on bone densities, volumes, and weights of many elements of several modern taxa.

A, the cross-sectional area appropriate to these expressions, is the area of the projection of the element on a plane perpendicular to the current direction. Thus the area varies with orientation and is usually at a minimum when the long axis is parallel to the current direction.

The drag coefficient, c_D, is a dimensionless parameter which varies with the shape, relative current direction, and Reynolds' number. The latter quantity is a measure of the relative amounts of viscous and inertial fluid force. Most bones will begin to move while within the range of Reynolds' numbers (10^3 to 10^5) through which drag coefficients tend to remain nearly constant. For purposes of determining initiation of movement, then, a

single drag coefficient should be applicable to bones of different sizes if they are similar in shape and orientation.   Initiation of movement may occur at lower Reynolds' numbers
for very small (e.g., limb bones smaller than those of *Rattus norvegicus*) or very low
density bones; these will have larger drag coefficients than their larger or denser
counterparts.   No direct measurements of drag coefficients for bones resting on a bed
have yet been made, but known coefficients for various bonelike shapes, adjusted to account
for effects of contact with a boundary (see Carty's data in Raudviki 1967, fig. 278), suggest that a value of about 1.5 should be approximately correct for most elements.

As defined above, the resistance coefficient, μ, may be either the rolling coefficient or the sliding coefficient, depending on the mode of movement.   Under the ideal
conditions, initial movement in a given case should occur in the mode associated with the
lowest coefficient.   Experimentally determined sliding coefficients for a few bones suggest that the value varies slightly with the shape and orientation of the bone, the normal
force, and probably with the nature of the substrate.   Most of the tested combinations
yielded coefficients between 0.75 and 0.90, with an average of about 0.85 for limb bones
on sand, but these values should be regarded as tentative.   The rolling coefficient μ_r,
also dependent on orientation, is equal to x_g/h_f or (h_g/h_f) tan α, where x_g is the horizontal distance from the center of gravity to the rolling fulcrum, h_g is the height of the
center of gravity, h_f is the height of the center of fluid force, and α is the angle of
the slope on which the bone will just begin to roll under its own weight.

The shear stress, τ, is the amount of tangential force per unit area exerted by a
moving fluid on the boundaries that confine it.   The local shear stress equals the product
of the density of the fluid (ρ), the acceleration due to gravity (g), the depth of flow
(D), and the slope of the energy grade line (S), which is also the slope of the fluid surface in equilibrium flow: $\tau = \rho gDS$.

The rate of increase of fluid velocity with height from the bed depends on shear
stress and on the von Kármán coefficient, K.   Under ideal conditions with a flat bed, K is
fairly constant, with a value near 0.4.   However, it appears to vary with the amount of
suspended sediment (Vanoni and Nomicos 1959), channel geometry (Rozovskii 1957), and bed
forms, especially large dunes (Nordin and Dempster 1963).

The effective diameter of the bed roughness elements, k_s, for a flat bed of uniform, spherical particles is simply the particle diameter itself.   For a flat bed of
nonuniform sediment, Einstein (1950) used the value $k_s = d_{65}$, where d_{65} is the particle
size for which 65% of the bed material is finer.   In natural channels with various bed
forms, Nordin and Dempster (1963) and Axelsson (1967) found the average values of k_s to be
about 400 times as great as d_{65}.   A relationship between k_s and size of bed forms has not
yet been determined.

In the force ratio equations (12) and (13), several parameters of the bone are explicitly stated, but it may not be obvious how the force ratio changes with size alone if
the other variables of shape, density, orientation, and flow conditions are constant.
Three variables, cross-sectional area (A), submerged weight (W_s), and height (c), are
size-related and all may be expressed, by allometry, as some exponential function of a
length measure, L.   Thus, $A \propto L^2$, $W \propto$ volume $\propto L^3$, and $c \propto L$.   Substituting these into
equation (13), we find $R \doteq Ac^{1/3}/W_s \propto L^2 L^{1/3}/L^3$, so $R \doteq 1/L^{2/3}$.   Since $L \propto V^{1/3}$, where V is
volume, $R \doteq 1/(V^{1/3})^{2/3}$ or $R \doteq 1/V^{2/9}$.   For a given kind of element, then, the force
ratio is approximately inversely proportional to the two-thirds power of its length or the
two-ninths power of its volume.

Under the ideal conditions, the state of movement is theoretically predictable by equations (12) or (13), if all of the variables are known.   These relationships should be useful for modern bone transport experiments and for fluvial taphonomic model studies involving established principles of river hydraulics. However, especially in the investigation of fossil deposits, the bed and hydraulic conditions of transport and deposition may not be known, or it may be desirable to determine only the *relative* transportability of different elements. Expressions suited to these potential applications may be derived from the general transport equations by assigning arbitrary, constant values to the flow and bed parameters.[3] Convenient values in the *cgs* system would be $\tau' = 1$ dyne/cm^2, $\kappa' = 0.4$, and $k_s' = 0.1$ cm. With these substitutions, equation (12) becomes

$$R' = 3.125 \; (C_D A/W_s \mu) \quad (1+1/20c) \quad [\ln(20c+1)+1.71]^2 + 1 \; - \; 1/5c. \tag{14}$$

Similarly, the approximation (13) becomes

$$R' \doteq 75 \; C_D A c^{1/3}/W_s \mu. \tag{15}$$

The value R' is thus a quantitative measure of predicted transportability, virtually independent of flow and bed conditions (except possible slight variation of μ_s with substrate) but subject to the same assumptions and approximations outlined earlier. Once the value of R' is established for a particular element and orientation, simple proportional and exponential relationships permit convenient conversion to R (for given flow and bed conditions), calculation of ideal critical shear stress (that needed to make $R = 1$), or calculation of R' for other elements of different size or density but the same shape and orientation.

Transport velocity as well as state of motion is important to an understanding of fluvial taphonomic systems and water-transported assemblages. Long-term velocity determines the relationship between distance of transport and extent of exposure to time-dependent destructive factors. The relationship of transport velocity to linear density and transport rate was discussed previously: for certain cases, this will determine the abundance of elements in secondary storage and permanent burial. Of particular paleoecologic importance is the expectation that the faster elements found in a type 2 assemblage most probably represent more recent biologic events which occurred at a given distance upstream, and more distant events which occurred at a given time. Inasmuch as long-term velocity depends on velocity during the short, seasonal periods of active transport, these short-term velocities deserve consideration.

The forces which act on a sliding bone are similar to those acting on a stationary one, except that the fluid force depends on the square of the *difference* between the velocity of the fluid and that of the bone. It can be shown, again from the balance of forces, that the mean sliding velocity, $\bar{v}_s$, is approximately predicted by

$$\bar{v}_s \doteq \bar{u}_c (1-1/\sqrt{R}), \quad R \geq 1, \tag{16}$$

where $\bar{u}_c$ is the mean ambient fluid velocity, that is, the mean velocity of the fluid layer between the bed surface and the height, c, of the bone. Forces on a rolling or

---

3.   This approach yields expressions which are strictly valid only for a bed whose roughness elements are of arbitrarily specified size. For other bed conditions, a small error, systematically related to element height, will be introduced, but in most cases this error will be negligibly small compared to those which will inevitably arise from real deviations from ideal conditions and from measurement uncertainty.

saltating element are more complex, but the transport velocity should still increase with R for values of R greater than unity.

Long-term transport velocity should also increase as R' increases above a critical minimum value, though bone parameters not included in R', and stream hydrology, will determine the exact nature of the relationship.

Experimental Tests of Transport Criteria

Two recent investigations of transport of bones by moving water, initiated by A. K. Behrensmeyer, provide quantitative data which can be used to test the validity of the threshold of movement criteria and transport velocity hypotheses outlined above. We will report the results of these studies in more detail in the near future; only that information most relevant to the present discussion is included here.

Flume Experiments

In the Spring of 1974, A. K. Behrensmeyer, J. T. Gregory, and I conducted a series of experiments to test the transport behavior of skeletal elements of several mammals under controlled conditions of flow and substrate in a laboratory flume. Boaz and Behrensmeyer (1976) conducted similar experiments with human skeletal parts using the same flume and substrate material.

We conducted the experiments in the 1 ft (.30 m) by 40 ft (12.2 m) recirculating flume in the Hydraulic Engineering Laboratory at the University of California, Berkeley. An artificially flattened bed of coarse sand (d_{50} = 1.4 mm, d_{65} = 1.5 mm) permitted experimentation within a broad range of hydraulic conditions without the complications of moving sediment and changing bed forms. Velocity profiles, water depth (approximately 15 cm) and surface profiles, discharge, and temperature were measured before the experiments and monitored during the runs. The rat, rabbit, dog, sheep, and goat bones used were weighed in water and in air at the times of the experiments so that bulk densities could be determined. Travel times of several runs through a segment of the flume yielded mean transport velocities for each transportable element. Shear stresses, calculated from water surface slopes and hydraulic radii for the various runs, ranged from 2.0 to 8.7 dynes/cm^2.

Light-shadows cast by the bones and measured on millimeter-ruled paper provided accurate measurements of cross-sectional area. The tangent of the roll angle was taken without correction as an estimate of the rolling coefficient. Sliding coefficients were not measured, but were consistently assigned the value of 0.85, no doubt introducing some error into the force ratio calculations. As no independent measurements of drag coefficients for skeletal parts are available, the estimated average value of 1.5 (discussed previously) was applied throughout. The diameter of the bed roughness elements was taken to be 0.15 cm., the 65 percentile diameter of the bed material. The value of von Kármán's K, calculated from our measured velocity profiles, was consistently close to 0.4, so this value was used throughout subsequent calculations.

Observed transport behavior, plotted in figure 9.5 against element height and force ratio (calculated from eq. [12]), shows a reasonably coherent pattern, considering the probable errors of estimation discussed above. Most of the discrepancies would tend to be reduced or eliminated by adjusting the drag and resistance coefficients in what would seem to be the appropriate directions. The apparent widening of the zone of discontinuous movement with decreasing element height is probably due to obstruction by individual grains or minor bed irregularities and to the greater sensitivity of small elements to small-scale turbulence.

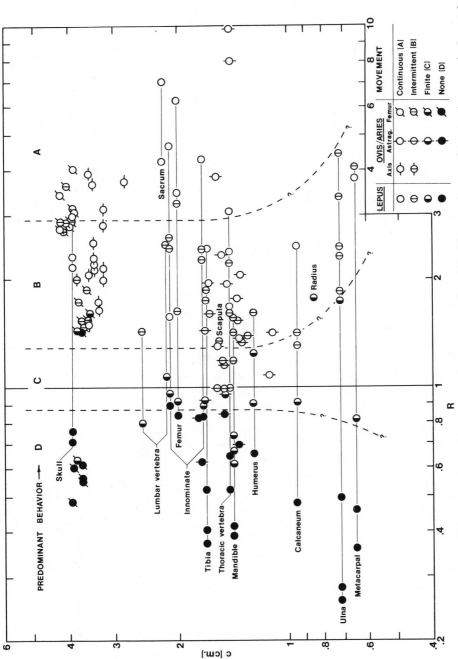

Figure 9.5. Flume observations of bone transport behavior plotted against predicted fluid-to-resistance force ratio, R, and element height, c. Movement is predicted for elements having values of R greater than 1. *Lepus* (jackrabbit) elements are labeled on diagram; *Ovis/Capra* (sheep or goat) elements are identified by spurs on symbols. Different values of R and c for a given kind of element reflect different orientations and flow conditions as well as individual variation. "Finite" movement (behavior category C) means that the element stopped after short initial movement.

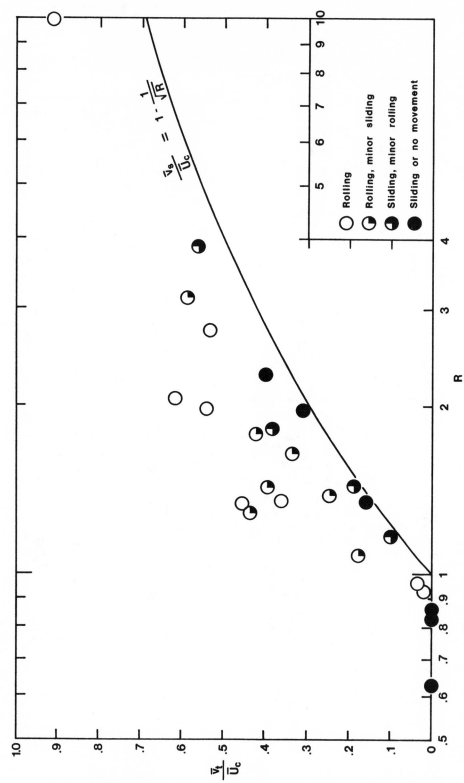

Figure 9.6. Mean observed transport velocity, $\bar{v}_t$, expressed as a fraction of calculated mean ambient fluid velocity, $\bar{U}_c$, is plotted against predicted force ratio, R, for *Ovis/Capra* astragali in flume. Curve indicates predicted velocity ratio as a function of R for sliding elements. Rolling elements are expected to have higher velocities.

Mean transport velocities of sheep/goat astragali, measured during three flume runs at different shear stresses, are plotted against R in figure 9.6. Transport velocities are expressed as decimal fractions of the calculated mean ambient fluid velocities, $\bar{U}_c$, to correct for expected velocity differences due to height variation. Velocities of continuously sliding astragali agree well with those predicted by equation (16). Sliding was found to be more common for the larger elements. Continuously rolling astragali moved at higher velocities than sliders with the same calculated R value, probably because less energy was dissipated by friction. Elements which alternated between rolling and sliding exhibited intermediate transport velocities.

Elements other than astragali show similar patterns of increase in velocity with R, but their patterns differ in detail from that for the astragali, indicating that shape factors not included in the R parameter have some effect on transport velocity.

River Experiments

In 1974, A. K. Behrensmeyer and I began a study of bone transport in the East Fork River in Western Wyoming. The study area is the site of several concurrent investigations of river geometry, hydraulics, and sediment transport being conducted by L. B. Leopold, his students, and the U. S. Geological Survey.

At the site of the experiment, the East Fork is a perennial, meandering river with a sandy, gravelly bed. Dunes reaching about 15 cm in height are common during flood flow, but parts of the channel bed are scoured to a basal gravel mostly composed of clasts less than 5 cm in diameter (Lisle 1976). The river has an average width of nearly 20 m and an average depth of 1.2 m at bankfull stage. Nearly identical maximum flood peaks during the 1974 and 1975 seasons exceeded bankfull stage, and discharges reached 45 m^3/sec both years (E. D. Andrews, personal communication).

Before the annual snowmelt floods of 1974 and 1975, measured and numbered bones were arrayed at surveyed sites on the stream bed. At the low-water stages after the floods, we returned to the area to visually relocate the bones, map their positions, and note orientation, damage, and nearby features.

During flood stage, when most bone transport occurred, the channel, flow, sediment transport, and bed conditions were neither constant nor similar to the conditions assumed in the derivation of the transport criteria. Nevertheless, within the relatively narrow range of height of the bones used in the experiment, entrainment and flood-season transport distance should be related to the relative transportability parameter, R'. To test this expectation, I estimated R' values for a representative sample (33) of the 42 relocated elements from the group of 120 assorted domestic cattle bones placed in 1975.

Submerged, soaked weights, W_s, were estimated according to the equation W_s = .562 W_d - .181 V, where W_d is the dry weight and V is volume, measured before the experiments. This equation was derived from a linear regression equation of soaked vs. dry sensity measured for seventeen assorted bones. Allometric equations, determined for each kind of element, provided a basis for determination of cross-sectional areas from measured heights and widths.

At the experimental site, the bones had been laid parallel to the current direction, and most of the transported ones probably moved by sliding with their long axes parallel to the current (except metatarsals; see below). The cross-sectional area appropriate to the R' calculation, then, is that which corresponds to parallel orientation, and μ is the sliding coefficient, assumed equal to .85, except as noted below. Values for R' were then calculated according to equation (15).

Transport distances were measured from plane table and tape-and-Brunton maps made before and after the flood season, and from enlarged, low-altitude aerial photographs on which the positions of relocated bones had been plotted in the field.

Figure 9.7 shows observed transport distances plotted against the estimated R'

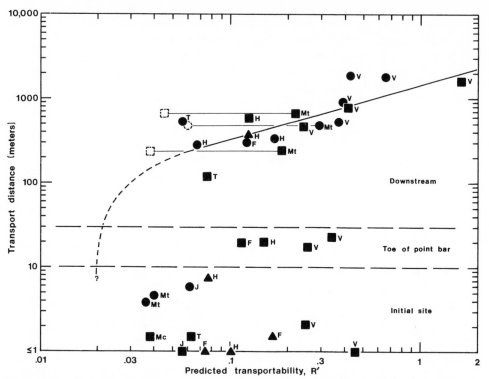

Figure 9.7. Observed one-season transport distance plotted against estimated transportability parameter, R', for domestic cattle bones visually relocated after exposure to 1975 flood on East Fork River. Horizontal dashed lines separate natural assemblages of elements, those which remained on the bar where they were initially placed or within 10 m along the channel, those which were trapped at the toe of the point bar site, and those which moved past the area influenced by features of the initial site. Oblique line is regression curve fitted to solid downstream data points. Little or no transport is expected for elements with R' less than about .02, as indicated by the dashed extension of the regression curve. Open, dashed symbols are plotted at R' values corresponding to parallel-to-current orientation for downstream metatarsals; solid symbols indicate more probable perpendicular orientation during transport. Triangular symbols denote elements initially placed at site A; square symbols, site B; and circles, site C. Letters identify element; F, femur; H, humerus; J, jaw; Mc, metacarpal; Mt, metatarsal; T, tibia; V, thoracic vertebra. Note increase in proportion of transported elements (downstream) vs. untransported elements (initial site) with increasing R'.

values. Although the body of data is not large, some significant trends are apparent. These can best be discussed in terms of the fluvial model and classification described above.

In 1975, nearly identical samples of forty assorted elements were initially placed on each of three sites; a midchannel sand and gravel bar (site A), a sandy point bar (site B) about 50 m downstream from site A, and the channel bed (site C) opposite site B. These samples may be regarded as type 1A assemblages in primary storage. The type 1B assemblages which remained in primary storage at the sites after the flood exhibited a marked bias toward low transportability,[4] but the separation is not complete; a few elements with high R' values remained at the site, obstructed by vegetation or apparently stabilized by scour or partial burial during the rising flood stages. Flood waters carried the complementary type 2A assemblages from sites A and B across the steep, prograding downstream ends of the bars to areas of low shear stress, where parts of the assemblages (type 2B) remained buried at least until the falling stage of the flood. The type 2A assemblages which escaped these local traps were then redistributed downstream.

The transported assemblages include some elements which should be among the most difficult to transport, implying that shear stresses locally exceeded those necessary to entrain even the least transportable cow elements. This is consistent with estimated mean critical R' values of .015 to .02, calculated from mean peak bed shear stress of 265 dynes/cm^2 and k_s equal to 5 cm (approximate diameter of basal gravel clasts) and 15 cm (approximate height of dune crests), respectively.[5]

Once entrained, the more transportable elements would be expected to travel farther during a flood event for two reasons: (1) they are exposed to supercritical conditions for a longer period of time, and (2) they will move faster at each instant within the supercritical period. The river observations are consistent with this expectation.

Values of R' appropriate to transported elements depend on mode of movement and orientation during transport. Observations of behavior in flumes (Boaz and Behrensmeyer 1976, and our own) suggest that smaller, symmetrical elements with low rolling coefficients tend to roll with their long axes perpendicular to the current, while others tend to slide in parallel orientation. Of the transported elements included in figure 9.7, only the metatarsals fall in the former category, so these were assigned R' values corresponding to perpendicular rolling. Initiation and cessation of significant movement, however, is probably determined by R' corresponding to the static-stable (parallel) orientation.

The upper part of figure 9.7, showing the data for the downstream sample, exhibits a general trend of increasing transport distance with increasing R'. A linear regression fitted to the logarithms of the two variables yields a correlation coefficient of .77. The corresponding regression equation in exponential form is thus transport distance = 1450 $R'^{0.65}$, where transport distance is in meters.[6]

Several factors are no doubt responsible for the considerable scatter of the data points about the general trend line. Both random and systematic errors enter into the estimation of R'. Most of the scatter, however, is almost certainly due to randomly distributed obstructions and unpredictable temporal and spatial variations in flow and bed conditions along individual transport paths. These factors would cause an original association, even of identical elements, to become spread over progressively longer stream

---

4. Both the amount of bias and the percent of elements transported from the sites are greater than suggested by figure 9.7. Seventy-three percent of the elements in the original sample are not included in the figure, and most of these are easily transportable elements not found at the carefully searched and excavated placement sites. About 12% of the initial sample was relocated at the sites.

5. Peak shear stress was estimated by extrapolation to peak discharge of an exponential regression curve fitted to data (Lisle 1976) on shear stress vs. discharge for the reach including the B and C sites. Dune heights, also recorded by Lisle (1976), were measured during moderate flows using a sonic sounder in the reach just upstream from site A.

6. This should not be taken as a generally applicable relationship. For any given case, the curve relating mean transport distance to R' will depend on the frequency distribution of shear stress with time and on the history of sediment movement and bed morphology during the flood event. In fact, I can find no a priori reason that the curve should follow a simple exponential law; it almost certainly would not for values of R' closer to the critical value.

segments as their mean transport distance increases. (This dispersion is masked by the logarithmic distance scale in fig. 9.7.) The probability of their reassociation within a given interval of time and distance decreases correspondingly. Conversely, if an association of disarticulated elements does occur in a transported assemblage, the elements probably originated at different times and distances upstream, and hence represent different individuals. At least, the number of individuals represented will probably be closer to the number of specimens than to a minimum number calculated from the most common element.

For the purpose of testing the predictive value of R', the important points are as follows: (1) the scatter of the downstream data for each kind of element is centered very close to the same curve; and (2) the ratio of transported to untransported elements exhibits a consistent increase with R'. This suggests that R' adequately describes element transportability in common terms, at least within the size and shape ranges of the included elements, and within the present limits of resolution.

Surveys made seven months apart between the 1975 and 1976 floods revealed virtually no transport of bones except for one atlas which had moved 43 m farther; most of the bones probably moved only during the few days of highest flow. The distance indicated for each element in figure 9.7, however, represents its total transport for the year, and so is numerically equal to its mean one-year transport velocity. We have yet to determine whether these velocities are truly representative of long-term transport velocities in this river, but our present information suggests that they are better than order-of-magnitude estimates.

Downstream Loss

Although entrainment and transport comprise the focal center of fluvial taphonomic processes, the phenomena of destruction and permanent burial set the limits on the extent of possible fluvial dispersal. From the viewpoint of the assemblage in transport, both permanent burial and destruction have the similar effect of reducing the linear density and transport rate as distance from the source increases. It would be useful to know in more detail how these processes individually and jointly affect the relationship between distance and abundance.

There is as yet little direct information about the rate of destruction as a function of distance of stream transport, but several pieces of evidence offer indirect support for the following hypothesis:

Hypothesis 1: The probability of destruction of a given stream-stransported element is uniformly distributed with respect to distance from the point of initial entrainment measured along a uniform channel reach.

Observations which tend to corroborate this are as follows: (1) Elements carried nearly 2 kilometers in the East Fork River showed virtually no abrasion or other transport-related damage, though minor damage occurred during bank storage, due to weathering and trampling. (2) Between transport intervals, elements may remain in any of a broad range of microenvironments, and only a small percent occupied exposed surfaces in the East Fork River. (3) Rate of destruction is very sensitive to microenvironmental factors (see papers by Behrensmeyer and Dechant Boaz, Coe, Hare, and Hill, this volume), especially degree of burial (Gifford, this volume). Together, these suggest that transport distance is effectively limited by random accidents of exposure to destructive agents during bank storage, and, at most, weakly limited by factors correlated with transport distance.

A similar hypothesis for the uniform probability of permanent burial can also be proposed. This hypothesis is even more probably valid than that applying to destruction because it is very unlikely that local circumstances leading to burial are in any way correlated with time or distance of transport.

Hypothesis 2: The probability of permanent burial of a given transported element is uniformly distributed with respect to distance from the point of initial entrainment measured along a uniform channel reach.

A "uniform channel reach," for these purposes, is one in which the potential burial and destructive agents are either uniformly distributed or common and randomly distributed with respect to time and to distance along the reach, and the most probable long-term transport velocity of each element is the same for each segment along the reach.

These two hypotheses each have consequent corollaries which carry significant taphonomic and paleoecologic implications.

Corollary 1a (2a): The logarithm of the probability that an element, once entrained, will reach at least a given distance along a uniform reach downstream from its point of initial entrainment without being destroyed (permanently buried) decreases linearly with the distance.

Corollary 1b (2b): Given only that an element has reached a point within a uniform, unbranched channel which is capable of transporting it, the logarithm of the probability that it was first entrained at least a given distance upstream from that point decreases linearly with the distance.

The mathematical form of both of these corollaries is $\ln P = -\lambda x$, or $P = e^{-\lambda x}$, where P is the probability associated with the distance, x, downstream or upstream; and $\lambda = \lambda_D = \lambda_B$ is the unit loss ratio, that is, the probability of destruction (λ_D) or permanent burial (λ_B) per unit length of channel. The form of the equation is analogous to that for radioactive decay, and we can calculate a half-distance, x_H, analogous to the half-life of radioactive elements: $x_H = (\ln 2)/\lambda = .693/\lambda$.

Long-term transport velocity relates the unit loss ratios λ_B and λ_D (probabilities per unit distance) to the transition probabilities (per unit time) for destruction and burial from the combined fluvial transport and secondary storage states of the model in figure 9.2. Thus the problems of testing the general model and the above hypotheses, and, if these are valid, quantifying the included variables, are all interrelated and could best be attacked through combined investigations involving model studies and detailed, multifaceted studies of contemporary natural taphonomic systems.

Examples of Applications

The intent of this paper is to develop a few implements with which some kinds of taphonomic problems can be attacked. The potential results or applications are no more determined by these implements than is the shape or use of a structure determined by a builder's tools. Nevertheless, some specific examples may help to illustrate potential applications and bring out additional points.

Stream-Delta System

     Consider a portion of a stream channel terminating in a lake delta distributary complex. A simple model of this partial system might be constructed of three consecutive segments as shown in figure 9.8: an upstream segment (I) with only nonfluvial input at a

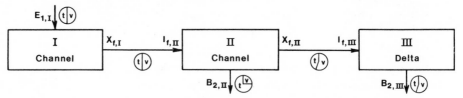

Figure 9.8. Simplified three-segment model for transport and burial within a stream channel and lake delta system. Divided circles indicate relative transition rates for teeth, t, and faster vertebrae, v. Transition rates and transport velocities for the two elements are assumed constant throughout, and their respective rates are equal until the assemblage reaches segment II. At equal transport rates, linear density is inversely proportional to velocity, so that as they enter segment II, the slower teeth have greater linear density than do the vertebrae. Burial rate in segment II, assumed proportional to linear density, is therefore greater for teeth, and the buried channel assemblage has a high t/v ratio. The transported assemblage, now slightly more deficient in teeth than in vertebrae, enters the delta segment (III). As this is a "dead end" to transport, the burial rates equal the import rates and the buried assemblage has a t/v ratio which is somewhat lower than that of the parent assemblage.

rate E_1, no burial or destruction, and fluvial export (X_f); a middle segment (II) with progressive burial (B_2); and the delta segment (III) with fluvial import and burial only. Let us follow the progress of two kinds of elements, say teeth and vertebrae, which we assume will differ in long-term transport velocity by a factor of three (vertebrae are faster). Assume these elements have equal, constant rates of initial entrainment in segment I. As there is no burial or destruction here, their export rates, $X_{f,I}$ (number exported from segment I per unit time), are also equal. As they enter segment II, their linear densities, δ_t, are inversely proportional to their mean long-term transport velocities, $\bar{\bar{v}}_t$, since $\delta_t \bar{\bar{v}}_t = I_{f,II} = X_{f,I}$, that is, on the average, there will be one-third as many vertebrae as teeth in an increment of channel length at any instant. If the rate of burial is proportional to linear density, a buried sample at the upstream end of segment II (a type 2B assemblage) will have three times as many teeth as vertebrae.[7] As the transported assemblage progresses downstream, the linear densities and transport rates of both elements decrease because of losses to burial, but those of teeth decrease somewhat faster than those of vertebrae. If segment II is short, or the burial rates are low, or (as is probably the case in a real channel) more vertebrae than teeth are destroyed in each unit transport distance, then the rates of export of the two elements may still be nearly equal. These rates are also the import rates for segment III. Within the delta segment, the rapid downstream decreases in slope and channel depth within the distributary channels cause shear stress to drop below critical values for both teeth and vertebrae, and they are quickly buried in the rapidly aggrading channels and foresets. The nearly equal import rates become the burial rates; the assemblage (type 2A) is virtually "reconstituted" within the delta system taken as a whole.

_____

     7. If the burial rate in segment II were related to transportability rather than linear density, similar results would obtain; teeth would be selectively buried and over-represented relative to vertebrae, but not necessarily in the same ratio as that produced by linear-density-dependent burial.

The foregoing example is analogous to the system which produced a number of the Lake Turkana (East Rudolf) fossil-bearing localities described by Behrensmeyer (1975). Although greatly oversimplified, the example may explain the relatively unbiased sample obtained from locality 103-0267, a deltaic distributary-beach deposit, in contrast to the tooth-rich samples yielded by nondeltaic channel deposits.

Drainage Network

Natural fluvial systems are not composed of single channels, of course, but of branching networks of channels of varied character. How might such a branching network sample the organisms within its realm and deliver the samples to particular points?

Taphonomically important aspects of streams in a given basin tend to vary systematically with the amount of water the channels carry—that is, with increasing discharge downstream. Typical downstream changes in channel slope, depth, width, grain size, bedforms, seasonality of flow, and erosion/deposition tend to offset each other in such a way that the net effect on the rates of entrainment and loss of elements may be nearly the same throughout a drainage network if these elements can be transported in all parts of the system. Much more investigation is needed to confirm, modify, or refute this tentative conclusion, but for the moment let us assume its validity; the rate of initial entrainment, E_1, per unit of channel length and the combined unit loss ratio for destruction and burial, λ, are uniform throughout the basin for a given kind of element. The uniform unit loss ratio also carries the assumption that hypotheses 1 and 2 (stated earlier in this paper) are valid.

Consider a small channel increment, segment I_1, at a distance x_I above a sample point at segment III, as shown in figure 9.9. New elements of a given kind entrained within I_1 constitute the input to the next reach, segment II_1, which extends to the sample point. After exponential losses in the intervening distance, the elements which were entrained in segment I_1 are delivered to the sample point at a rate proportional to $e^{-\lambda x_I}$. All other increments (I_2, I_3 ... , I_M) of the same length and upstream distance from the sample point, but on other channels, contribute to the sample point at similar rates. The total contribution from all such increments is proportional to M, the number of channels which exist at that distance, times the incremental contribution per channel, or $Me^{-\lambda x_I}$. Approaching the sample point from upstream, $M \to 1$, $x_I \to 0$, and $Me^{-\lambda x_I} \to 1$. Hence, if any upstream increment of distance has a value of $Me^{-\lambda x} > 1$, it will contribute more elements to the sample than will an increment of the same size immediately above the sample point.

I am aware of no presently available data which can be used to test this prediction or its basic assumptions. Wolff's late Pleistocene fossil data compared with Recent distributions and population densities of the same species (Wolff 1971, 1973, 1975) indicate that species which were probably abundant in the community proximal to the sampled deltaic deposits are underrepresented relative to some members of the distal grassland and hillslope communities. This may be due to a combination of factors: hydraulic sorting (as Wolff suggests), proximity of habitats to channels, greater total drainage area occupied by distal communities, and greater relative contribution per unit drainage distance for the distal communities as predicted by the above model.

Summary

Organisms and their remains undergo four basic kinds of change which fall within the realm of taphonomy. The extreme versions of these changes, death, destruction, import-

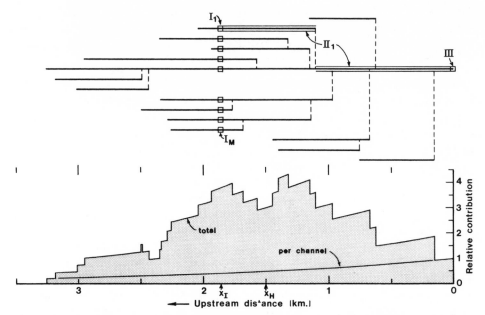

Figure 9.9. Hypothetical relative contribution of a given kind of element to a sample point as a function of up-drainage distance to points of initial entrainment. Uniform rate of entrainment per unit distance and uniform loss ratio (λ = 4.62 x 10^{-4} m^{-1} or x_H = 1.5 km) for all channels are assumed. Upper part of figure is a straightened representation of a small drainage basin near Moraga, California. After losses from transport through the intervening distance (segment II$_1$), each small channel increment such as I_1 contributes elements to the sample point (segment III) at a rate proportional to $e^{-\lambda x_I}$ (the height of the "per channel" curve at distance x_I in the lower diagram). The total contribution from all such M segments existing at that distance is proportional to $Me^{-\lambda x_I}$, the height of the "total" curve. The model suggests that, in a transported sample, members of communities approximately one or two half-distances (x_H) upstream from the sample point may be more abundantly represented by the given element than are members of the community in the immediate vicinity of the sample point.

export, and permanent burial, may be discretely defined and compatibly placed within a common conceptual framework based on gains and losses to a segment (or a mosaic of transport-linked segments) of the earth's surface and near-surface. The agents collectively responsible for each kind of change determine the probability of occurrence, within a given time interval, of the discrete change for each organism or part affected. This basic structure may be elaborated or extended to construct models commensurate with taphonomic systems or subsystems of any temporal or spatial scale.

An expanded version of the segment model facilitates a more detailed examination of the multiple taphonomic effects of moving water and its naturally associated agents. The possible compound and complementary transport histories of assemblages lead to a genetic classification which is correlated with the nature of hydraulic bias likely to be imposed on each assemblage.

Quantitative relationships derived from basic hydraulic and physical concepts and tested in a laboratory flume and a natural river satisfactorily predict some aspects of the transport behavior of bones within certain ranges of imposed conditions. Tentative, semiquantitative relationships between transport distance and destruction and burial may simplify the modeling of these processes, but need further testing. Incorporated within the segment model, these relationships begin to link important physical taphonomic processes to descriptive parameters of fluvial systems, and thus to established concepts of sedimentation and fluvial geomorphology.

Some of the ideas included in this paper may be immediately useful in the taphonomic interpretation of certain fossil assemblages, especially those affected by fluvial transport. The broader goal, however, has been to initiate consideration of a systems approach to terrestrial taphonomy—an approach which seems necessary at the present time when we are passing from a trickle of data into a rapidly widening river of information, without a map or a paddle.

Acknowledgments

Financial support for the flume and river studies cited here was provided by the Archeological Research Facility at the University of California, by grants to Anna K. Behrensmeyer from the Miller Research Foundation and the Leakey Foundation, and by Joseph T. Gregory's research allotment from the University Committee on Research, University of California.

The staff of the Hydraulic Engineering Laboratory, University of California, Berkeley, generously allowed access to their equipment and proffered the technical advice and assistance which made our flume experiments possible. Dr. Luna B. Leopold encouraged us to undertake the experiments at the East Fork River and offered helpful comments on an earlier version of this paper. He and his students, especially Thomas P. Lisle and E. D. Andrews, provided us with direct assistance, equipment, and invaluable river data. I am grateful to Anna K. Behrensmeyer and Joseph T. Gregory for allowing me to cite unpublished data from our jointly conducted experiments.

The Wenner-Gren Foundation, the Burg-Wartenstein staff, and especially Lita Osmundsen, Director of Research, deserve special acknowledgment for making this conference possible and for providing the unique physical and intellectual ambience conducive to a very productive exchange of ideas. I finally wish to thank all of the other conference participants for their many constructive comments and suggestions which influenced the ideas in this paper.

10. FUNCTIONAL ANATOMY AND TAPHONOMY

Alan C. Walker

Introduction

Deductions about the functional anatomy of extinct primates have to be based on an understanding of the anatomy of living species. The sort of samples of fossil bones that primate paleontologists deal with are often treated as though they were collections of museum bones of recent species that have some parts missing and others broken. In fact, an understanding of those taphonomic processes that have led to the preservation and discovery of any particular fossil assemblage can help the functional anatomist. On the other hand, a knowledge of the functional anatomy can help elucidate some of the taphonomic processes. The following two brief examples will, I hope, make this clear.

When an anatomist deals with a fossil bone it is rare that the preserved bone will be exactly as it was in the animal. The skull of an *Australopithecus* that has been punctured by carnivores, squashed by compressing cave sediments, broken by mineworkers, and cleaned of matrix in a laboratory is not easily comparable with a macerated skull of a recently dead human. What can be seen and measured on the recent specimen is often not the same as that on the fossil, and the differences must be taken into account. To take the other example, when dealing with the fundamental properties of bones as sedimentary particles and accounting for their hydraulic behavior, it is clear that their basic properties of shape, density, bone organization, and suchlike are products of the animal's functioning anatomy. An awareness of that anatomy can help in the understanding of the accumulation of bone assemblages. With these two examples in mind, I want to suggest some approaches that might spur on further efforts to integrate the two disciplines to their mutual advantage.

Problems Facing the Functional Anatomist

Does the Anatomist Have the Right Bones?

It is common practice for the excavator of a site to sort the excavated bones in a preliminary way in the field or laboratory before distributing them to each relevant specialist. Sometimes mistakes occur and, since the specialist is usually unfamiliar with groups other than his own, he may not recognize the mistake. In this way crocodile femora have been described as hominoid clavicles (Le Gros Clark and Leakey 1951), lateral toes of *Hipparion* called *Australopithecus* clavicles (Boné 1955), crocodile naviculocuboids called *Palaeopropithecus* capitates (Sera 1935), and so on. It is no accident, perhaps,

182

that mistakes occur most frequently in human and primate paleontology, because every scrap is seen as important and the anatomists are sometimes unfamiliar with other orders of mammals, let alone other classes of vertebrates. As a general rule, the smaller the fragment, the greater the chances of mistaken identity. The chances are especially high, it seems, if the bone to be examined is presented together with a set of similar bones. The two curved phalanges from the Olduvai hand bone set offer a slightly curious corollary to this. These two bones were immediately sorted from the rest of the assemblage on their individual age. The remainder of the bones were juvenile and these were adult. All the bones were taken, and in no way unreasonably, to belong to the same species. These two bones were singled out for special study in a preliminary investigation by Oxnard (1973), who made a study of stress distribution by using a photoelastic method applied to a simplified copy of one of them. He found that there were indications that the phalanx came from a hand that was habitually used in climbing and hanging. Day (1975) has recently suggested that these two bones are not hominid at all, but rather belonged to a giant colobine monkey. This might, in fact, be so and would bear out Oxnard's analysis. The problem is, however, that identification as hominid has been made and it will be necessary to demonstrate clearly to which other taxon they belong in order to refute their hominid status. In this case there are no hand bones of the contemporaneous giant colobines for comparison, and it would be unwise to sweep the Olduvai bones under some "nonhominid" carpet—if only for the very good reason that to do so might raise the possibility of removing evidence that supports an alternative hypothesis. It appears that only the discovery of new fossil hand bones will help in answering this particular problem.

Apart from bones finding their way to the wrong specialist, it may also be true that a large number of unrecognized primate fossils still remain in museum collections. This is because the time involved in sorting through boxes of bone scraps and unidentified fragments is too long for most visiting scientists to expend.

How Does an Anatomist Deal with Isolated Limb Bones to Determine Relative Limb Proportions?

It often happens that a species is known by many isolated limb bones but not by a single articulated skeleton. Yet the relative proportions of the limbs and parts of limbs are extremely valuable in deducing presumed locomotor habits. Rather than take a case from the hominoids, where there are great difficulties, the example of an extinct genus of Madagascan lemur, *Archaeolemur*, will show the problem clearly. The two presently recognized species of subfossil *Archaeolemur* are known from several isolated limb bones and one relatively complete skeleton. There is a fair size difference between the skulls of the two species. The individual limb bones vary in size from small to large and, apart from the partial skeleton that can be used as a test case, are not certainly associated. An indication of the variation can be gained from table 10.1.

The standard way of expressing limb proportions is by three basic indices, the brachial (radius x 100/humerus), crural (tibia x 100/femur), and intermembral (radius + humerus x 100/tibia + femur). A collection of limb bones of the living genus *Lemur* demonstrates the nature of the relationships (table 10.2). The indices from these animals are as follows: brachial 106.3 (100.8-109.3), crural 95.0 (91.4-98.4), and intermembral 69.9 (67.7-71.6). The proportions, being closely related to locomotor abilities, are relatively constant between animals, so that the following highly significant (p < .001) correlations are found:

Table 10.1

Numerical Data for Fossil Limb Elements of the Madagascaran ·Subfossil Archaeolemur

Bone	N	Average Length (Millimeters)	Range	s.d.	V
Humerus	8	135.5	126.0-156.0	10.43	13.61
Radius	8	145.0	137.0-160.0	8.29	8.55
Femur	11	165.5	144.0-195.0	16.26	22.41
Tibia	9	150.7	142.0-162.0	2.28	5.19

Table 10.2

Numerical Data for Limb Elements of the Living Genus Lemur

Bone	N	Average Length (Millimeters)	Range	s.d.	V
Humerus	16	82.5	71.5- 89.5	4.96	6.01
Radius	16	87.7	75.9- 94.0	5.79	6.61
Femur	16	124.9	108.5-138.0	8.64	6.92
Tibia	16	118.6	102.0-128.0	6.96	5.87

   correlation of humerus on radius lengths, $r = 0.9330$,

   correlation of femur on tibia lengths, $r = 0.9497$,

   correlation of total forelimb on total hind limb lengths, $r = 0.9152$.

Because of these high correlations, the ratios of the means are good approximations of individual proportions.

   There is no reason to doubt that *Archaeolemur* had similar constraints on its limb proportions and, in fact, the calculated ratios of means match the values of the one skeleton extremely closely (table 10.3). The only method I have found of incorporating the sample data to give estimates of the standard error of ratios of means needs some further

Table 10.3

Limb Proportion Indices Calculated for Archaeolemur

Index	Calculated from Means	Associated Skeleton	Longest Bones	Shortest Bones
Brachial	107.0	106.2	102.6	108.7
Crural	91.1	91.1	83.1	99.3
Intermembral	88.7	88.7	88.5	91.9

estimate of the relationship between the variables. I have included in the table indices calculated using the longest and the shortest bones. Because of high correlations between the lengths of limb elements, these, too, are reasonable estimates. Such a procedure might be very appropriate when dealing with related species of dissimilar sizes, where

allometric considerations might rule out close correlations, or where samples are affected
by taphonomic factors such as hydraulic sorting.

In conclusion, it does seem that the ratios of means can provide accurate proportions,
given adequate samples and no selective preservation.

How Can an Anatomist Deal with the Problem of
Distortion in Fossils?

There are several types of distortion commonly present in fossil mammals, and all
present problems to an anatomist. In the skull, individual bones or parts of bones may
be displaced and recemented by matrix. Sometimes it is possible for the preparator
to correct this effect by removing matrix and realigning the parts. If this can be done,
the accuracy of the result depends upon the skill of the preparator and the difficulties
of the particular preparation. Sometimes this is impossible to carry out on the original,
but it might be possible to correct graphically or on accurate casts. The graphic method
can be used in simple cases and, in those cases where bilateral symmetry is present, can
be checked against an undistorted side. A method of using thin, accurate epoxy resin
casts is also possible. The cast is cut along the line of folding or splitting and can
then be realigned. In such cases there are usually checks on accuracy of realignment
provided by anatomical necessities, such as midline relationships or limits on joint
movements. Similar solutions could be found for correcting broken and recemented limb
bones.

The most severe problem is that of plastic deformation, where the bone has warped
during fossilization. It may be impossible, in such circumstances, objectively to re-
create the original condition. Detailed measurement of distorted adjacent sedimentary
clasts followed by a reversal of that distortion in the fossil might, I suppose, just be
feasible, but it is doubtful that a confident reconstruction can be made from information
of that sort. The best solution, probably, is to assess the direction and magnitude of
distortion relative to some known biological constraint, such as bilateral symmetry. This
is best carried out graphically. Figure 10.1 is an example of a plot made to show
distortion in calvarium KNM-ER-1470, an early hominid from East Lake Turkana, Kenya (Day
et al. 1974). This stereographic projection was made using a stereoplotting craniostat
(Oyen and Walker 1977). This gadget records points in terms of angulation and distance
from a chosen center (three-dimensional polar coordinates). It consists of an open,
rotatable circle to which is perpendicularly attached a 180 degree arc of the same
diameter. A movable, retractable, calibrated needle is attached to the arc so that it
moves radially. Through rotation of the circle and radial movement of the needle sliding
within the arc, any point can be reached that is within the hemisphere described by the
circle and arc, and that point can be defined in terms of azimuth, elevation, and radial
distance. The specimen is mounted within the hemisphere and centered about any chosen
point, which may be on the surface, within, or outside the specimen. Recorded results
can be handled in the same way as any other coordinate measurements, but stereographic
projections in two dimensions can also be made. Any point can be plotted using a meri-
dional stereographic net.

The calvarium KNM-ER-1470 still has its sagittal points (glabella, bregma, lambda,
external occipital protuberance) aligned. These are distributed along the 90 degree
meridian in figure 10.1. Homologous points were identified on the left and right sides
and their coordinates taken and plotted. Points on the left are shown by crosses, those
on the right by circles. The right side has been moved downwards and forwards relative

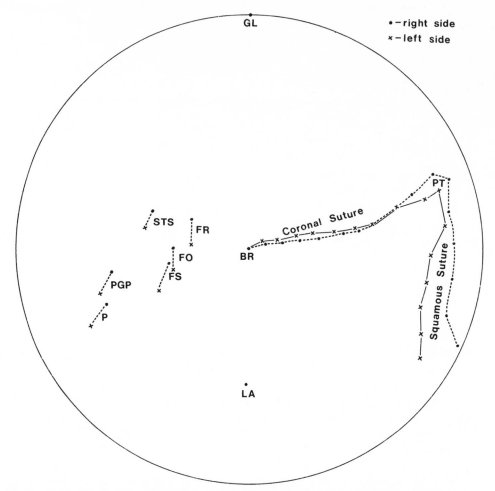

Figure 10.1. Equal-angle stereoplot of KNM-ER1470, showing the symmetry of coronal suture and assymetry of squamous suture. The right squamous has been depressed inferiorly and anteriorly. All temporal markers show right side has moved anteriorly, despite the fact that the midline structures are maintained as sagittal. Abbreviations: GL = glabella; BR = bregma; LA = lambda; PT = pterion; P = porion; PGP = postglenoid process; FS = foramen spinosa; FO = foramen ovale; FR = foramen rotundum; STS = sphenotemporal suture.

to those on the left (or the left side upwards and backwards). On the temporal bone, the position of porion has moved most and indicates a rotation of one of the temporals by movement along the squamous suture. This part of the skull is useful for diagnostic purposes, and Wood (1976) has already used measurements of this region to show that 1470 should be considered as *Homo*. I have pointed out (Walker 1976) that on this feature 1470 could be either *Homo* or *Australopithecus*, depending on which side is taken! The stereoplots show movement in one of two directions. The skull shows that the left temporal is not overridden on the parietal and that the right temporal is the displaced one. Here is a clear case where taphonomic distortion has biased taxonomical assessment (and also functional considerations, since the temporomandibular joint is on the temporal bone).

In the case of 1470, the asymmetry is measurable and can be seen to be the result of compressive forces during fossilization. What conclusions would be arrived at by someone studying the endocast and looking for signs of cerebral lateralization?

In general, distortion is not, by chance, of a simple type, and if the only plane of symmetry is no longer correct, then it may be impossible to decide more than the general degree and direction of distortion. This is the case in the cranium of *Dryopithecus (Proconsul) africanus* (Le Gros Clark and Leakey 1951). In such cases of badly squashed crania (Olduvai Hominid 24, many from Swartkrans, etc.) the only useful measurements are arcs along the bone surfaces. None of the chords will be accurate, some being stretched to equal the arcs and others compressed to zero. Thus, if all arc measurements could be taken, many severely distorted specimens could be added to our samples.

The innominate SK50 from Swartkrans has suffered both plastic deformation and displacement of parts, and it is very difficult to decide just what could possibly be measured on it. The circumference of the acetabulum can be determined (ignoring a 10 mm wide matrix-filled crack that traverses the joint surface) and the diameter calculated. Most of the rest of the specimen is traversed by cracks and faults. The difficulty of even graphically restoring the shape is made worse by the faults that run through the inner cancellous bone, so that a fault starting on one side runs through the inner bone and reappears elsewhere. Perhaps reconstruction of plastic casts will help resolve the minor controversies surrounding this specimen, which is one piece of evidence for differences in gait between different species of hominid.

How Can an Anatomist Make Use of Fragmentary Remains?

Fragmentary fossils make up most of our sample. For both taxonomic and functional anatomical purposes, they are usually used to gather some idea of variability. It frequently happens that fragmentary remains are not sufficient to be taxonomically useful, i.e., there is not enough preserved of diagnostic value. Their use in samples can be misleading, therefore, since parts of one species can be added to another, which makes testing the differences between sample means difficult, to say the least.

The only way of making comparisons that are genuine is to take into account only those parts that remain and not reconstructions. If valid comparison is to be carried out, then this can be thought of as breaking a cast of a more complete specimen to match the parts preserved on the fragment, rather than reconstructing the fragment and comparing the reconstruction with the more complete part.

Fragmentary remains often offer the anatomist a chance to examine internal structures that are not possible to see on a whole bone, such as trabecular organization, developing tooth germs, cortical bone thickness, and suchlike. In this way, they are most useful adjuncts to more complete material. They must, however, be treated with caution to avoid mistaken identity.

How Can an Anatomist Find Out the Internal Features of Fossils?

The outside of a specimen has only so much information for the anatomist. In the case of long bones especially, it is now clear that the organization of the trabeculae in cancellous bone is the key to a more complete functional understanding. Studies that are made on only the outer shape and size, such as all present multivariate statistical studies (see especially those of Oxnard [1975], who has pioneered such studies on primates), will give information about joint excursions, joint pressures, and mechanical lengths, but not about the real load-bearing tissue. This is the cancellous bone found in the ends of long bones and beneath all load-bearing joints. It is *theoretically* possible that two bones that are identical in surface shape and size might be accepting very different stresses during life, by virtue of trabecular organization and remodeling. The work of Pugh, Rose, and Radin (1973) is of great significance in that it demonstrates the enormous capacity that the large surface area of trabecular bone has for remodeling in response to applied stresses.

How can this be examined and quantified in a fossil? So far the organization of trabecular bone has been studied by destructive techniques that are obviously not applicable to fossil material. Radiographs are of some limited use, and integrative techniques that present an overall impression of trabecular density are available. These include the use of density-gradient photographic representation (Konermann 1971), and it is possible to take into account differential mineralization of fossils to prepare relative-density pictures. It might be possible to use microradiography to record trabecular organization, but the method would be tedious and exhausting and would require large comparative samples in order to decide the significance of results from fossils.

The resolution available from normal tomography is not usually high enough for most fossil specimens, but when allied with xerography it may prove useful in the future.

I have experimented with computer-generated tomography, using the CATSCAN machine at Massachusetts General Hospital and the CATSCAN Laboratory. Tests on oreodont skulls show that it is possible to gain information about the size and shape of matrix-filled cavities such as sinuses, orbits, and the braincase, but that any irregularities of mineralization can obscure the resulting scan. These machines were developed to record irregularities in brain anatomy and so are optimally useful for soft tissue of the size of a human head. If the computer could be reprogrammed or the pelicle size reduced, this technique might be a useful tool in the future for paleontologists.

How Can an Anatomist Examine the Submicroscopic Detail on Bone or Tooth Specimens?

Normal light microscopy is often inadequate for resolving details on fossils. The scanning electron microscope can give high resolution of details of structures such as tooth enamel and wear scratches on teeth. The fossils, however, are often much too large to fit in the scanning chamber and are often in institutions where there is no such machine. I have developed a method of replication using cold-cure silastomers and epoxy resins that preserves good detail up to between 1,500 and 2,000 times magnification (figs. 10.2a-10.2f). Results on fossils, replicated fossils, and recent species are proving to be very exciting, and it is now possible to distinguish between forest and open-country herbivores, between browsers and grazers, and between carnivores that chew bones and those that do not (figs. 10.3 and 10.4).

Hoeck (1975) has demonstrated seasonal dietary differences in two sympatric species of hyrax, *Heterohyrax brucei* and *Procavia johnstoni*. In this case the diets were carefully counted in the field and aspects such as size of the animal, masticatory mechanics, and tooth morphology, as well as local climate and food plants, were all controlled. Dr. Hoeck has kindly given me teeth from the animals he studied, and examples of the results can be seen in figures 10.5-10.14. The grass-eating episode in *Procavia* leaves the enamel and dentine very finely scratched, so that complete ablation of dentine tubules and enamel prisms occurs. In both species, prismatic outlines and tubular detail are restored during the browsing period. This method may now be applied to extinct species that are presumed to have grazed and browsed, and a matrix of wear and diet is being prepared against which fossils can be compared.

ACTUAL TOOTH

REPLICA

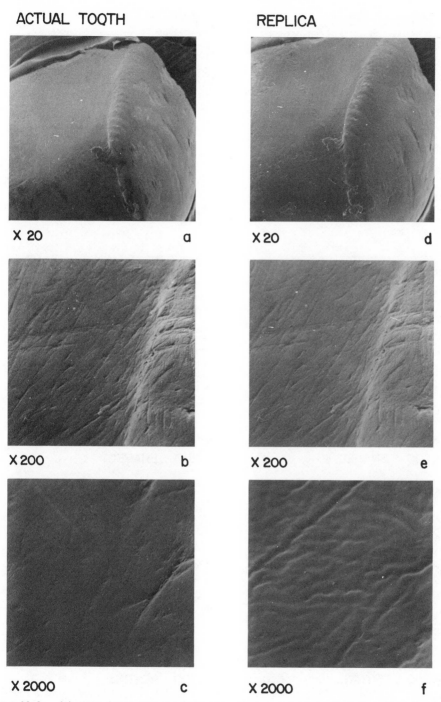

X 20   a

X 20   d

X 200   b

X 200   e

X 2000   c

X 2000   f

Figure 10.2.   (a) Wear facet of baboon lower canine; original tooth X 20.   (b) Enamel/
dentine junction of same baboon lower canine; original X 200.   (c) Enamel/dentine junction
of same baboon lower canine; original X 2000.   (d) Wear facet on baboon lower canine;
replica X 20.   (e) Enamel/dentine junction of same baboon lower canine; replica X 200.
(f) Enamel/dentine junction of same baboon lower canine; replica X 2000, showing breakdown
of silastoma fidelity at high magnification.

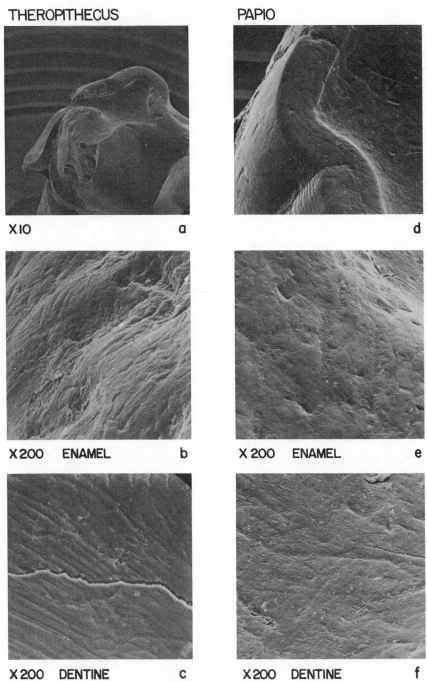

THEROPITHECUS

PAPIO

X 10   a

d

X 200   ENAMEL   b

X 200   ENAMEL   e

X 200   DENTINE   c

X 200   DENTINE   f

Figure 10.3. (a) Low-power magnification of lower molar of *Theropithecus gelada;* replica
X 10. (b) Enamel of same tooth; replica X 200, showing coarse grooves lined with spiral
scratches. (c) Dentine of same tooth; replica X 2000, showing parallel fine scratches.
(d) Low-power magnification of lower molar of *Papio sphinx*; replica X 20 of enamel/
dentine junction. (e) Enamel occlusal surface of same tooth; replica X 200, showing small
pits and 10 μ wide scratches. (f) Dentine of same tooth; replica X 2000, showing occasional scratches and solitary pits.

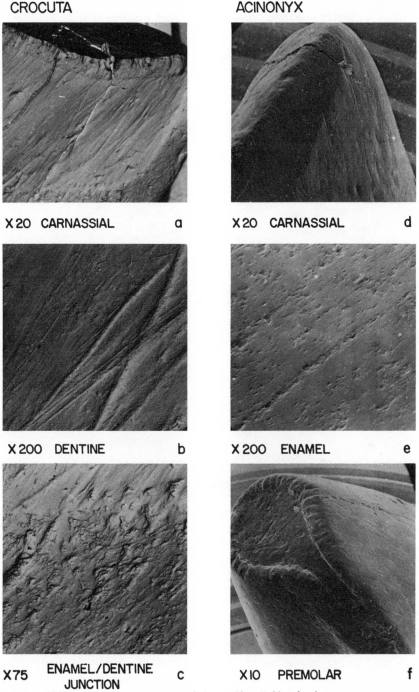

CROCUTA

ACINONYX

X 20   CARNASSIAL   a

X 20   CARNASSIAL   d

X 200   DENTINE   b

X 200   ENAMEL   e

X 75   ENAMEL/DENTINE   c
        JUNCTION

X 10   PREMOLAR   f

Figure 10.4.   (a) *Crocuta crocuta* carnassial; replica X 20, showing many coarse scratches
scoring enamel (*top*) and dentine surfaces.   (b) *Crocuta crocuta*, dentine of same tooth;
replica X 200, showing scratches of varying width from .5 mm to 10 μ.   (c) Enamel/dentine
junction of same tooth; replica X 75, showing portions of enamel removed by flaking.   (d)
Acinonyx jubatus carnassial; replica X 20.   (e) Enamel of carnassial wear plane of same
tooth; replica X 200, showing linear plucking of enamel prisms.   (f) *Crocuta crocuta* premolar at low power; replica X 10, showing crushed dentine surface resulting in "piecrust"
enamel wear.

FIGURE 5. SEASONAL DIFFERENCES IN FOOD DIFFERENCES IN 2
SPECIES OF HYRAX (HOECK, 1975)

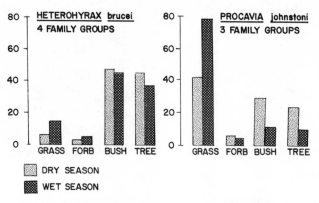

Figure 10.5.   Seasonal differences in food preferences in two species of hyrax (from
Hoeck 1975).

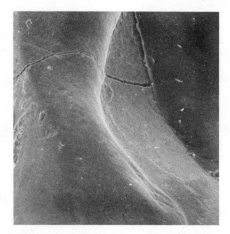

Figure 10.6.   *Heterohyrax brucei* molar, enamel/dentine junction; replica X 100, showing
extremely smooth enamel and dentine.

A cautionary note must be placed here. Care must be taken to avoid mistaking artifacts for natural wear. Reworking or natural sandblasting of teeth as well as the use of
Airbrasive developing tools can replace or confuse natural tooth wear with spurious
scratches.

How Can an Anatomist Deal with Unusual Changes in the Anatomy That Occurred before Death?

Repaired bone fractures, results of diseases that affect teeth bones and joints,
and unevenly erupted teeth that lead to unusual occlusion are frequently found in the
fossil record. The temptation to invoke pathology to explain unusual morphology must be
resisted. Clear demonstration of an abnormal condition is needed to avoid the sort of
thinking that led to the original Neanderthal fossil being considered a pathological
human. On the whole, unusual features in skeletal parts obscure normal anatomy and render
specimens less useful. Inferences from the abnormalities can be helpful, however, in

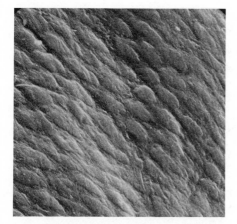

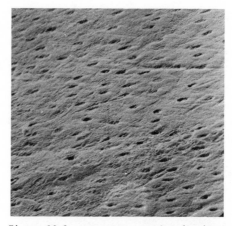

Figure 10.7. *Heterohyrax* molar enamel; replica X 1500 showing detailed polishing of enamel prism outlines.

Figure 10.8. *Heterohyrax* molar dentine; replica X 1500, showing detailed polishing of dentinal tubules and peritubular sheath material.

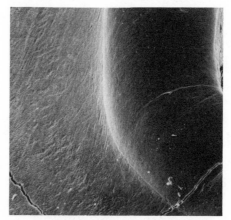

Figure 10.9. *Procavia johnstoni* molar, enamel/dentine junction at lower power (dry season specimen); replica X 100.

Figure 10.10. *Procavia* molar, enamel/dentine junction, at low power (wet season specimen); replica X 100, showing marked scratching of both enamel and dentine.

determining whether other conclusions about life habits match the conditions for the onset of the abnormalities. For instance, the manner in which a limb bone was broken may point to the way in which normal stresses were applied during life.

How Can an Anatomist Deal with Bone Characteristics Acquired between Death and Burial?

Damage by predators, scavengers, weathering, and transport are frequent causes of missing information in fossils, although sometimes internal anatomy is exposed that would normally not be available. The conflicting interpretations of fractures and breaks usually are set in a discussion of predatory behavior of hominids. Often claims are made to support cannibalism and hunting that are based on fractures caused by other taphonomic processes. One hardly expects forensic scientists to know what fossils are really like, and to ask them what caused crushing of a skull is inviting an answer based only on analogies with their postmortem-room findings.

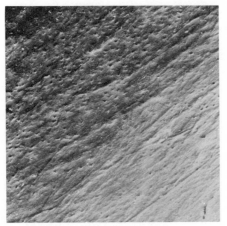

Figure 10.11. *Procavia* molar, dentine (dry season specimen); replica X 500, showing persistence of dentinal tubular sheath structure despite some scratching.

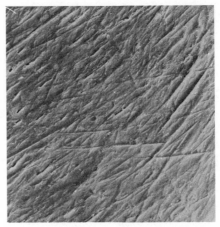

Figure 10.12. *Procavia* molar enamel (wet season specimen); replica X 500, showing obliteration of tubular structures by marked subparallel scratching.

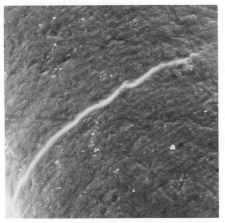

Figure 10.13. *Procavia* molar enamel (dry season specimen); replica X 500, showing some enamel prism relief.

Figure 10.14. *Procavia* molar enamel (wet season specimen); replica X 500, showing obliteration of all enamel prism outlines by massive subparallel scratching.

Artifacts due to erosion by chemical weathering and by temperature changes frequently affect teeth, causing pitting and cracking. Teeth left in their crypts on a land surface putrefy and "explode," as in KNM-ER-1171B (Leakey and Walker 1973), whereas those that were placed in a sedimentary environment soon after death may be perfectly preserved (e.g., Olduvai Hominid 39). Olduvai Hominid 34, from a channel deposit in Bed IV, was not accepted as hominid for some years because of extensive surface erosion by particle abrasion (Day and Molleson 1976). The radiographic appearance showed the femur clearly to be hominid, despite the extensive loss of surface bone that included enlargement of the nutrient foramen.

The relative positions of the associated parts of fossil skeletons can also be of help to the functional anatomist, for drying tendons and ligaments pull the bones into positions that may be characteristic of a habitual locomotion. The *Dryopithecus (Procon-*

sul) africanus forelimb from Gumba, Rusinga Island, had its hand bones in a position such that all the phalanges were folded back on the metacarpals. This is a death posture, as Napier and Davis (1959) noted, that is characteristic of many primates, but it is not typical of digitigrade primates such as baboons. This supports the anatomical conclusion that a digitigrade posture was not the habitual one for this species.

The extent of the paranasal sinuses in the maxillary and facial fragments of *Dryopithecus (Proconsul) major* from Moroto (Pilbeam 1969) can be determined easily because of a pink staining on the yellow bone which came about, presumably, by differential filling by sediment types and/or different times of decay of soft tissue. This probably can be taken to indicate that the antrum of the maxillary sinus was small in diameter, and not large, as in species that thermoregulate mainly by panting.

Can an Anatomist Use Paleoecological Findings to Support Functional Anatomical Conclusions?

There are many instances in the literature where anatomists have extended their findings in the light of sedimentary information or evidence from other fauna or flora. The most amusing instance is that of Sera (1935), who decided that, because some specimens of lemur were found in lake beds, the animals must have been aquatic. Of course, in practically every case of fossil primates, they ended in a sedimentary environment in order to be fossilized, but this in itself says little about their behavior when living. Details of paleogeography, fauna and flora, and climate are important only as a framework in which to set the suite of behaviors that can be deduced from the anatomy and to examine hypotheses concerning adaptation in relation to the rest of the community. To take a rather crude example, if someone decided on the basis of functional anatomy that early hominids were rather like orangs and retained climbing abilities, then it would be appropriate *afterwards* to ask whether suitable habitats were available for climbing in. Similarly, if a certain dietary habit is deduced, then details of available resources and competitors would be of great interest. However, the presence in a fossil site of abundant remains of types of edible plants or animals should not color an anatomist's functional assessment. The taphonomic history of each faunal and floral element has to be determined before community structures can even be guessed at.

What Can the Taphonomist Learn from the Functional Anatomist?

Assemblages of vertebrate bones are concentrated or scattered by many phenomena, but the functional anatomy of some concentrating and scattering agencies (such as owls, carnivores, crocodiles, and hyenas) sets limits for their activities. In a similar way, the functional anatomy of species preyed on or scavenged from determines bone, tendon, and ligament combinations that are, to a certain extent, predictors of how predators or scavengers can work on the bones or the order of joint disarticulation during transport.

Taking the first cases first, a knowledge of the anatomy of owls and the process of pellet formation might begin to be applied to certain owl-pellet concentrations in the fossil record with a view to determining owl species sizes and prey proportions. Brain's excellent studies (e.g., 1976) on the methods employed by carnivores might be extended to identify anatomical correlates of the behaviors that could be identified in extinct carnivores. It is perhaps not enough just to learn what behaviors occur and to quantify them. Integrated studies of the behavior and anatomy should provide a very powerful tool for the elucidation of ancient bone collecting and scattering activities, especially for the more ancient sites in which exact modern analogues are lacking.

The ways in which the functional anatomy of living species may determine how their bones can be scattered or destroyed are many. A few examples will help to make my point. Species with no clavicles have forelimbs that are held to the thorax only by muscle and tendon. This makes detachment of the whole forelimb easier than the hind limb, whether by carnivore, scavenger, or water action. The anatomy of the bovid shoulder is the ultimate cause for the usual disproportion of distal and proximal ends of the humerus in bone assemblages, and is a reflection of the different mechanical demands on shoulder and elbow joints; all other taphonomic factors are dependent on this. Densities of bone as they appear to a paleontologist may be misleading. Giraffe crania and mandibles, for instance, are extremely light, but this would not be noticed after mineralization and filling by matrix. Again, functional considerations might help correct any possible errors in taphonomic interpretation that could arise from using fossils alone, and can certainly help in the choice of modern equivalents of extinct species for such experimental work as flume studies.

Summary

An understanding of taphonomic processes, including biases introduced by researchers themselves, seems important for interpretation of the anatomy of fossils that are in anything but perfect condition. Knowledge of the functional anatomy of collectors and scatterers of bones can aid in understanding of their behaviors and, together with information about the anatomy of prey species, can help in evaluating the significance of taphonomic features of bone assemblages.

11. TRACE ELEMENTS IN BONES AS PALEOBIOLOGICAL INDICATORS

Ronald B. Parker and Heinrich Toots

Introduction

The mineral phase of vertebrate skeletal tissue (bone, dentine, and enamel) consists of apatite, a calcium phosphate mineral. Biologically formed apatite always contains significant amounts of several minor elements in addition to the essential elements of the apatite structure. The content of these minor elements differs between different species and also between individuals and populations of the same species. These variations are caused by various biological and environmental factors. Thus, the composition of skeletal apatite offers clues about the biology of the animal.

The composition of the mineral part of vertebrate skeletons undergoes postmortem changes, mainly diagenetic changes after burial. These changes affect the content of most of the elements contained in bones and teeth. Uptake of fluorine, for example, during fossilization has been well known for over a century (Middleton 1844). Strontium is one of the few elements not affected by diagenetic processes.

If one wants to draw conclusions about the biology of a fossil vertebrate from the composition of its bones and teeth, then one must first separate the biological factors from the diagenetic effects. Then, when one has determined what the original lifetime (antemortem) content of an element was, one can try to account for the biological factor or factors controlling the content.

It is our aim in this article to show the methods by which we can recognize diagenetic change and separate its effects from original composition of a bone or tooth, and give the theoretical reasons why we have confidence in these methods. This will be followed by an account of the biological factors that control the incorporation of a minor element into living bones and teeth. We will conclude with a discussion of specific elements and their biological significance.

A Few Remarks about Scientific Method

The history of our investigations pointedly demonstrates the need to approach a new subject methodically. Our progress was slow until we learned to consider all possible factors and sort out the variables.

Analyzing a randomly, or at least an uncritically, selected set of samples will produce rather noisy data. Such data may give valuable clues to profitable lines of research, but will rarely if ever solve any problems. Our first results were of this nature

197

and now have only historical interest (Toots 1963). Investigations by other workers into the chemistry of fossil bones have been discontinued before proceeding beyond this point.

The experimental sciences rely on the controlled experiment in which all factors that could influence the outcome are kept constant except for the one variable that is under investigation. In geology and paleontology it is not possible to perform experiments in the strict sense. However, one can at least approximately satisfy the reasons behind the controlled experiment by selecting samples so that they differ only in the one aspect to be studied.

For example, in the study of the significance of strontium in fossil bone, Toots and Voorhies (1965) used only bones from a single quarry site. This insured that all of the animals studied were of the same geologic age and had lived in the same complex of environments in the same broad depositional basin. Thus, the background level of strontium was probably uniform throughout the area from which the fossils were derived. Therefore, any differences in the strontium content of the fossils could only be due to biological factors. Fossils of the same species were tested to check whether age of the individual had any influence on the strontium. Different parts of the skeleton were compared. When none of these variables was found to have an effect, and strontium content was found to show only minimal variation within a species, comparisons between species could be made. The differences found were fairly small, but highly significant.

Unless other factors are kept constant, merely finding a correlation between two variables is not necessarily meaningful. Grossly erroneous conclusions can be drawn. For example, Pflug and Strübel (1967) found a positive correlation between geologic age and strontium content of fossils. Statistically, the correlation is good; however, they based their conclusion on specimens which are not comparable. They compared late Cenozoic terrestrial vertebrates with Cretaceous marine vertebrates. The real cause and effect in this case is that the marine environment provides much higher environmental strontium levels than does the terrestrial environment.

In all studies of bone and tooth minor elements, natural variation must be taken into account and reliance on comparisons of single analyses must be eschewed. One must compare sets of data, not individual analyses. Precision is influenced by the natural variations, ranging from slight to very large, within and between samples. In addition, the laboratory procedures involved in the analysis can introduce variation into the data. Any given analytical method has inherent limits of both sensitivity and reproducibility, and a method must be chosen which is capable of providing a high degree of reproducibility at trace levels. Also, variation can be introduced early in the procedure by unsuitable specimen preparation.

Standard statistical tests should be made to estimate what number of replications is appropriate to the task at hand and should be used to compare data sets. Simply making a very large number of analyses is no guarantee that the means of the data sets will be meaningful in geological or biological terms. It is not the absolute number of samples, but the care with which they are selected and analyzed which determines the success of an investigation. As with all scientific investigations, liberal application of common sense goes a long way in selecting appropriate methods.

Diagenetic Changes

After burial, vertebrate skeletons undergo a variety of chemical changes, the magnitude of which is a function of the chemical environment in the burial site, the properties of the enclosing sediment, and the properties of the hard tissue itself. Rates of diagenetic change differ enormously from place to place and from tissue type to tissue type so that any attempt to relate the composition of the fossil to its age is doomed to failure (Parker, Murphy, and Toots 1974; Parker and Toots 1976a). In any case, some elements in hard tissue are characteristically depleted. Very few elements are unaffected, though there are exceptions such as strontium.

Additions

Enrichment during diagenetic change or fossilization is characteristic of fluorine, silicon, manganese, iron, yttrium, and a host of other elements. The enriched elements may be subdivided into those that are present in minerals filling voids in the tissue and those that become combined in the crystal structure of the mineral apatite of the bone or tooth.

An element which is present in a void-filling mineral leaves no cause for doubt about its diagenetic origin. Silicon is present as silica and various aluminosilicates such as feldspar and clay minerals (Parker and Toots 1974). Manganese is present as various simple and complex oxides, the latter also containing barium and lead (Parker and Toots 1972). Iron is present as oxides and sulfides. Other common pore-filling minerals include calcite ($CaCO_3$) and gypsum ($CaSO_4 \cdot 2H_2O$), and, locally, barite ($BaSO_4$). The localization of elements in void fillings can be demonstrated readily by use of an electron microprobe analyzer or scanning electron microscope with X-ray spectrometer (Parker 1967; Parker and Toots 1970; Parker and Toots 1974). Minerals present in large pores can sometimes be observed microscopically and occasionally identified by means of X-ray diffraction.

Fluorine and yttrium are localized in the apatite of the solid bone, as demonstrated by the electron microprobe studies cited previously. Because of their electrochemical properties fluorine and yttrium preferentially replace hydroxyl and calcium, respectively, in the apatite structure. The concentration of these elements in fossil bone and tooth is several orders of magnitude higher than in unfossilized material.

Losses

Depleted elements include sodium, magnesium, chlorine, and potassium (Parker, Toots, and Murphy 1974; Wysoczanski-Minkowicz 1969). However, magnesium and potassium can be reintroduced into pore spaces in diagenetic minerals. We have not found any secondary sodium-bearing minerals to date.

Differences among Tissue Types

Bone, dentine, and enamel differ greatly from one another in their physical properties. Bone is highly porous and permeable and is composed of submicroscopic crystallites which consist of only a small number of unit cells. Enamel of teeth, in contrast, has no pore spaces to speak of and is essentially impermeable unless fractured. Further, enamel apatite crystallites are two orders of magnitude larger than those of bone (Brudevold and Söremark 1967). Dentine is intermediate, more closely resembling bone.

As a consequence of these differences in properties, diagenetic processes operate much more slowly in enamel than in either bone or dentine. Chemical differences among the three tissue types are small in modern animals. The amount of magnesium in the dentine of modern animals can exceed that in bone or enamel by a factor of 2. For other elements the differences are considerably smaller (of the order of 10% in the case of strontium, an amount which can be of significance). Diagenetic changes commonly result in enrichments of an order of magnitude or more, and depletions by a factor of 3 or more in bone and dentine.

When we compare the three tissue types from the same fossil, the composition of the enamel most nearly reflects the primary composition of the skeleton. Thus, comparisons of the three tissue types from the same individual are an excellent indicator of which elements are gained, are lost, or remain stable during diagenesis (fig. 11.1).

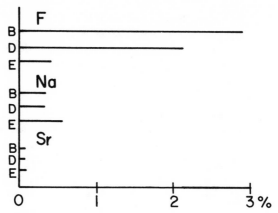

Figure 11.1. Comparison of F, Na, and Sr in bone (B), dentine (D), and enamel (E). Specimen is a *Subhyracodon*, S. Torrington, Wyoming (No. HT-154).

Correcting for Diagenetic Loss

In order to correct for diagenetic removal of an element one must have an independent method of quantitatively estimating the degree of diagenesis. Diagenesis is only incidentally a time-dependent process, so geologic age is of little or no use in this regard. One can, however, use enrichment or depletion of an independent element as a measure of degree of diagenesis. The indicator element should occupy a site type in the apatite structure rather than just be present in a void-filling mineral. In addition, the structure site must not be the same one as that occupied by the leached element of interest.

For example, fluorine would be a poor choice for studying chlorine loss inasmuch as they both occupy the hydroxyl site. Similarly, yttrium would be an unwise choice for examining sodium loss because they both, at least in part, occupy the calcium site.

The indicator element should also be one which is taken up or lost in proportion to the amount of diagenesis without high sensitivity to environmental conditions such as pH, Eh, and the like. Also, if the indicator element is one which is added during fossilization, it must be one which would be expected to be present in small amounts in most underground waters.

For correction of sodium loss, fluorine appears to meet all the criteria. It occupies a structural site in the apatite which is not the same one occupied by sodium. Its behavior in solution is independent of Eh and should be insensitive to pH in the hydroxyl-replacing reaction, and it is a widespread ion in subsurface waters.

In theory there should be a high negative correlation between a depleted element and its enriched indicator. In practice, high correlations are found only when the sample set has a wide range in the amount of diagenesis. The latter is true for the specimens of fossil *Equus* which we have studied (fig. 11.2) for which sodium is the depleted element and fluorine is the indicator.

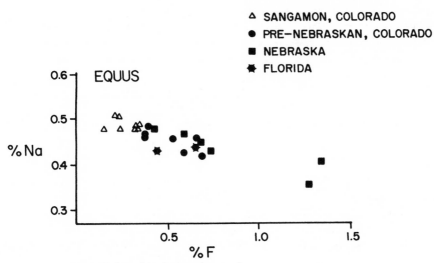

Figure 11.2.   Sodium and fluorine in *Equus* enamel.

A regression line can be fitted to data such as the Na and F values for *Equus* enamel by assuming that error exists in both values and calculating a line that minimizes departures in both—a reduced major axis line (Till 1974, p. 99). Such lines for *Equus* and other data sets with high correlation coefficients intersect near 3.8% fluorine and 0.1% sodium. The 3.8% fluorine value agrees with a theoretical fluorapatite and the 0.1% sodium value agrees with values for the most highly leached bone we have analyzed. As a consequence we have adopted 3.8% fluorine and 0.1% sodium as a theoretical value towards which Na-F diagenesis tends.

The use of the theoretical point allows us to fit reasonable lines to data sets in which there is low spread in fluorine content combined with a high scatter in sodium content. An example of such a case is found for the titanotheres, animals with very thick impermeable enamel which is therefore resistant to diagenetic change (fig. 11.3).

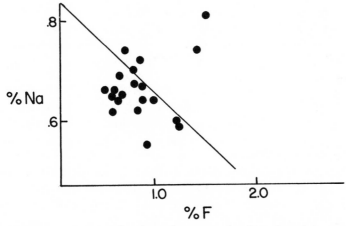

Figure 11.3.   Sodium and fluorine in titanothere enamel. Line forced to pass through F = 3.8%, Na = 0.1%.

The reason for fitting lines is to allow us to assign a corrected sodium value to each specimen which approximates life sodium levels. Fluorine is present in very small amounts in enamel of modern teeth (about 0.02%). We have arbitrarily corrected to a normalized sodium value for 0% fluorine by projecting each data pair parallel to the calculated regression line slope. Such is our correction for diagenetic loss.

Factors Controlling Amounts of Minor Elements in Skeletons of Living Mammals

The levels of minor elements in skeletal tissues in living mammals are a function of several major factors. These factors are physiological, crystal chemical, dietary, and environmental. In any given instance a single factor may dominate or several may have influence.

Physiologic Factors

By physiologic factors we mean that the internal mechanisms of the animal exercise major control over the amount of the element in the body tissues. Elements which would be expected to be physiologically controlled are those which are important to the functions of a living mammal such as the so-called cation body electrolytes—sodium, magnesium, and potassium. In a healthy animal, levels of these elements in blood and tissue are regulated within narrow limits. For example, levels of the alkalies sodium and potassium control the osmotic pressure balances within the body of a mammal. Potassium is largely intracellular and sodium is largely extracellular (i.e., in the blood). We assume that body levels of functionally important elements are related to levels in any tissue, including bones and teeth.

Crystal Chemistry

Inasmuch as bones and teeth are composed of crystalline hydroxyapatite, the concentration of minor elements—impurities, as it were—will also be influenced by crystal chemical restrictions. There are two sorts of restrictions which should be considered.

The first of these might be called the upper limit restriction. An example will illustrate the point. In any apatite crystal there are a finite number of hydroxyl sites. Thus the amount of fluorine which can be accommodated in those sites has an upper limit—about 3.8% as previously mentioned.

The second restriction is that thermodynamic factors controlling the stability of the crystal can set an upper limit to the amount of substitution of one element for another in a crystal structure. Thus, for strontium, which can occupy the calcium site, there is a maximum Sr/Ca ratio which can be built into a stable crystal. Even assuming overwhelming amounts of strontium, the Sr/Ca ratio should not exceed the maximum ratio.

Having given these two rules, we should point out apparent exceptions. Both exceptions relate to the size of the apatite crystals. Crystals in enamel are large relative to those in dentine or bone, large enough to behave chemically like large crystals. In contrast, the crystals in bone and dentine are very small indeed, containing only a few unit cells. Such tiny crystallites result in a large surface area for a given volume of apatite. The large surface exposes excess ionic charges which are not satisfied by neighbor ions. Thus, both cations and anions can become bonded to the surface in apparent violation of the structural limits and the partitioning coefficient. This is a likely explanation for the fact that fluorine amounts to as much as 4.2% in some dentine and the fact that strontium is slightly more abundant in bone and dentine than in enamel.

Intake of the Elements

There are two major sources from which elements enter the animal during life—food and water. A third source which may be of local importance is salt licks. A source of the same nature is the "mineral supplement" often made available to domestic animals. In marine animals seawater is the most important source of trace elements. In contrast, terrestrial vertebrates obtain only small amounts of trace elements from water in comparison with intake via food, with the exception of animals that take in water of unusual composition, such as saline water from springs or closed drainage basins. Different plants and different parts of the same plant tend to concentrate different elements in varying degrees so the animals' diet choice is of major importance in determining body levels of many elements. The same generalities apply to animals as to plants, and there are systematic differences between animals and plants. For example, animals contain considerably more sodium than plants (McDonald et al. 1966).

Environmental Abundance of Elements

The abundance of elements in both the life and burial environment of an animal will have pronounced effects. Environmental abundance is a particularly influential factor in the case of elements which are not physiologically regulated by the animal, such as strontium. The abundance of strontium in the vegetation of a given area is related to the amount of strontium in the environment. For elements which are regulated physiologically, only extreme departures from normal levels will be reflected in the animal's tissues. For example, we find abnormally high concentrations of sodium in modern mammals from the East African Rift Valley, an area noted for waters rich in sodium carbonate and bicarbonate.

Significance of Specific Elements

Strontium

Strontium is a particularly useful paleobiological tool in that it appears to be unaffected by diagenesis over a wide range of conditions. This fact is well illustrated in figure 11.1, where strontium is seen to be nearly equal in the three tissue types, compared with sodium, which is leached, and fluorine, which is added.

Strontium level does not appear to be regulated physiologically, and while there is some discrimination against strontium in favor of calcium (Odum 1957; Thurber et al. 1958), tissue levels do reflect intake levels. Thus, strontium levels in bone and teeth may be used to draw dietary and environmental inferences. In the case of land mammals, and probably for all land vertebrates, food is the primary source of strontium and the strontium levels provide a powerful tool for estimating diets. In contrast, the strontium content of aquatic vertebrates is largely determined by the composition of their environmental water (Rosenthal 1963).

Plants contain different amounts of strontium. Menzel and Heald (1959), studying crop plants, found that legumes contain more strontium than grasses. Toots and Voorhies (1965) concluded that succulent, leafy forage contained more strontium than grasses, and showed that browsing and grazing animals could be distinguished by their tooth and bone strontium content (table 11.1).

Food chain fractionation effects are also found. The loss of strontium at each level of the food chain is relatively small, with the resultant need for high-precision analytical procedures. The strontium retained by herbivores is largely in bone and tooth.

Table 11.1

*Concentration of Strontium in Bones of Various Pliocene Vertebrates (ppm)**

Bone Source	Number of Specimens	Mean Concentration	Standard Deviation	Coefficient of Variation
Carnivores	4	477	15	3.1
Merycodus				
Mandibles	10	523	27	5.1
Mature metacarpals	10	526	34	6.5
Immature metacarpals	10	529	28	5.3
Total	30	526	29	5.5
Pliohippus	10	552	35	6.4
Hypohippus	10	630	37	5.8
Testudo	10	636	24	3.7

*From Toots and Voorhies (1965).

Thus, animal flesh (muscle) contains low strontium, and carnivores can be recognized from the low strontium content of their hard tissues. It should also be possible to separate flesh-eating carnivores from those which also eat bone. Note that there is a discrimination against strontium entering herbivore bone as mentioned above so that even carnivores which consume large quantities of bone would still ingest less strontium than a herbivore.

In the case of animals of omnivorous habit such as man, it is possible to draw inferences regarding the relative amounts of plant and animal food consumed. Brown (1974) has provided an instructive example, relating bone strontium content to diet and thence to social status in a prehistoric human community.

Practical applications of strontium to dietary problems must take into account the fact that the strontium content of a given fossil is a function of two variables of equal importance. One of the variables is the dietary intake, and the other is the overall strontium level, or, more precisely, the Sr/Ca ratio, in the environment. In order to determine one variable the other must be kept constant. As mentioned previously, Toots and Voorhies (1965) have discussed this problem and collected their material to keep environmental factors constant for their sample set. Also, their fossil collecting site represents part of a wide alluvial plain of sediments derived from a distance source (Voorhies 1969). Brown (1974) used material from a single site also.

Environmental background effects can be larger than differences due to dietary habits. For example, herbivores from Agate, Nebraska, contain 3-4 times as much strontium as the samples from Knox County, Nebraska, reported by Toots and Voorhies (1965). In other localities, the strontium levels fall into a lower range than in Knox County.

Inasmuch as the environmental background level serves as a starting point for fractionation along the food chain, any inferences regarding feeding habits can be obtained only by comparing animals from the same habitat. Even then, the relative strontium content does not provide us with a unique solution to the problem of an animal's diet. However, just as anatomy, especially dental morphology, puts restraints on the interpretation of strontium data, so the strontium data in turn place restraints on interpretations from anatomy.

For example, the mean strontium concentration in bones of *Diceratherium* from the Agate quarry, Nebraska is 0.16%. Bones of *Moropus* from the same locality contain 0.12% mean strontium (the means are unequal at $P > 0.9999$). Since the highest strontium levels characterize browsing herbivores feeding on soft, leafy vegetation, we interpret *Dicera-therium* as a browsing herbivore. The dentition of *Moropus* eliminates the possibility that it could have been a carnivore or grazing herbivore. The idea that *Moropus* may have fed on roots and tubers therefore seems to be the most plausible interpretation.

Sodium

Sodium is an important element in the physiology of vertebrates, and its amount is regulated by physiological processes. Different genera of mammals have different sodium levels, reflecting different physiological adaptations. Sodium is leached during the fossilization process (Parker et al. 1974), and a correction is therefore needed to estimate life levels. Fortunately, in tooth enamel, the leaching is slow and systematic and is accompanied by a diagenetic enrichment in fluorine. The negative correlation between sodium (losses) and fluorine (gains) is surprisingly good (fig. 11.2) and is normally independent of local environmental conditions. We use the correction method to normalized "life levels" discussed previously. The sets of normalized values can be treated by standard statistical tests.

In the marine environment, an abundant supply of sodium is available in the water. As evolution of vertebrates generally progressed from marine carnivore to terrestrial carnivore to terrestrial herbivore (Bakker 1975), the physiology had to become adapted to progressively reduced sodium intake. Plant tissues contain much less sodium than animal flesh (McDonald et al. 1973), and the advanced herbivore must adapt to sodium retention rather than sodium elimination in order to maintain a satisfactory alkali balance. We theorize that effective sodium conservation permits lower sodium levels in body fluids, which is reflected in the sodium content of the tooth enamel. One would expect highest sodium in marine mammals with progressively smaller amounts in carnivores, omnivores, and primitive herbivores, and least of all in advanced herbivores (table 11.2).

Table 11.2

Sodium in Enamel (Normalized Values Given for Fossils)

	Mean Weight (Percent)
Desmostylus, fossil	0.91
Canis latrans, modern	0.86
Ursus americanus, modern	0.74
Subhyracodon, fossil	0.70
Equus, fossil	0.52

The highest sodium content, as expected, is found in the marine mammal *Desmostylus*. We have no data on fossil carnivores, but modern coyote and bear fit our model as expected. *Subhyracodon*, an early representative of the rhinoceros family, contains less sodium than the carnivores and omnivores, and *Equus*, a highly advanced herbivore, contains the least.

The resolution of the method is far better than can be appreciated from these comparisons. We have been able to trace gradual changes in the sodium metabolism of horses as reflected in the sodium content of enamel from the primitive *Hyracotherium* through several intermediate genera to the advanced *Equus* (fig. 11.4). All genera in figure 11.4 are in the direct line of descent of horses, with the exception of the *Hipparion* horses, which are a side branch on an evolutionary level intermediate between *Parahippus* and *Pliohippus*. It might be noted that *Hyracotherium* is a very primitive

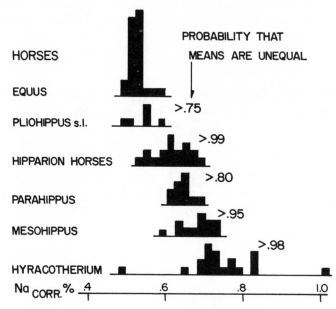

Figure 11.4.   Change in sodium content of enamel during the course of evolution in the horse lineage from *Hyracotherium* through intermediate genera to the advanced *Equus*.

herbivore which may have been transitional between omnivores and herbivores. A more detailed review of sodium in mammals will be presented in a forthcoming publication.

Evolution of physiological adaptations does not necessarily parallel evolution of skeletal parts and dentition. Proboscidians are highly advanced in the evolution of their dentition but are primitive in their sodium metabolism. Because of the latter fact, distribution of elephants in modern Africa is closely correlated with high environmental sodium levels (Weir 1972), and elephants are known to depend on food that is particularly rich in sodium (Dougall et al. 1964; Laws et al. 1975). This is also reflected in the high sodium levels of fossil probiscidians (normalized mean Na percentage = 0.68) and may be a factor in the unusually high rate of extinctions of probiscidians during the Pleistocene. One should keep this factor in mind when theorizing about the extinction of such high-sodium groups of mammals as titanotheres or oreodonts, as an animal with a poorly developed sodium metabolism would be more vulnerable to environmental stress, and at a disadvantage in competition with better-adapted animals (Parker and Toots 1976b).

Sodium levels in enamel show little variation within any species or genus or, occasionally, family within a wide range of environments. However, in hypersodic environments, the sodium intake may be so large that abnormal enrichment of sodium in the skeletal tissues results. A good example is seen in the present-day East African Rift Valley, as previously mentioned. Our method of sodium correction may not work in abnormal environments of fossilization. A preliminary survey of carnivores from Rancho La Brea, California suggests rapid depletion of sodium without the usual degree of enrichment in fluorine. Sodium is apparently readily leached, but the surrounding bitumen seemingly inhibits circulation of fluorine-bearing waters.

Other Elements

Potassium and magnesium are important elements in the physiology of mammals and have the potential to be valuable paleobiological indicators. We have identified potassium and magnesium minerals in the pore spaces of fossil bones, and this creates difficulties in correcting to life levels. Therefore, any work on these elements will have to be limited to use of nonporous enamel. We are just starting to explore the usefulness of potassium and magnesium, and a great deal more work needs to be done before we can reach any conclusions (Parker and Toots 1976*c*).

Life levels of fluorine and yttrium in living mammals are insignificant, especially in enamel, compared to diagenetic gains. We see no potential for paleobiological use of these elements, but they may prove to be useful in purely geochemical or geological studies.

A number of other elements have been found in fossil bone. Most of them occur in secondary minerals deposited in pore spaces (Al, Si, S, V, Mn, Fe, Se, Ba, Pb, U). Others (As, La, and the rare earths) appear to substitute into the apatite structure and require further investigation.

12. ORGANIC GEOCHEMISTRY OF BONE AND ITS RELATION TO THE

SURVIVAL OF BONE IN THE NATURAL ENVIRONMENT

P. E. Hare

Introduction

The degree of preservation of bone in the natural environment after the death of
an animal is a function of time and the environment. Given a constant environment, the
chemical and physical changes in bone can be used in many cases as an index of the
relative age (i.e., age since death) of the bone. On the other hand, if this age of a
bone or fossil is known, it should be possible to use the degree of chemical and physical
preservation to determine something about the history of the diagenetic environment. It
is of course first necessary to evaluate the environmental factors and determine their
effects on the chemical and physical changes in bone. Some of the environmental factors,
such as temperature, moisture conditions, pH and Eh (oxidation-reduction), may be evalu-
ated by reacting bone samples under controlled conditions in the laboratory.

After a brief discussion of the structure and chemistry of modern and fossil bone
material some preliminary laboratory simulation experiments will be described. These
should not be considered as final and definitive but rather as an initial attempt to
separate some of the numerous variables characterizing the environment. Hopefully,
comparison of field data with the laboratory simulation data will show up weaknesses in
the experiments and indicate future directions to pursue to achieve progressively closer
approximations.

Living Bone: Description

The bone in living animals is a heterogeneous dynamic system made up of a complex
and intimate mixture of organic and mineral components. Small crystals of hydroxyapatite and bundles of collagen fibers are bound together by an amorphous "cement" consisting of organic and inorganic ingredients.

The structural unit of bone is the osteone, a cylindrical, branching structure
of the order of 100 µ containing one or more small blood vessels. Since osteones
generally have their long axes parallel with the long axis of the bone, a transverse
section of bone reveals the osteones as circular patterns called Haversian systems.
These structures are often preserved in fossil bone.

The cellular components are distributed within the osteones and are involved in
the dynamics of bone formation (osteoblasts) and destruction or resorption (osteoclasts).
Osteocytes are cells differentiated from the osteoblasts and apparently are able to perform both deposition and resorption of bone matrix and general bone maintenance.

During the lifetime of an animal there is a continuing remodeling of the osteones in bone. Cavities are formed by the osteoclasts and new bone matrix deposited by the osteoblasts. At any given time a bone will have osteones in all stages of development. Some osteones will show incomplete mineralization and a relatively large vascular canal, others will be densely mineralized, and still others in the process of disappearing. In general, as the biological age of the animal increases, the density of the bone tends to increase—i.e., the ratio of mineralized components to organic matrix increases. This increased density might be expected to enhance the stability and preservability of mature bone as compared to that of less dense bone material. Laboratory experiments could readily be performed with bone of varying density as well as from different species to evaluate possible differences in stability to leaching effects, etc. (For more information on bone structure and chemistry, the reader is referred to Pritchard 1972; Ortner 1976; Halstead 1974.)

Organic Matrix Composition of Modern Bone

The organic matrix makes up approximately 20-25% of the dry weight of bone, with "inorganic" mineral material making up the balance. If a bone is left in a solution of a weak acid the mineral material will dissolve and there will be left a pseudomorphic model ("ghost") of the bone formed by its organic matrix. Ninety percent of the dry weight of this organic matrix is accounted for by the protein collagen with the balance made up of noncollagenous proteins (some of which contain carbohydrates) and a number of other organic materials not all of which have yet been characterized.

Since collagen makes up such a large fraction of the total organic matrix it is not surprising that much of the analytical effort in characterizing the organic matrix of bone has gone into the analysis of collagen from various sources. The relative amino acid composition of collagen extracted from bones of a number of different species is summarized in figure 12.1. Differences in the composition of collagen from different species of vertebrate bones are small but measurable. In fact, collagen from noncalcified tissues, such as skin and tendon, shows little difference from bone collagen (Piez and Likens 1960; Herring 1972).

In table 12.1 and also shown in figure 12.1 is the total amino acid composition of a bovine bone. A fragment of the bone was dissolved and heated in 6N HCl to hydrolyze the proteins and free amino acids. Again, the relative amino acid composition fits into the range for collagen extracted from bone and shows that collagen is so abundant in bone that the noncollagenous proteins present do not significantly alter the collagen pattern, a pattern that is characterized by a high glycine content, relatively high proline content, and the presence of hydroxyproline and hydroxylysine. The relative amino acid composition of a noncollagenous protein fraction from bone left after the extraction of collagen is also shown in figure 12.1. This fraction contains no hydroxyproline or hydroxylysine. It has relatively much less glycine and proline and more glutamic acid and aspartic acid than does collagen. The total bone amino acid composition includes this noncollagen protein, but because collagen is so abundant the overall bone amino acid composition resembles collagen.

Fossil Bone

Unlike modern bone, fossil bone shows extreme variations in amino acid composition. In general, for a constant environment, the older the fossil bone the less organic

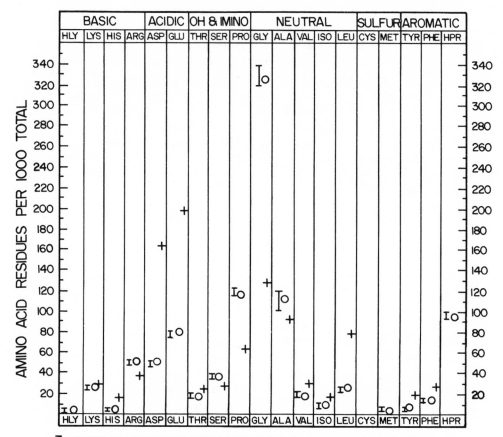

Figure 12.1. Relative amino acid composition of modern bone material.

matrix it contains and consequently the fewer total amino acid residues per gram of
bone. The relative amino acid composition also shows extreme variations, ranging from
collagen-like patterns to noncollagen amino-acid patterns. Table 12.2 shows the
relative amino acid compositions of some selected fossils compared to modern bone for
reference. A nitrogen content of about 4% for modern bone corresponds to around 2500
nanomoles (1 micromole = 1000 nanomoles) of amino acids per milligram of bone (table
12.1). As long as the amino acid content of bone remains above about 250 nonomoles or
so (~ 0.4% N) the relative amino acid pattern resembles that of collagen, with high
amounts of glycine and substantial amounts of proline and hydroxylproline. Below about
50 nonomoles of amino acids per mg of fossil bone the relative amino acid pattern looks
very different from that of collagen. Some workers have considered this dissimilarity
to modern bone as evidence of contamination (Wyckoff 1972); however, it seems probable
that at these low levels of amino acids in bone in many cases we are seeing indigenous
and highly resistant noncollagen components that were masked by the high concentrations
of collagen (Hare 1974).

Table 12.1

Amino Acid Analysis of 1 mg of Modern Bovine Bone (nitrogen content = 4%)

Amino Acid	Symbol (Fig. 12.1)	μ Moles/mg Bone	Residues per 1000
Hydroxyproline	(HPR)	0.253	100.0
Aspartic acid	(ASP)	0.127	50.1
Threonine	(THR)	0.047	18.5
Serine	(SER)	0.088	34.7
Glutamic acid	(GLU)	0.197	77.7
Proline	(PRO)	0.276	108.9
Glycine	(GLY)	0.835	329.5
Alanine	(ALA)	0.284	112.1
Valine	(VAL)	0.054	21.3
Methionine	(MET)	0.009	3.6
Isoleucine	(ISO)	0.029	11.4
Leucine	(LEU)	0.072	28.4
Tyrosine	(TYR)	0.016	4.3
Phenylalanine	(PHE)	0.035	13.8
Histidine	(HIS)	0.010	3.9
Hydroxylysine	(HLY)	0.011	4.3
Lysine	(LYS)	0.062	24.5
Arginine	(ARG)	0.134	52.9
Total		2.539*	1000.0

*I.e., about 2500 nanomoles/mg bone.

Table 12.2

Amino Acid Composition of Some Selected Fossil Bones Compared to Modern Bone

	Modern Bone	Egyptian 25th Dynasty (~2,000 yr)	Egyptian 12th Dynasty (~4,000 yr)	California Early Man (^{14}C = 4,990)	San Diego No.18402 (~6,000 yr)	Maryland Pleistocene (~80,000 yr)
OH-Pro	100	99	63	--	90	97
Aspartic	50	34	160	230	66	48
Threonine	19	3	3	19	16	22
Serine	35	11	8	23	32	37
Glutamic	78	84	180	224	82	73
Proline	109	114	80	25	110	124
Glycine	330	362	288	200	334	340
Alanine	112	134	110	78	117	111
Valine	21	26	20	47	21	18
Methionine	4	0.2	4	8	6	4
Isoleucine	11	10	9	19	8	10
Leucine	28	29	25	38	23	23
Tyrosine	4	--	--	--	--	2
Phenylalanine	14	--	--	--	20	12
Histidine	4	--	--	--	2	2
OH-Lys	4	--	--	--	2	2
Lysine	25	41	22	33	34	28
Arginine	53	53	28	36	40	48
Nanomoles/mg	2500	2300	140	2.3	200	1250
D/L Aspartic Acid	0.05	0.56	0.25	0.40	0.22	0.20

Racemization of Amino Acids in Fossils

All of the amino acids found in proteins except glycine can exist in one of two possible optical isomeric forms designated the D- and L-amino acid. An equal mixture of the D and L isomers shows no optical activity (rotation of polarized light) and is often referred to as a racemic mixture. Normal acid hydrolysis of proteins yields L-amino acids with traces of the D-amino acids. Prolonged hydrolysis in 6N HCl eventually produces a racemic mixture of amino acids. Base hydrolysis of proteins causes nearly complete racemization of the amino acids and is usually avoided as a sample preparation technique.

The presence of increasing amounts of D-amino acids in progressively older fossils was first noted for a series of molluscan shells (Hare 1969). The first use of the extent of amino acid racemization for estimating the age of fossil was also on calcareous shell material (Hare and Mitterer 1969). The various amino acids found in proteins racemize at different rates, with aspartic acid fastest and valine and isoleucine the slowest. Aspartic acid racemization has received wide attention as a tool for dating bone (Bada et al. 1975), particularly since it was applied to the dating of San Diego Man at around 50,000 years B.P. (Bada et al. 1974). I believe this date is grossly in error. One of the assumptions involved in arriving at this date is that temperature is the only environmental factor that is important (Bada et al. 1975). Laboratory simulation experiments (Hare 1974) suggest that water distribution in the environment may be equally as important as temperature in determining the extent of amino acid racemization, and, indeed, in the entire diagenesis process. At present I feel it is premature to use amino acid racemization alone as an age-dating tool for bone material. It has a great deal of potential, however, and, used along with data on free and total amino acid concentrations and relative amino acid compositions, it can be helpful in determining something of the diagenetic history of the fossil.

Laboratory Simulation of Protein Diagenesis in Fossils

Soon after Abelson first reported the presence of amino acids in fossil material, he showed by heating amino acids in water at elevated temperatures that some amino acids should be stable enough to persist through geologic time (Abelson 1956). The Arrhenius equation (see Appendix A) relates the rate constant and the absolute temperature, and is valid for many chemical reactions.

The initial chemical reaction in the diagenesis of proteins in shell or bone is the relatively slow hydrolysis of the peptide bonds between adjacent amino acids in the protein chain. This hydrolysis lowers the molecular weight of the protein and increases its solubility. Some amino acids, such as valine and isoleucine, form more stable peptide bonds than do others such as serine, alanine, and glycine (Hare et al. 1975).

Water is necessary for the hydrolysis of the peptide-bound amino acids and for racemization. The indigenous water in bone, or even water vapor in the atmosphere, is usually sufficient for these reactions to occur. If excess liquid water is present, varying amounts of leaching can occur in addition to hydrolysis and racemization.

Modern bone generally contains 7-10% indigenous water. This water is reversibly bound and can be removed by heating to 110° C. The dried bone will gain the water back again if cooled to room temperature in the open atmosphere. If a sample of bone is heated to remove water, then sealed in a glass tube before it can gain back the lost water, the protein in the bone will not undergo the usual reactions of hydrolysis and racemization when the sample is heated at an elevated temperature (Hare 1974).

Bones from the La Brea tar pits in Los Angeles, California, may have had a somewhat anhydrous diagenetic history, with hydrocarbons displacing the indigenous water in the bones and then sealing them against further moisture from the outside.

In another experiment bones were sealed in tubes with only their indigenous water present or in other cases only water vapor present ($Na_2CO_3 \cdot 10H_2O$ gives up H_2O when heated and provides a convenient source for water vapor). These samples approximate a

closed system in which essentially all the original protein material is retained within
the bone but is extensively altered by reaction with water. Racemization reactions in
such a closed system follow a predictable rate law for first-order reversible reactions.

In most natural environments, however, water is present in excess, at least
periodically, and this results in the leaching of soluble components. To simulate this
possible effect, bone samples were heated in various amounts of water. In one series of
experiments the surrounding water was changed frequently so that the bone fragments were
in contact with fresh water most of the time. The simulated diagenetic changes resulting
from the bone-heating experiments varied dramatically according to whether or not water was
present and how much. Table 12.3 (from Hare 1974) summarizes the results of heating

Table 12.3

Effects of Water on Leaching and on the Apparent Racemization of Amino Acids in Bone
Samples Heated with and without Water (3 Days at 157¼ C)

Water Conditions	% Nitrogen in Bone Fragment	% Leached	% Racemization (for Isoleucine)
1) Anhydrous (heated to remove indigenous H_2O)	4.0	0	0
2) Indigenous H_2O of H_2O vapor (no excess liquid H_2O for leaching)	4.0	0	95
3) Liquid H_2O (10:1 ratio of H_2O to bone)	0.6	85	56
4) Excess liquid H_2O (> 1000:1 ratio of H_2O to bone)	0.04	99	25

modern bone fragments with varying amounts of water for the same length of time and at
the same temperature. The sample heated under anhydrous conditions looked nearly the
same as the unheated fragment, with no nitrogen loss and little or no racemization of the
amino acids. In contrast, the sample heated in indigenous water or water vapor, although
it showed amounts of nitrogen similar to modern bone, showed the highest degree of amino
acid racemization for any of the experimental simulation systems. This is the closed
system model in which the racemized amino acids and peptides from the protein breakdown
are retained in the bone material.

With liquid water present there are further dramatic effects due to leaching.
When the ratio of water to bone is 10 to 1, around 85% of the protein is leached out
of the bone, whereas in the experiment in which the water was changed frequently, the
amount of protein leached out was 95%. As more protein material is leached out, the
proportion of D- to L-amino acids (percent racemization) in the bone fragment diminishes
because more of the D-amino acids are leached out into the water. This was confirmed
by determining the extent of racemization of the amino acids in the water used in
leaching. The reason relatively more D-amino acids are leached out of bone is that the
free amino acids and peptides formed from the hydrolysis of the bone proteins are more
racemized than the residual protein. During leaching these free amino acids and paptides
are more readily leached than the more intact residual proteins which still contain
mostly L-amino acids.

Even though the temperature and time of heating were identical in all of the above
experiments, the results show a wide range of D/L amino acid ratios. Using only the D/L
amino acid data, it would be easy to wrongly conclude that the samples had been subjected

to different temperatures and/or heating times. The situation is analogous for amino acid dating on natural samples and shows the need for using other parameters in conjunction with racemization data.

Kinetics of Protein Breakdown and Leaching of Bone

Fragments of modern bovine bone varying in size from 0.2 mm to 10 mm diameter were weighed out into glass tubes. In one series of tubes only water vapor was allowed to contact the bone fragment, whereas in the other series a large excess of water was added. The tubes containing bone and only water vapor showed a uniform increase in the D-amino acid content as well as in the proportion of free amino acids. There was little difference in reaction rates among the various-sized fragments.

In the tubes with excess water to simulate leaching effects, the various-sized fragments showed some considerable differences in apparent reaction rates. As might be expected, protein loss was more rapid for smaller bone fragments than for larger fragments. For a given size bone fragment, the protein loss curve appears to follow a curve with three distinctly different rates (slopes)(fig. 12.2). The initial part of the curve is very slow, probably because the collagen in the bone fragments is still of high molecular weight and relatively insoluble. The second segment is relatively rapid and grades into a slower third segment. Although the total amino acid content continues to diminish, the relative amino acid composition of the bone during this third segment remains relatively constant and is completely unlike collagen. Collagen is leached out during the relatively rapid leaching shown by the second segment of the curve. Apparently this rapid leaching is initiated not at the beginning of the heating experiment but at some later time when the collagen has broken down to a series of more soluble lower-molecular-weight polypeptides (subprotein units). The samples heated in water vapor alone show only relatively slight changes in overall amino acid content (or nitrogen content), or in relative amino acid composition. They do, however, show substantially higher degrees of racemization because the breakdown products which are more racemized are not leached out. The difference for some amino acids (e.g., isoleucine) can be more than an order of magnitude (Hare 1974).

Figure 12.3 shows the aspartic acid racemization data. Again the curve is not a simple straight line. A slow initial racemization rate where the collagen is still largely insoluble is followed by a relatively faster rate segment in which the collagen is being leached out, revealing the noncollagen protein that appears to be more resistant and better retained in the bone structure than collagen. The racemized products of the noncollagen protein are not leached out nearly as fast as the collagen-breakdown products.

Implications for Amino Acid Racemization Dating

It has been assumed, in using amino acid racemization rates for age dating, that the L-amino acids in bone follow the rate law for first-order reversible reactions. This is strictly true only for a closed system in which the reaction products are not leached out. Since the racemization curve for continuously leaching bone does not follow a straight-line relationship, it follows that estimates of age based on racemization data along may be grossly in error if the sample used for calibration and the sample to be dated are on different segments of the racemization curve (fig. 12.3). For example, if the calibration sample (^{14}C dated) happens to be on the slow initial curve, the age determined for a sample that may be on the steeper curve will be overestimated. If both

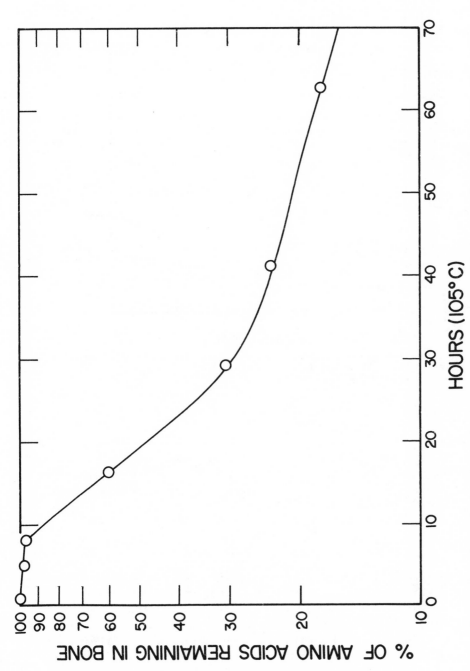

Figure 12.2. Three-segment curve showing changes in the rate of loss of amino acids from bone fragments (2-3 mm) under continuous leaching conditions.

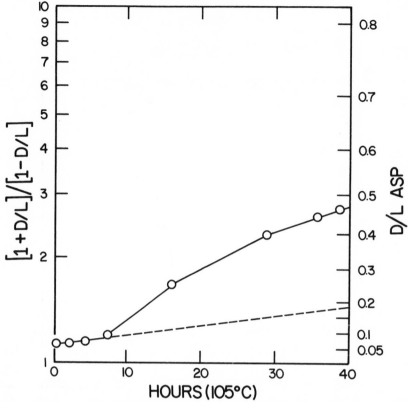

Figure 12.3. Curve of aspartic acid racemization in bone fragments under continuous leaching conditions. Lower dashed line represents the extrapolated expected rate based on the slow initial racemization occurring during the initial collagen-gelatin transition.

the calibration and sample to be dated are on the same segment of the curve; then a reasonable age estimate may be possible.

Probably few of the amino acid racemization dates published to the present time are reliable. Some are possibly off by an order of magnitude. However, in spite of the problems involved in calibration, a useful tool will be developed that will make it possible to determine the age of a fossil and something about its diagenetic environment. This will be done by combining data on racemization reactions with other reactions having different activation energies (see Appendix A).

Comparison with Natural Samples

Most samples of fossil bone from natural environments seem to follow a diagenetic pathway similar to the continuous leaching curve in the simulation experiments. There are, of course, an infinite number of possible pathways between what would seem to be the two limiting pathways represented by the water vapor curve and the continuous leaching curve. The proportion of time a sample is exposed to or immersed in liquid water would no doubt make the diagenetic pathway a good deal more complex than the segmented curve shown in figure 12.2. Adequate data on bones from known natural environments are not yet available for comparison with the simulation models.

Since most bones from natural environments show diminished amounts of protein material, it seems reasonable to conclude that some leaching has occurred. The absence of significant amounts of free amino acids from fossil bones is probably due to the fact that they have been leached out. In the water vapor experiment there was a continuous increase with time of the level of free amino acids from the protein breakdown. No significant levels were found in any of the continuously leached samples. The rate-determining step appears to be in the protein breakdown rather than the leaching of the breakdown products. This would suggest that even occasional or periodic leaching conditions would adequately remove the breakdown products formed by the reaction with water vapor during the "dry" period. It may be possible to check this on bones in the field toward the end of an extended period of drought and after the next period of rainfall.

If most bones in the natural environment follow a curve approximating the continuous leaching model, then the diagenesis of protein or nitrogen loss or even racemization will not follow a simple straight line. The initial slow segment of the continuous leaching curve may extend for thousands of years or only a few years, depending mainly on environmental factors. Samples from the same archaeological sites sometimes show variable amounts of protein remaining even in samples of supposedly the same age. Is this due to microenvironmental effects within the site, or to species effects? In a sample of Egyptian bone material from the 12th Dynasty (3800-4000 B.P.)(table 12.2) there is a profound difference within a single section of a bone fragment. A sample from the outside edge of the bone shows only 20-25% of the amino acids recovered from a sample near the center only a few millimeters away. The degree of aspartic acid racemization was also significantly different and would be equivalent to several thousand years' difference in age if used for amino acid racemization dating without additional data.

A sample of bone from the 20th to 25th Dynasty (~2000 B.P.) of Egypt shows even a higher D/L aspartic acid ratio than the older sample. Again, using only the racemization data would yield an unrealistic result. The 20-25th Dynasty sample appears to have had very little if any leaching. There are significant levels of free amino acids present, and the overall recovery of amino acids indicates over 90% of the original material is still in the bone. Much of the protein, however, has broken down to smaller and more soluble peptide products. This sample appears to have followed the water vapor model curve of the simulation experiments where the racemized products are retained in the bone structure. A small amount of intact collagen was isolated from several hundred milligrams of sample and found to have almost the same amino acid composition as modern collagen (figure 12.1), but the total amino acid composition of the bone itself shows some changes compared to modern bone, particularly in the presence of some oxidation products of amino acids indicating oxidation conditions at some point in its diagenetic history.

Application to Survival of Bone

The survival of bone in many natural environments is a function of the mechanical strength and hardness of bone, which, in turn, is a function of its microstructure. Both the organic matrix and the mineralized components are essential in maintaining the physical properties of bone (Ascenzi and Bell 1972). If either the organic matrix or the mineralized components are partially removed or weakened, the hardness and strength of bone will decrease.

Although little or no quantitative data are available on the strength and hardness

of fossil bone, a general qualitative observation is that bone strength and hardness continues to decrease until the bone disintegrates or, if in a suitable environment, is
mineralized with silica and/or other materials.

Qualitative observations during the simulation experiments showed that, as water
reacted with the protein in the bone fragments, the fragments became progressively chalkier
and easier to break apart. Samples that had been leached extensively were generally easy
to crush and cut. In the early stages of the reactions where collagen was still present,
the fragments when dissolved in acid would show the intact pseudomorphic ghosts of the
bone fragments. Bone strength and hardness appeared only slightly less than that of fresh
bone material. As the reactions progressed, the pseudomorphic ghosts looked progressively
less intact until there was no longer any pseudomorph left—only a few scattered fragments
of organic material. At this stage there was substantially less strength and hardness left
in the bone fragment. The fragments were somewhat chalky and easily crushed with the
fingers.

The sample of 2000 year old bone from Egypt (table 12.2) showed only limited
mechanical strength in spite of the fact that it had over 90% of the amino acids and
nitrogen of modern bone. When the sample was dissolved in acid there were only scattered
fragments of organic matrix left. Most of the amino acid material must have been in the
form of soluble peptides and free amino acids rather than intact collagen fibers.

Fossil bone samples of pre-Pleistocene age frequently show much greater strength
and hardness than younger fossils. In these cases the microstructure of bone has been
strengthened, or, in some cases, completely replaced by mineral matter deposited from
ground water. Structural details of the original microstructure of the bone are often
preserved, in contrast to the almost complete lack of preservation of any of the organic
matrix components.

In considering the role of the bone organic matrix in the potential physical and
chemical preservability of bone in various natural environments, there are many questions
for which present data are not adequate to supply satisfactory answers. For example,
here are just a few:

1. What effect does alternate wetting and drying or cooling and heating have
(dew or rain followed by hot sun)?

2. What effect does the deposition of a carbonate caliche-type deposit in and
around bone have on the organic matrix of bone and its potential reaction with water?

3. What effect does the composition of the ground water have on leaching organic
and mineral components?

4. What about the role of microorganisms and other biological factors?
(Marchiafava et al. [1974] describe the action of fungus on the dissolution of buried
bone. Biological activity must be an important factor to consider.)

5. What are the diagenetic effects on other components of the organic matrix
of bone such as the sugars?

Conclusion

It seems to me the best approach to investigation of these and other, related
questions is a combination field and laboratory study. After bone assemblages are collected in the field with different species and bone types represented and from as many
different environments as possible, they can be compared to modern bone materials that
have undergone various laboratory simulation treatments. Many of the chemical changes

observed within minutes and hours in the laboratory treatments at relative high temperatures (75°-150° C) should be observable within a few years at environmental temperatures (15°-20° C).

The mutual feedback of laboratory and field data should make it possible to refine the future questions as well as the methods to help answer them.

Appendix

Arrhenius Equation

Most chemical reactions are affected by temperature changes in a predictable manner. The Arrhenius equation relates the rate constant, k, in the first order rate expression $\ln A/A° = -kt$, to the temperature, T:

$$\ln k = - E_a/RT + C \tag{1}$$

where E_a is the activation energy, R is 1.987 calories per mole, C is the integration constant, and T is the absolute temperature.

Instead of the rate constant, k, it is possible to use the equivalent expression $k = 1/t \ln A°/A$ and relate the time for a specified amount to react such as the half-life or any other specified amount. If two reactions have appreciably different activation energies, it is possible to solve the two equations for both time and temperature.

Part 5

PALEOECOLOGY

In this section we come to the ultimate aims of work in taphonomy and paleoecology —the reconstruction of whole paleocommunities and the study of their changes through time. The papers provide us with examples of how taphonomic factors affect paleoecological interpretations and how faunas are used to construct habitats and overall characteristics of the paleoenvironments. Along with these examples, we are introduced to some new insights and approaches in paleocommunity analysis.

Klein provides us with examples of the problems encountered in using real data in paleoecological reconstruction, and much of his discussion is important in demonstrating the limitations of the fossil record. The subject is faunal analysis of South African fossil assemblages, particularly those associated with archeological sites. Klein demonstrates the value of the comparative approach, using different sorts of bone accumulations which allow him to specify the bone-collecting agent and draw conclusions concerning its predatory and bone-transporting behavior. These conclusions bear on the question of early human behavior in South Africa and elsewhere. Klein points out that subrecent or historical faunal assemblages in which aspects of taphonomic history are known can provide a valuable basis of comparison for the more distant past.

Vrba examines the extent to which a fossil fauna can be used to interpret past environments and herbivore habitats. She analyzes the Bovidae, the dominant group of large herbivores in the later Cenozoic of Africa and an important faunal component in Neogene fossil assemblages in general. She has assembled much important comparative data on the recent ecology of bovids, and shows how this can be used to infer differences in past environments by analogy with the present. This approach has been widely used by vertebrate paleontologists studying various time periods on different continents, and Vrba's demonstration of its strengths and weaknesses is of general importance to paleoecology. She provides interesting additional comments and conclusions regarding hominid paleoecology in the South African cave localities. Her work demonstrates the problems of separating time and ecology in a region where there is no independent stratigraphic control on the time succession of faunas.

Van Couvering's paper serves to illustrate many of the points on theory, method, and interpretation in paleoecology which are made by Olson in his introductory treatment of taphonomy and paleocommunity analysis. We see that the basis for interpreting the habits and habitats of Miocene to Pleistocene faunas in Africa is almost entirely biological, that is, based on the ecology of modern faunas and the assumption that close relatives will be more similar in their ecological preferences than distant onces. The habitats of the extinct species or genera are in turn used to indicate dominant plant community types, such as forest and savanna, and their change through time in response to environmental pressures. Van Couvering postulates gradual stages in the evolution of a new community type over some 10 million years, and this contrasts with the concept of rapid transitions between successive communities presented by Olson. There is added interest in the fact that human ancestors were participants in the gradually changing community proposed by Van Couvering. She bases her conclusions on a relatively new method of analyzing the fossil record which has potentially wide application. This method illustrates one of the possible means of overcoming taphonomic biases in faunal samples: the use of large samples of localities and faunal lists in which these biases will tend to cancel out. Assembling the right kinds of samples for paleoecological reconstruction is obviously of great importance to the future of large-scale paleocommunity analysis.

13. THE INTERPRETATION OF MAMMALIAN FAUNAS FROM STONE-AGE

ARCHEOLOGICAL SITES, WITH SPECIAL REFERENCE TO SITES IN

THE SOUTHERN CAPE PROVINCE, SOUTH AFRICA

Richard G. Klein

Introduction

Although paleoanthropologists have long appreciated the value of mammalian fossils
as indicators of geological age and of past environments, detailed studies of archeological
faunal assemblages are remarkably rare. In the past, most studies were done by paleontologists who were not particularly interested in the archeological implications of the
material. Most often they produced a simple species list, sometimes only rough or partial,
which was appended to an archeological report. Often, little or no quantitative information was presented on species frequencies, and all the faunal materials were treated as
forming a single assemblage, with no regard for the separate levels from which they may
have come.

Although archeologists today are almost universally aware of the interpretative
potential of archeological faunal remains, relatively few archeology training programs
include meaningful exposure to the practice and problems of faunal analysis. This is
unfortunate because it means that relatively few prospective archeologists are being encouraged to become faunal specialists. Even more importantly, it means that many archeologists do not learn how to minimize breakage and sampling bias in recovering faunal
materials, and they often unwittingly bring excavations to a close before a statistically
meaningful faunal sample has been accumulated. Inadequate size is one of the two most
important obstacles to the detailed interpretation of many archeological faunal samples
today. The other is a shortage of complementary samples suitable for the comparisons
that allow the analyst to sort out the effects of culture, environment, and sedimentary
context on any given sample.

There is probably no part of the world where the basic interpretative requirement—
a large number of different kinds of sites, each providing a large quantity of faunal
debris—has been fully satisfied. Within sub-Saharan Africa, reasonably intensive recent
and continuing research at Upper Pleistocene and Holocene sites in the southern part of
the Cape Province, South Africa (fig. 13.1 and table 13.1) has brought this area far closer
to the ideal than any other region. For this reason, and also because I have conducted
most of my own research on southern Cape faunal materials, I have chosen to use examples
from this area to illustrate the interpretative potential of stone-age archeological
faunas.

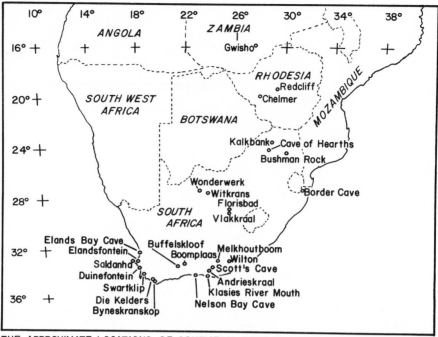

THE APPROXIMATE LOCATIONS OF SOUTHERN AFRICAN SITES MENTIONED IN THE
TEXT. THE SOUTHERN CAPE IS ROUGHLY THE PART OF SOUTH AFRICA SOUTH
OF 32° S. lat.

Figure 13.1.   The approximate locations of Southern African sites mentioned in the text. The Southern Cape is roughly the part of South Africa south of 32° S latitude.

Sources of Bias: Excavation, Sieving, and Sorting

In an ideal excavation, every bone would be recovered and its stratigraphic provenience and three-dimensional coordinates would be recorded. In some open-air sites (for example, Duinefontein 2 in the southern Cape) where bones and artifacts appear to be scattered across buried surfaces, where the density of materials is not especially great, and where fragmentation is relatively limited, a good excavation may approach the ideal. However, even in the best open-air excavations, some small bones will only be recovered in the process of sieving. In many cave sites, including all the southern Cape examples with which I have been concerned, relatively intensive occupation and/or relatively slow sedimentation has led to very high bone densities and substantial fragmentation. In these cases, in order to obtain sufficient samples in a reasonable amount of time, it has generally been necessary to rely on sieves for the recovery of most bones (and artifacts, which are also usually present in high density and are often very small). At southern Cape sites, most often two tiers of sieves have been used, the upper with a mesh of 6-12 mm, the lower with a mesh of 1-2 mm. The principal exceptions to this rule are the sites at Klasies River Mouth where only larger mesh screens were generally employed. At most sites, vigorous shaking was used to speed the passage of sediment through the sieves. At Nelson Bay Cave and Die Kelders, however, where water was readily available, the matrix was often washed through. This had the dual advantage of reducing damage to the bones and increasing their visibility. Among the sites considered here, therefore,

Table 13.1
Principal Southern Cape Mammalian Faunas Analyzed by the Author and Used as the Bases for Examples in this Paper

Site	Age of Faunal Materials (and Nature of Supporting Evidence)	Cultural Affiliation	Excavator or Collector	Principal Publications
Melkhoutboom Cave	Ca. 15,500 to ca. 2,500 B.P. (^{14}C)	LSA	H.J. Deacon	H.J. Deacon (1972,1974)
Scott's Cave	Very late Holocene (^{14}C)	LSA	H.J. and J. Deacon	H.J. and J. Deacon (1963); H.J. Deacon (1967); Klein and Scott (1974)
Klasies River Mouth Caves	Earlier Upper Pleistocene and Middle to late Holocene (^{14}C and geological inference)	MSA, LSA	J.J. Wymer and R. Singer	Wymer and Singer (1972); Klein (1974a, 1975a, 1976b).
Nelson Bay Cave	Ca. 18,500 to ca. 1,500 B.P. (^{14}C)	LSA	R.R. Inskeep and R.G. Klein	Klein (1972a,b,1974a); Butzer (1973)
Boomplaas Cave	Upper Pleistocene and Holocene (^{14}C and geological inference)	MSA, LSA	H.J. Deacon and M. Brooker	H.J. Deacon and M. Brooker (1976)
Buffelskloof Shelter	Late Upper Pleistocene and Holocene (^{14}C and geological inference)	LSA	H. Opperman	Opperman (1978)
Byeneskranskop Cave	Ca. 12,500 to ca. 2,000 B.P. (^{14}C and archeological inference)	LSA	F.R. Schweitzer	Schweitzer and Klein (in prep.)
Die Kelders Cave	Earlier Upper Pleistocene and late Holocene (^{14}C and geological inference)	MSA, LSA	F.R. Schweitzer	Schweitzer (1970,1974, 1975); Schweitzer and Scott (1973); Tankard and Schweitzer (1974); Klein (1975a)
Swartklip 1	Earlier Upper Pleistocene (^{14}C and geological inference)	None	Principally Q. B. Hendey and R.G. Klein	Hendey and Hendey (1968); Klein (1975b)
Duinefontein 2	? Later Middle Pleistocene (paleontological inference)	MSA	R.G. Klein and L. Stoch	Klein (1976a); J. Deacon (1976)
Elandsbay Cave	Terminal Pleistocene and Holocene (^{14}C)	LSA	J.E. Parkington	Parkington (1972,1976); Parkington and Klein in prep.)

Note: For locations of the sites, see Fig. 13.1. Under "Cultural Affiliation," MSA = Middle Stone Age, LSA = Later Stone Age.

sampling bias, particularly against small bones, is probably least at Nelson Bay and Die Kelders, and only the Klasies sites present a special problem in this regard.

Beyond sieving, another potential source of bias in archeological faunal samples is sorting. Generally speaking, the accuracy of sorting increases with the experience of the sorter. In my own case, I have attempted to maintain comparability between assemblages I studied as a beginner and ones I analyzed several years later by resorting (and reidentifying) substantial portions of the earlier assemblages, including all the cranial material. For most of the southern Cape assemblages discussed below, I either did all the sorting myself or had the opportunity to check and correct any preliminary sorting done by others. The principal exceptions are the Klasies River Mouth sites where the bone was sorted into identifiable and unidentifiable components in the field, and the unidentifiable pieces were discarded. However, the field sorting at Klasies was generally done by persons with experience in bone identification, and I doubt that it seriously biased the Klasies identifiable bones with regard to ones in the assemblages over which I have had more complete control.

Estimation of Species Frequencies

The most widely used numerical data in archeological faunal analysis are estimates of species frequencies. Several different methods of making estimates are in use, and the relative merits of these have been discussed by various authors (for example, Shotwell [1955, pp. 330-32]; Chaplin [1971, pp. 63-72]; Payne [1972, pp. 68-71]). The method I prefer involves calculation of the minimum number of individuals by which a species is represented in a bone assemblage. My purpose in examining various methods here is to justify this preference.

1. Frequency Estimation Based on the Number of
    Bones Assigned to a Species

The main advantage of this index of species frequency is that it is relatively easy to compute. It has several disadvantages: (1) It will overemphasize the importance of species whose skeletons contain more bones. Carnivores and primates, for example, generally have more bones than ungulates, and suids generally have more bones than bovids or equids. (2) It will overemphasize the importance of a species for which nearly every skeletal element is diagnostic as opposed to a species for which only certain bones are positively identifiable. In most African archeological sites, for instance, all hippopotamus bones are immediately assignable to the single, locally recorded species, but many bovid postcranial bones, especially fragmentary ones, cannot be unequivocally assigned to one bovid species versus another of similar size. (3) It will overemphasize the importance of a species whose skeleton tended to reach a site intact as opposed to one for which only selected body parts were ordinarily brought back. More generally, it will overemphasize the importance of a species that is well represented by all its diagnostic skeletal elements as opposed to one that is well represented by only a few of them. (4) It will overemphasize the importance of a species whose bones tend to frag-ment into readily identifiable pieces. With regard to the bovids, for example, the jaws of larger species are more prone to fragmentation than those with smaller species. Since even isolated bovid teeth are usually specifically identifiable, species indices based on simple bone counts will tend to overemphasize the importance of larger bovids.

Perkins (1969) has suggested a method of correcting bone counts to eliminate the biases I have listed. Basically, it consists of dividing the total number of bones as-signed to a species by the number of its diagnostic skeletal elements that occur most commonly in a site. For the Turkish Neolithic site of Çatal Hüyük, for example, Perkins found that most cattle bones belonged to just thirty body-part categories (various limb and foot bones), while most horse bones belonged to an even smaller number—just sixteen categories (different foot bones only). He therefore obtained species frequencies for cattle and horses by dividing the total number of bones assigned to each by 30 and 16,

respectively. The major drawbacks I see to the use of Perkins's formula are that the
number of diagnostic elements to use in the divisor must be arbitrary to some extent and
the species frequencies that result are not necessarily comparable among sites. In fact,
on close examination, I think Perkins's method is actually a crude and inadvisable way of
calculating "minimum numbers" estimates like those discussed below.

2. Frequency Estimation Based on the Minimum Number of Individuals by Which a Species is Represented

The minimum number of individuals by which a species is represented in a bone
assemblage is the number that is sufficient to account for all the bones assigned to the
species. Procedurally, the minimum number is arrived at by sorting all bones assigned
to a species among appropriate body-part categories. The number of bones in any given
category—for example, the number of left distal humeri or right second lower molars—
is the minimum number of individuals represented by that body part. The minimum number
of individuals represented by the most abundant body part is also the minimum number of
individuals by which the species as a whole is represented.

As indices of species frequencies, minimum numbers counts are superior to simple
bone counts because they are relatively insensitive to the various factors (listed above)
that máy bias bone counts. However, they are generally more difficult to calculate. And
in the small samples usually generated by archeologists, minimum numbers estimates may be
significantly altered by the precise method used to calculate them. Further, they can be
seriously affected by the number and quality of different provenience units recognized by
the excavator of a site.

For most archeological bone assemblages, especially smaller ones, the larger the
number of categories into which the analyst sorts bones, the higher the minimum individual
counts that will result. I present two major examples. (1) A higher number will generally
emerge if the analyst separates rights from lefts for any body part category and estimates
the minimum number from the larger sum. The alternative is to divide the total number of
bones in the category by an appropriate figure (two for most skeletal elements). The difference between a minimum individual count obtained by prior sorting into lefts and rights
and one gotten by after-the-fact numerical division is likely to be especially significant
in small samples. Since most of the samples I have dealt with are relatively small, I
have almost always sorted body parts between lefts and rights. (2) A higher number will
generally result if the analyst matches complementary bones to determine if they could
have come from the same individual. As an example, a species may be represented by ten
left distal humeri and six rights. If systematic matching suggests that three of the rights
came from animals not represented among the lefts, then the best estimate of the minimum
number of individuals would be thirteen. Matching by size involves a considerable amount
of subjectivity and I have generally avoided it. However, I have attempted matching
when sex or age criteria are available. For example, for any given species, I have
generally assumed that fused and unfused epiphyses of the same body part (e.g., the
proximal femur) must come from different individuals.

The way in which the number and quality of provenience units recognized by an
excavator may influence minimum numbers counts as illustrated by the following example.
A buffalo phalanx from one layer of a site and a buffalo tooth from a second layer will
constitute a minimum of two animals if the layers are truly separate, but need indicate
no more than one animal if the stratigraphic distinction between the layers turns out to
be meaningless. In practice, it is probable that most excavators recognize fewer closed

provenience units than actually exist (in fact, recognition of the most basic units, representing discrete occupations of only a few days or weeks, is probably impossible at most sites). For most sites this means that, barring countervailing sources of bias, minimum individual counts will be underestimates. This is particularly true for sites where bone samples are small.

In sum, the main disadvantages of minimum individual counts are that they are relatively difficult to calculate and they may be seriously affected by the precision of excavation and sorting. I believe these disadvantages are minor compared to the disadvantages of simple bone counts. Further, they may be largely offset in detailed reports by listing the frequencies of all body parts for each species and specifying precisely how minimum individual estimates were obtained.

3. Other Methods of Frequency Estimation

In recent times, almost every stone-age site has been excavated with respect to a horizontal grid of squares laid down over the surface of the site. This opens up the possibility of estimating the frequency of a species by the number or percentage of squares in which its bone occur, analogous to the method used by Behrensmeyer (1975) to establish taxonomic frequencies in her study of the taphonomy and paleoecology of vertebrate assemblages from the Koobi Fora Formation east of Lake Turkana (Kenya). Alternatively, a kind of "minimum individuals" estimate might be derived by assuming that bones found a certain number of squares apart probably came from different animals. Both methods of using squares obviously involve assumptions about the degree to which bones of a single animal become dispersed across the surface of a site. Empirical verification of these assumptions would be difficult, if not impossible, in most instances. Additionally, the results of both methods can be substantially altered by arbitrary changes in the size of the squares. For these reasons, I think dispersion-based frequency estimates are generally inferior to conventional "minimum individual" ones. One obvious exception is where dispersion itself is a matter of interest (as in Behrensmeyer's study). Another possible exception, which perhaps deserves exploration, is in cases where extreme fragmentation makes it impossible to obtain conventional minimum numbers greater than one or two, while straight bone counts provide enormously inflated numbers. In such instances, dispersion-based frequency estimates might be the only practical ones.

The Interpretation of Body-Part Counts

The frequencies of different body parts by which a species is represented can often provide clues as to how the species was butchered or utilized by prehistoric people. Even more importantly, body-part frequencies ultimately underlie the species frequency estimates which provide the basis for most paleoecological and cultural/behavioral interpretations of archeological bone assemblages. This means that it is vital to have at least some understanding of the factors that affect body-part frequencies.

In most archeological bone assemblages, for any given species the frequencies of different skeletal elements show at least some significant departures from the frequencies in which they would be represented in complete skeletons. It is now widely accepted that such anatomically unexpected frequencies result from the fact that not all skeletal elements are equally susceptible to destruction or transportation. Most assemblages have been subject to destructive forces such as butchering, chewing, digesting, leaching, burning, and trampling, or to the withdrawal or additon of bones, for example, by flowing water, scavengers, porcupines, or people.

The overwhelming majority of bone assemblages I have examined are from human occupation sites, mostly caves, where there is little or no reason to suppose that flowing water played a role in bringing about body-part frequency discrepancies. That narrows the field, but still leaves a number of important destructive and transportational factors to consider. In most assemblages, in fact, it seems likely that several factors have operated together to bring about body-part frequencies, and it can be very difficult to separate their effects. The most fruitful approach is the controlled comparison of frequencies of different body parts for which size, density, or some other factor varies in a known way, or the comparison of frequencies of the same body parts among species with different characteristics (size, behavior, etc.), or the comparison of frequencies among sites that have been subjected to different agencies of accumulation, withdrawal, or destruction. The purpose of this section is to illustrate the utility of comparisons, using southern Cape assemblages as examples. As in other regions, the examples are limited mainly by the limited number of comparably analyzed assemblages and by the small size of many of them.

1. Frequency Discrepancies That Reflect Differential Bone Durability

Brain (1967, 1969a) has argued that the occurrence of various body parts in disproportionately high frequencies is to be expected in most bone assemblages, if for no other reason than that some bones are more resistant to destruction than others. In particular, Brain has suggested that bones which are relatively dense and which undergo epiphyseal fusion fairly early would tend to outlast bones which are less dense and which take longer to undergo epiphyseal fusion. This means that bovid distal humeri could be expected to outlast proximal humeri, proximal radii distal radii, proximal femora distal femora, and distal tibiae proximal tibiae. This is precisely what Brain found to be the case in a moderately large sample of goat bones he collected from around Hottentot villages along the Kuiseb River in South West Africa. Table 13.2 shows that by and large the same pattern characterizes a representative sample of southern Cape archeological sites and a local nonarcheological site as well. The fact that the "expectable" pattern is not present to the same degree at all sites and is sometimes even reversed simply indicates that relative bone durability is not the only factor influencing body-part frequencies.

Under most circumstances, teeth are among the most durable body parts, and in large assemblages, they almost always provide the highest minimum counts. However, because teeth are made up of several different substances, they are particularly susceptible to destruction (shattering) in the presence of intense heat. The uppermost deposits of Boomplaas Cave provide a case in point. These deposits were intensely baked by the repeated combustion of sheep dung floors. For the most part, teeth are represented only by tiny fragments, while other body parts, though often incredibly deformed, remain more or less identifiable. This has led to a tooth to nontooth ratio which is uniquely low in my experience.

2. Operation of the "Schlepp Effect"

In most archeological sites, striking disproportions in body-part frequencies are more common for larger wild ungulates than for smaller ones. This is presumably because the hunters tended to carry smaller animals home intact, while they butchered larger ones at the place of the kill, taking away only selected parts. Perkins and Daly (1968) have labeled this relationship between larger animal size and a greater tendency to

Table 13.2

Minimum Numbers of Bovid Individuals Represented by Limb Bone Epiphyses in Various Southern Cape Sites

	Nonarcheological/ Earlier Upper Pleistocene Site	Middle Stone Age/ Earlier Upper Pleistocene Sites		Later Stone Age/Later Upper Pleistocene and/or Holocene Sites		
	Swartklip 1	Klasies 1	Die Kelders 1	Elandsbay	Byeneskranskop	Die Kelders 1
Humerus						
Proximal	6	20	47	19	9	68
Distal	53	87	94	34	23	125
Radius						
Proximal	37	63	43	33	21	67
Distal	39	41	33	11	13	48
Femur						
Proximal	10	49	65	19	15	75
Distal	27	46	34	15	10	76
Tibia						
Proximal	21	32	34	8	8	64
Distal	57	93	37	24	22	75

Note: For references on the sites see table 13.1. The assemblages for tabulation were chosen for their relatively large size and for the fact that the same methods of sorting and of calculating numbers were used for all of them.

selective body-part representation the "schlepp effect," from the German verb "to drag." With regard to the ungulates found at Suberde, an early Holocene hunters' camp in Turkey, they showed that wild oxen (*Bos primigenius*) were relatively poorly represented by leg bones versus foot bones, especially in comparison with smaller ungulates. Their interpretation of this was that the hunters discarded oxen leg bones at the place of the kill, after stripping them of meat. The feet may have been left as handles in the skins, which in turn were perhaps used as containers for transporting the meat back to camp. Alternatively, the feet may have been especially valued as sources of sinews for sewing. This leads to the point that some bones at a site may be represented in disproportionately high frequencies because they had direct utilitarian value as tools or as raw material for tools.

Table 13.3 suggests that, as at Suberde, there is a tendency in southern Cape sites for larger ungulates to be disproportionately better represented by foot bones (versus leg bones). The tendency is not equally clear-cut for all sites, indicating the operation of other factors in determining body-part frequencies. Interestingly, table 13.3 suggests that the single nonarcheological occurrence which was included, the probable carnivore accumulation from Swartklip 1, deviates from the pattern that characterizes the archeological sites. Perhaps the reason is that in comparison with hominids, carnivores would be more likely to consume most of the meat of any animal where it was killed. A "schlepp effect" would therefore probably operate more with respect to individual bone size than to animal size, and in fact, the Swartklip assemblage does seem to show a strong selection for body parts about the size of a reedbuck skull or a wildebeest limb.

3. Hominids versus Other Bone Accumulators

There are several bone assemblages of late Pleistocene or Holocene age in the southern Cape which are accompanied by few or no artifacts, suggesting that hominids

Table 13.3

Minimum Numbers of Different-Sized Bovids Represented by Limb Bones (Humerus, Radius, Ulna, Femur, and/or Tibia) and Foot Bones (Carpals/Tarsals Through Terminal Phalanges) at Various Southern Cape Sites

Minimum Number of Individuals Represented by:	Small	Small-Medium	Large-Medium	Large
	Nonarcheological/Earlier Upper Pleistocene Site Swartklip 1			
Limb bones	10	7	39	5
Foot bones	12	11	40	6
	Middle Stone Age/Earlier Upper Pleistocene Sites Klasies Cave 1			
Limb bones	24	18	21	41
Foot bones	20	16	29	65
	Die Kelders 1			
Limb bones	87	11	8	2
Foot bones	43	11	14	9
	Later Stone Age/Later Upper Pleistocene and/or Holocene Sites Elandsbay Cave			
Limb bones	36	2	-	1
Foot bones	53	8	6	6
	Byeneskranskop 1			
Limb bones	18	2	6	1
Foot bones	26	11	22	19
	Die Kelders 1			
Limb bones	108	2(+14)*	1	4
Foot bones	127	5(+12)*	3	9

*The numbers in parentheses under small-medium bovids for Die Kelders 1 Later Stone Age levels are estimates of the minimum numbers of domestic sheep represented by limb bones and foot bones. Sheep have been separated out since the "schlepp effect" which the table is intended to illustrate may operate differently on domestic animals than on wild ones.

Note: In the context of the table, small bovids include *Cephalophus, Sylvicapra, Raphicerus, Oreotragus,* amd *Ourebia;* small-medium ones *Pelea, Redunca fulvorufula, Antidorcas, Tragelaphus scriptus,* and *Ovis aries;* large-medium ones *Redunca arundinum, Hippotragus, Damaliscus, Alcelaphus, Connochaetes,* and *Tragelaphus strepsiceros;* and large ones *Taurotragus, Syncerus, Pelorovis, Bos,* and *Megalotragus.*

played little or no role in accumulating them. Hendey and Singer (1965) reported on one such assemblage from Andrieskraal 2 in which 385 (60%) of the 640 identifiable bones exhibited the shallow, flat, parallel superficial grooves characteristic of porcupine gnawing. Porcupine quills were also present, and Hendey and Singer concluded that porcupines were the primary bone collectors at the site. They did not establish any body-part frequencies which might be peculiar to porcupine-derived accumulations, nor have I been able to isolate any in a reexamination of their data. Brain (1968, p. 131) points out that the clearest indication of a porcupine accumulation is a high percentage of gnawed bones, averaging around 70% in collections with which he was familiar. The highest percentage of gnawed bones I have encountered in an archeological fauna is 0.5% for the Middle Stone Age assemblage from Klasies River Mouth Cave 1. This suggests that porcupines are unlikely to have contributed significantly to any of the known southern Cape archeological assemblages. The extent to which they may have removed bones, possibly selectively, is impossible to establish with the data on hand.

Among the other nonarcheological bone assemblages in the southern Cape, only one has been examined in detail. This is the assemblage from Swartklip 1 which I believe is most economically explained as a carnivore (hyena ?) accumulation (Klein 1975a). In commenting on table 13.3, I have already noted that the "schlepp effect" may have operated in a different way for Swartklip than for the archeological sites. Table 13.4 suggests a

Table 13.4

Minimum Numbers of Bovids Bontebok (Damaliscus dorcas) Size or Larger Represented by Cranial and Postcranial Material at Various Southern Cape Sites

Minimum Number of Individuals Represented by:	Swartklip	Klasies	Die Kelders (MSA)	Elandsbay	Byeneskranskop	Die Kelders (LSA)
Cranium	44	292	61	15	85	14
Postcranium	48	151	40	12	41	12

Note: For the geological ages and cultural affiliations of the sites, see tables 13.1, 13.2 or 13.3.

further point in this context, namely that larger bovids tend to be better represented by postcranial material versus cranial material at Swartklip than at the archeological sites. I suspect the contrast would be even clearer if I could present the data by species for each site. Unfortunately, the archeological material is generally too fragmentary to allow easy taxonomic separation of postcranial bones derived from bovids of similar size.

It is not clear to me why larger bovid cranial material should be relatively so common the the archeological bone assemblages. Perhaps the selection of jaws and teeth for raw material or tools is a major part of the answer. In any case, the relatively smaller amount of larger bovid cranial material at Swartklip seems to me to fit into a wider pattern in which all bones heavier or bulkier than a wildebeest or kudu limb bone are underrepresented, suggesting that the collector presorted bones on the basis of size. In this context, it is relevant that insofar as cranial material of large bovids and of comparably sized or larger animals (equids, rhino, and hippo) is represented at Swartklip, it derives almost exclusively from young to very young animals, while the postcranial bones come largely from adults. This is not generally the case at the archeological sites, where both the cranial and postcranial material of large ungulates is more evenly distributed between juveniles and adults.

Swartklip further contrasts with the archeological sites in containing a very high proportion of identifiable long bone shafts without epiphyses (versus identifiable epiphyses with or without attached lengths of shaft). I believe this reflects the fact that hominids are more inclined than carnivores to enter a bone through the shaft (versus the ends).

Finally, the Swartklip 1 body parts are much less highly fragmented than those in most of the archeological assemblages. However, I think the factors involved may be largely postdepositional. In the hominid sites, the bones are always most fragmented where arti- factual or sedimentological evidence indicates occupation was very intense or sedimenta- tion was very slow. Bones are least fragmented, sometimes only to the extent of the Swartklip example, where (as in the lowermost levels of Die Kelders or in the earliest Later Stone Age occupation of Nelson Bay Cave) there is evidence that occupation was only sporadic or sedimentation was relatively rapid. In sum, unless postdepositional factors can be controlled for, I do not believe that degree of bone fragmentation by itself can serve as a good criterion for separating hominid and carnivore-derived bone accumulations.

4. "Occupation Sites" versus "Kill Sites"

The archeology of American Indian bison (*Bison* spp.) hunters suggests clear con- trasts between the frequencies of body parts found in village or occupation sites and those occurring in kill or kill/butchery sites (compare, for example, White 1953, 1954a, 1954b on occupation sites with Wheat 1972 or Frison 1974 on kill sites). The occupation sites are generally deficient in bones of the axial skeleton, especially vertebrae, which seem most often to have been left at the kill sites. For most southern Cape archeological sites, which are caves, body-part frequency information is obviously not critical to determining whether they were kill sites or not. The single principal exception so far is the open-air station of Duinefontein 2 where excavations have revealed Middle Stone Age artifacts and bones scattered over what appear to be buried land surfaces (Klein 1976a). Table 13.5 shows that the body-part frequencies of large-medium and large bovids in

Table 13.5

Minimum Numbers of Large-Medium and Large Bovids Represented by Different Sections of the Skeleton at Duinefontein 2 and Klasies River Mouth Cave 1

The Minimum Number of Individuals of Large and Large-Medium Bovids Represented by:	Duinefontein 2 Level 2	Klasies 1 MSA All Levels
Cranium	4	232
Vertebrae	8	24
Ribs	3	No data
Scapula and/or pelvis	7	59
Humerus, radius, ulna, femur and/or tibia	3	60
Metapodials, carpals, tarsals and/or phalanges	3	86

Note: For the purposes of the table, large-medium bovids include *Redunca arundinum, Hippotragus, Tragelaphus (strepsiceros), Damaliscus, Alcelaphus,* and *Connochaetes.* Large bovids include *Syncerus* and *Taurotragus.*

Duinefontein 2, level 2, contrast significantly with those in the bone assemblage from the Middle Stone Age levels of Klasies River Mouth Cave 1 (Klasies 1 was chosen for comparison because it is the only like-aged southern Cape archeological site with a high frequency of large-medium and large bovid remains). Unfortunately, the Duinefontein 2 sample is small, reflecting the small scale of the excavations so far (the biggest sample is from level 2), but the data in table 13.5 clearly suggest a much higher relative frequency of axial bones, particularly vertebrae, than at Klasies 1. In combination with the relatively low fre- quency of artifacts in Duinefontein 2, level 2, and a plot indicating that bones found near one another often came from the same animal, the body-part frequencies constitute evidence that Duinefontein 2/2 was probably a "kill site" rather than a camp site.

5. Body-Part Representation in Garbage Heaps
   and Living Areas

As a result of the "schlepp effect," which pertains mainly to larger animals,
especially ones killed some distance from a base camp, relative body-part frequencies may
differ between spatially distinct kill/butchery sites and base camps. For smaller
animals or for larger ones killed nearer home (creatures which might be brought back more
or less intact), butchering might be expected to produce spatial segregation of different
body parts within the confines of the base camp. The Holocene (Wilton Culture) deposits
of Nelson Bay Cave provide an excellent test case for this proposition. To begin with,
of the two most common mammal species in these deposits, one, the Cape fur seal, was
almost certainly acquired very nearby, while the other, the grysbok, was probably also
taken in the immediate vicinity and, additionally, is very small. Furthermore, the
pertinent Nelson Bay deposits may be fairly clearly divided between two types, which I
think are reasonably interpreted as "garbage heaps" and "underfoot (= living area)
deposits." These two types were interstratified in the excavations and sometimes could
be seen intergrading laterally. More extensive intergradation would probably be found
if the relatively small excavations were substantially enlarged. The two types of
deposits are (1) "garbage heaps"—shell middens in which whole or nearly whole shells are
the principal constituents, followed by crushed shell, mineral material, vertebrate
remains, and artifacts, in roughly that order; and (2) "underfoot deposits" or "occupation soils"—brown, highly organic loams with a great deal of crushed shell, high frequencies of vertebrate remains and artifacts, numerous traces of hearths, and only small
quantities of whole shell.

Table 13.6 shows that the relative frequency representation of different body

Table 13.6

*Frequencies of Seal (Arctocephalus pusillus) and Grysbok (Raphicerus melanotis) Foot
Bones in the Wilton "Middens" and "Soils" of Nelson Bay Cave*

The Minimum Number of Individuals Repre-sented by:	Seal		Grysbok	
	"Middens"	"Soils"	"Middens"	"Soils"
The most common body part	48	18	83	15
Foot bones	46	5	47	5

Note: Chi-square (seal) = 5.56 (p = .02-.01); chi-square (grysbok) = 0.94 (p = .5-.3).

parts of the fur seal is not the same in the two kinds of Nelson Bay deposits. Foot
(flipper) bones are relatively more common in the middens. The same sort of contrast may
characterize the grysbok, although the available data do not permit statistical validation.
The occurrence of disproportionate numbers of seal and possibly also grysbok foot bones
in the supposed garbage heaps probably indicates that the relatively meatless feet of these
animals were frequently discarded in the butchering/food preparation process.

6. Disproportions in Small Mammal Body-Part Frequencies

Since the "schlepp effect" does not operate to the same degree on small animals as
on large ones, there is a greater tendency for the different skeletal elements of smaller
animals to be represented in roughly their anatomically expectable proportions in archeological sites. Most of the discrepancies in small mammal body-part frequencies encountered
in southern Cape bone assemblages are probably attributable to such relatively uninterest-

ing factors as differential bone durability and sieving and sorting biases. Perhaps the
most interesting exception concerns the frequencies of mole rat skeletal elements in the
Middle Stone Age levels of Die Kelders Cave 1. Table 13.7 shows that mole rats were

Table 13.7

*Extent to Which Mole Rats (Bathyergus suillus) are Represented by Foot Bones in the Middle
Stone Age Levels of Die Kelders Cave 1*

The Minimum Number of Mole Rats Represented by:	Levels with > 50% Mole Rats on Minimum Individual Counts	Levels with < 50% Mole Rats on Minimum Individual Counts
The most common body part	1704	142
Foot bones	361	86

Note: Chi-square = 53.58 ($p < .001$).

significantly underrepresented by bones of the foot, while the fact that the degree of
underrepresentation varies significantly between levels suggests that relative bone
durability and sieving and sorting factors cannot be the sole cause. I think the best
interpretation of the available facts is that mole rats were hunted at least in part for
their pelts from which the feet were removed. Mole rat fur is of high quality and only
the relatively small size of the species has saved it from exploitation by commercial fur
interests. The especially low incidence of mole rat foot bones in levels where the mole
rat is particularly common (constituting 50% of more of the mammal individuals mole rat
size or larger) may indicate that these levels were more or less specialized fur-trapping
occupations. Levels with lower frequencies of mole rats but higher incidences of their
foot bones may represent less specialized occupations where relatively more attention was
paid to the value of mole rats as food sources.

7. Artifacts and Body Parts

In a multivariate statistical analysis of materials from the Spanish Acheulean
site of Torralba, Freeman (in press) has demonstrated the possibility of uncovering significant associations between different kinds of stone tools and various skeletal elements.
This has obvious implications both for identifying the functions of stone tool types and
for isolating specific activity areas within sites. So far no study of this kind has been
attempted in the southern Cape (or to my knowledge anywhere in Africa), though a definite
potential for it exists, particularly at open-air sites like Duinefontein 2.

The Interpretation of Age and Sex Distributions

The age and sex composition of a group of individuals of the same species in an
archeological site can provide important information on how people utilized a particular
species—whether, for example, predation was selective or opportunistic or whether the
species was domesticated. In certain instances the age composition of a fossil animal
population can also be used to determine the likelihood that people only occupied a site
seasonally.

Generally speaking, sex is more difficult to establish than age. For some species,
e.g., fur seals, most adult body parts can be readily sexed on size, since females are so
much smaller than males. For most species, however, sexual dimorphism is more subtle, and
only select body parts may be easily sexed. For example, in the most common species in

southern Cape sites, various members of the Bovidae, frontlets are by far the most useful parts for sexing. Depending on the species, female frontlets either lack horn-cores altogether or bear ones that are significantly less robust and sometimes differently shaped than male horn-cores. Unfortunately, frontlets and especially horn-cores are usually not found intact in high frequency, at least in part because they are not very durable.

Aging is generally more feasible, mainly because the most useful elements, the teeth, are among the most durable and most common body parts in most archeological assemblages. However, in many assemblages, including nearly all those I have worked with, the occurrence of isolated teeth introduces a complication. The larger species in particular tend to be represented by isolated teeth, rather than by whole or near-whole jaws. This means that there is always a possibility that the same individual will be aged and counted more than once because it is represented by more than one tooth. I have attempted to resolve this difficulty by systematically observing the state of eruption and wear on various teeth (P_4's, M_1's, M_2's, etc.) in complete jaws belonging to appropriate comparative specimens. These observations then form the bases for matching isolated teeth from any given species in order to reconstruct the original jaws from which the teeth probably came. For various reasons, however, especially the fact that tooth wear is continuous and that judgments on the amount of wear are subjective, the age distributions I have derived are arbitrary to some extent. I regard them only as broad approximations, useful mainly in comparisons with other, similarly derived distributions.

1. The Nature of Predation

The clearest instance of "selective predation" in any bone assemblage I have examined is in the nonarcheological assemblage from Swartklip 1. Here it appears that all but one or two of the seventeen springbok (*Antidorcas australis*) represented were probably adult males. I know of no observations of predation on modern springbok that may be used for comparison. However, Kruuk (1972, pp. 92-93) and others have collected relevant data regarding predation on the Thomson's gazelle (*Gazella thomsoni*), a close relative of the springbok in East Africa. These data indicate that disproportionately high numbers of adult male "tommies" are taken by hyenas and other predators. I therefore regard the high incidence of adult male springbok at Swartklip as consistent with the apparent carnivore origin of the bone accumulation.

Among the archeological sites, the Middle Stone Age (MSA) levels of Klasies River Mouth provide an interesting example of apparent selective predation on certain age classes of giant buffalo (*Pelorovis antiquus*)(Klein 1975b). The overwhelming majority of giant buffalo in these levels were either advanced fetuses/newborn calves or full adults. There are very few older calves, probable yearlings, etc., particularly in comparison with the relative numbers of such intermediate age classes in the other large bovid species represented at Klasies, eland and Cape buffalo. Unfortunately, so far I have found no way to sex the majority of giant buffalo represented at Klasies. Still, the age distribution suggests to me that the Klasies MSA people perhaps preyed selectively on near-term pregnant females or even on females giving birth, thereby obtaining both cow and calf. Such a predation pattern might have initiated a long-term, progressive decline in giant buffalo numbers, such as is apparent upwards through the sequence at Klasies and subsequently in the Later Stone Age (LSA) levels of Nelson Bay Cave. Nelson Bay LSA levels dated between 12,000 and 10,000 years B.P. contain the latest record of the giant buffalo, after which the creature seems to have been extinct.

2. The Age-Sex Composition of Domestic Herds

In those areas of the world, especially South Western Asia, where local wild ungulates were domesticated, it can be difficult or impossible to separate the earliest domesticates from their wild ancestors on strictly morphological grounds. In such instances, the most reliable indicator of whether a species was wild or domestic may be the age-sex composition of the fossil herd in a site (see, for example, Perkins 1964; Hole and Flannery 1967; Ducos 1969). In the southern Cape, there is neither osteological/ morphological nor age-sex evidence that any of the local ungulates were ever domesticated, nor were any of them domesticated at time of historic contact. The only domesticated ungulates to appear in the local archeological or ethnohistoric record are sheep (*Ovis aries*) and cattle (*Bos taurus*), which lack local wild antecedents and were thus obviously introduced from elsewhere. Present evidence, especially from Die Kelders, Nelson Bay Cave, and Boomplaas, suggests introduction at about 2000 B.P., more certainly for sheep than for cattle.

Although the absence of wild predecessors for sheep ensures that all sheep bones found in southern Cape sites came from domestic animals, there is still the problem of determining whether the human occupants of the sites were herders or thieving hunter-gatherers. At Boomplaas, the sedimentary matrix provides more or less direct evidence for herding. It contains large amounts of burnt sheep dung, suggesting the site was used as a herders' kraal, at least during the last 1500 years or so. At Die Kelders and other relevant sites (especially Nelson Bay, but also Buffelskloof, Scott's Cave, Byeneskranskop, and Elandsbay), such independent evidence is lacking. However, the main sheep-bearing level at Die Kelders contains a sample of sheep bones sufficiently large and well-preserved to allow a search for clues in the age and sex composition of the fossil flock. My age and sex determinations suggested that the animals represented were over-whelmingly subadult males, followed by adult, generally quite old females, with very little else. This suggested to Schweitzer (1974) that the fossil flock was the product of systematic culling (elimination of reproductively unnecessary and potentially disruptive subadult males and females past peak fertility), which in turn would imply that the occupants of Die Kelders were herders, not thieves.

3. Seasonality of Occupation

For any given species within a site, a distribution of the ages of individuals at time of death which exhibits several distinct modes, each separated from the last by roughly the same interval, suggests (1) that the species had a restricted birth season, and (2) that the people who hunted it only occupied the site for a limited period of each year (not necessarily coinciding with the birth season). Archeologists are of course especially interested in establishing the second point.

For many of the mammalian species represented in southern Cape sites, observations on extant populations (or, in the case of extinct species, on populations of near-relatives) indicate that a more or less sharply defined birth peak may be expected if there is marked seasonal variation in precipitation and/or temperature. Seasonal contrasts characterize the southern Cape today and were perhaps even more marked at various times in the past. Unfortunately, most southern Cape faunal assemblages are too small or too highly frag-mented to allow clear-cut demonstration of multimodality in age distributions. This is all the more true because the probable occurrence of minor year-to-year variation in the duration and peaking of births or in the duration and timing of occupation would introduce sufficient "noise" to dampen out modes in small samples.

The problem of establishing analyzable age distributions is particularly acute for larger bovids, among which are some of the best candidates for demonstrating seasonal site occupation. To begin with, as I pointed out earlier, the teeth of larger bovids tend to occur as isolated specimens, rather than in complete or near-complete jaws. The age categories that result from the eruption and wear-state matching technique (described earlier as an obvious way of returning a sample of isolated teeth to the probable jaws from which they came) are of unequal chronological length. Some, especially the earlier ones, are probably much shorter than the interval between birth peaks or seasonal occupations; others are probably much longer. Obviously, the kinds of equidistant age clusters (modes) that are necessary to establish seasonal occupation cannot emerge from such an aging technique.

There are at least two alternative aging methods that are capable of producing the sorts of data from which seasonality of site occupation could be established. The first involves counting cementum annulations (see, for example, Chaplin 1971, pp. 84-85). However, cementum annulations have been established as reasonably reliable age indicators for only a few African bovids (for example for the springbok *Antidorcas marsupialis* by Rautenbach, 1970 and for the kudu *Tragelaphus strepsiceros* by Simpson and Elder, 1969). Additionally, the method requires special equipment, considerable practical experience, and a lot of time, especially if the object is to establish not just a few ages, but a statistically significant age distribution. It also results in the destruction of material. To my knowledge, it has not been applied to fossil bovid teeth anywhere in Africa and certainly not in the southern Cape.

The second method requires far less equipment, relatively little practice, and much less time. It is also not destructive. It relies on the assumption that seasonally restricted births and deaths, combined with relatively constant wear in a hypsodont ungulate, will leave regular, observable gaps in the frequency distribution of crown height measurements for any given kind of tooth (Kurtén 1953, p. 60; see also Voorhies 1969, pp. 28-30; Reher 1974). The clusters between the gaps are in effect the age-related modes from which seasonally limited occupation may be inferred. I have attempted to use this technique to establish seasonality from southern Cape bovid tooth samples, but so far I have not been successful. I think the principal reason is that the samples of measurable teeth at my disposal are generally too small.

Other than bovids, two species with clear potential for seasonal dating in southern Cape sites are the rock hyrax (*Procavia capensis*) and the fur seal (*Arctocephalus pusillus*). Both are seasonal breeders and are often very numerous in archeological faunas. Age determinations of hyrax, however, are complicated by the difficulty of consistently distinguishing between deciduous and permanent teeth (cf., Parkington 1972, p. 236). So far, the fur seal has provided the most persuasive evidence for seasonal occupation. In a pioneer study involving fur seal remains from the terminal Pleistocene and Holocene (Later Stone Age) levels of Elandsbay Cave, Parkington (1972, pp. 239-41) showed that virtually all the individuals represented were six months old or older. Since the overwhelming majority of fur seal pups are born in November (Rand 1956), Parkington hypothesized that Elandsbay was occupied primarily in the winter months (June-July and later).

A similar pattern characterizes the large late Holocene (Later Stone Age) fur seal sample from Die Kelders analyzed by Scott and myself. Animals younger than five-to-six months were apparently not represented, suggesting once again that winter occupation was the rule. The Die Kelders data are especially interesting because they suggest that there may have been some change in the duration of site occupation after the appearance of sheep. In levels lying below those in which sheep have been positively identified, the clustering of seal bones at or near the size of five-to-six-month-olds is very tight. In higher levels in which sheep bones occur and in quantity, the size range of seal bones upwards from those belonging to five-to-six-month-olds is significantly greater and there is an increase in the number of full adults. The implication may be that, prior to the introduction of sheep on a major scale, human populations were only at Die Kelders for a few months in winter, from perhaps May through July or August. Subsequently, with the appearance of sheep, they may have extended their stay until September or October, thereby obtaining a large number of different-sized pups and also more adults which tend to frequent rookeries to a greater extent as the breeding season approaches.

The Interpretation of Species Frequencies

Faunal analysts often have interpreted species frequencies within an assemblage without explicit comparisons to frequencies in other assemblages or by reference to some implicit or arbitrary standard. In many, if not most instances, interpretations achieved in this way are both limited and difficult to verify. As I hope the following examples will show, I think the most detailed and most satisfactory interpretations of species frequencies in any given assemblage are obtained through comparisons with other assemblages. The most fruitful comparisons are ones in which the cultural, paleo-environmental, and preservational factors that may cause frequency differences are reasonably well controlled so that the effects of any one may be perceived. Because of space limitations here, I have not included the detailed faunal lists on which the comparisons below are based, but most of these are either in press or published in references cited in the text. In all cases where I have based an interpretation on a contrast in species frequencies, the significance of the contrast may be demonstrated by chi square or some other appropriate statistic.

1. The Use of Mammalian Faunas in Dating
Archeological Events

Southern Africa in general and the southern Cape in particular lack materials suitable for radiometric dating techniques other than radiocarbon, while radiocarbon itself can generally only provide finite dates back to ca. 40,000-30,000 B.P. This means that the dating of some important archeological sites and events is at present largely dependent on a combination of geological and faunal evidence. This is true, for instance, for the end of the Acheulean and beginning of the Middle Stone Age, as shown specifically by the following example, based largely on the Middle Stone Age levels of Klasies River Mouth.

Butzer's unpublished evaluation of the Klasies sediments, supported by Shackleton's (unpublished) oxygen-isotope analyses of relevant marine shells and Bada and Deem's (1975) amino acid racemization dates, indicates that the Klasies Middle Stone Age levels span the Last Interglacial and the earlier portion of the Last Glacial, that is, roughly the .interval from 125,000 B.P. to perhaps 60,000-55,000 B.P. At the same time, the Klasies fauna contains only a small number of extinct species and only one extinct genus (*Pelorovis*, the "giant buffalo"), all of which appear to have survived until the end of the Pleistocene, some 12,000-10,000 years ago. Data from Nelson Bay Cave indicate that at least one other extinct genus (*Megalotragus*, the "giant hartebeest") survived to the end of the Pleistocene, while one or two additional extinct species not represented at Klasies occur at other Upper Pleistocene sites in the southern Cape. Overall, however,

the other relevant faunas, especially the earlier Upper Pleistocene ones from Die
Kelders and Swartklip, clearly support the conclusion based on Klasies, that the mammalian
fauna of the southern Cape has been essentially modern since the beginning of the Upper
Pleistocene (i.e., the beginning of the Last Interglacial).

2. Mammalian Faunas as Indicators of Environ-
    mental Change

Southern Cape archeological faunas reflect two kinds of environmental change especially clearly: (1) changes in the position of the coastline as a result of eustatic sea
level change during glacial/interglacial transitions, and (2) vegetational changes probably
also related in a general sense to shifts from glacial to interglacial intervals and vice
versa. Faunal changes probably will remain the principal source of evidence for regional
vegetational changes since many pertinent localities, including the more important archeological sites mentioned here, are disappointingly poor in pollen. Additionally, there is
currently no palynologist who maintains an active interest in the southern Cape.

Faunal changes between the late Pleistocene and early Holocene levels of Nelson
Bay Cave reflect both sea level and vegetational changes particularly well. The late
Pleistocene levels contain few or no remains of marine creatures because a drop in world
sea level of over 100 m and the gentle local inclination of the continental shelf had
displaced the coastline up to 80 km away. Marine fossils (including shells and bones of
seals, birds, and fish) first appear in the terminal Pleistocene levels, dated to roughly
12,000 years ago, when deglaciation was well underway and the consequent rise in sea
level had brought the coastline within fairly easy reach of the inhabitants of the cave.
In deposits younger than 12,000 years, remains of marine animals become increasingly important, reflecting at least in part the progressive rise in sea level which eventually
brought the coast to the foot of the slope directly below the cave. The terminal
Pleistocene-early Holocene rise in sea level is similarly recorded at other pertinent
coastal or near-coastal sites (Byeneskranskop and Elandsbay Caves).

With regard to terrestrial fauna, the late Pleistocene horizons of Nelson Bay
are heavily dominated by the remains of such grassland or savanna ungulates as wildebeest
(*Connochaetes*), bastard hartebeest (*Damaliscus*), and springbok (*Antidorcas*). Bones of
most of the grassland ungulates make their last appearance in deposits dated to between
12,000 and 10,000 B.P. (= the terminal Pleistocene). At the same time, remains of bush
and scrub creatures such as bushbuck (*Tragelaphus scriptus*), bushpig (*Potamochoerus
porcus*), and grysbok either appear for the first time or increase dramatically in frequency. This faunal change almost certainly reflects a dramatic reduction in open
grassland, with a concomitant increase in bush and forest, probably leading ultimately to
the development of the evergreen forest which dominated the area historically. A comparable
if generally less dramatic reduction in grassveld is also apparent in mammalian faunal
change at every other southern Cape site where substantial fossiliferous terminal
Pleistocene-early Holocene deposits are known (from east to west, Melkhoutboom, Boomplaas,
Buffelskloof, Byeneskranskop, and Elandsbay).

The fauna from the long Middle Stone Age sequence at Klasies River Mouth suggests
that the same kind of vegetational change which characterized the beginning of the
Holocene (= Present Interglacial), roughly 10,000 years ago, took place at the beginning
of the Last Interglacial, roughly 125,000 years ago. The oldest fossiliferous horizons
at Klasies, probably dating from the very beginning of the Last Interglacial, are
characterized by relatively high frequencies of alcelaphine antelopes and relatively low

frequencies of bushbuck and grysbok. Higher up in the sequence, in deposits of probable mid-Last Interglacial age, the frequencies of grysbok and bushbuck increase significantly, while the incidence of alcelaphines is substantially reduced. This suggests increasing scrub, bush, or forest, and decreasing grassland. Yet higher still, in deposits which perhaps data from the earlier part of the Last Glacial, the frequencies of grysbok and bushbuck decline, while the incidence of alcelaphines increases significantly. There is also a reduction in the quantity of seal, suggesting a fall in sea level.

At both Klasies and later sites such as Nelson Bay Cave, the significant changes in mammalian species frequencies seem to have taken place within culture-stratigraphic units, that is, more or less independent of changes in culture as reflected in stone artifacts. This suggests clearly that the responsible variable was environment and not culture, even though direct paleobotanical evidence to corroborate the inferred vegetational changes is still lacking. Environmental factors are further implied by the fact that the same pattern of faunal change characterizes sites which are very different in age and cultural affiliation, but which share a common environmental setting and a similar temporal position in the transition from a glacial to an interglacial.

At sites such as Florisbad (Dreyer 1938; Hoffman 1955; Meiring 1956), Wonderwerk Cave (Malan and Cooke 1941; Malan and Wells 1943), the Cave of Hearths (Mason 1962), Bushman Rock Shelter (Louw 1969; Brain 1969c), and Border Cave (Cooke, Malan and Wells 1945; Beaumont and Boshier 1972; Boshier and Beaumont 1972; Beaumont 1973) in the South African interior, where the sequences are long enough to reflect substantial environmental change, paleoenvironmental interpretation of the faunal data is hindered by small faunal samples, by the absence of information on species frequencies in different levels, or both. At Florisbad, the presence of giraffe (*Giraffa camelopardalis*) bones in one or more levels suggests that trees were once far more common than they were historically. Additionally, a palynological study of the Florisbad spring deposits by van Zinderen Bakker (1957) has uncovered long-term vegetational changes that are of sufficient magnitude to be reflected in the fauna. My unpublished analysis of the faunal material excavated by Beaumont from the Upper Pleistocene levels of Border Cave has revealed an inverse relationship between the frequencies of bushpig, buffalo, tragelaphine antelopes, and impala on the one hand and those of warthog, zebra, and alcelaphine antelopes on the other. Levels relatively rich in bushpig et al. alternate with levels relatively rich in warthog et al., probably reflecting long-term fluctuations in the amount of bush versus grass cover in the vicinity of the site. Similar faunal changes, probably reflecting long-term vegetational changes, also characterize the long Redcliff Cave sequence (Klein, in preparation).

Besides environmental changes through time, archeological faunas may reflect environmental differences at any one time. By and large, I have found the differences among areas in the past to be roughly comparable in direction, if not in extent, to the differences that existed among them historically. Thus, as in historic times, the earlier Last Glacial, Middle Stone Age levels at Die Kelders contain significantly fewer large mammals than the like-aged levels at Klasies River Mouth. Similar long-term continuity for a historic contrast is suggested by the persistently lower frequency of warthog and equids in southern Cape sites than in contemporaneous ones to the north.

3. Archeological Faunas as Indicators of Sub-
    sistence Change

One of the most interesting features of the southern Cape archeological record is the evidence it contains for long-term subsistence change. The Middle Stone Age levels of

Klasies River Mouth, dating from the Last Interglacial and the earlier part of the Last
Glacial, contain abundant remains of seals, penguins, and shellfish, constituting the
oldest recorded instance of the systematic exploitation of marine resources. Outside of
the southern Cape, only the less substantial evidence from the Haua Fteah, Cyrenaica
(McBurney 1967) and Devil's Tower, Gibraltar (Garrod et al. 1928) approaches the Klasies
evidence in age. Within the southern Cape the antiquity of marine resource exploitation recorded at Klasies is supported by open-air Middle Stone Age shell middens of
probable Last Interglacial age recently discovered near Saldanha Bay (Avery and Klein,
unpublished) and also by the occurrence of abundant seal and penguin bones in the earlier
Last Glacial, Middle Stone Age levels of Die Kelders Cave 1.

      One curious feature of earlier Upper Pleistocene Middle Stone Age levels at
Klasies River Mouth and Die Kelders is the rarity or absence of bones of fish and flying
birds. This is in distinct contrast to the situation in comparable Later Stone Age
(terminal Pleistocene and Holocene) sites where amounts of seal and penguin bones comparable to those in the relevant Klasies and Die Kelders levels are accompanied by abundant
bones of fish and of flying birds (cormorants, gannets, etc.). I have hypothesized elsewhere (Klein 1974a, 1975b) that active fishing and fowling may have been beyond the
technological capabilities of Middle Stone Age peoples. Just when active fishing and
fowling evolved may be undeterminable, since the most pertinent sites, dating from the
middle part of the Last Glacial, may all be under water.

      A contrast between the Middle Stone Age levels of Klasies and Die Kelders on the
one hand and southern Cape Later Stone Age sites on the other is also apparent in terrestrial faunal remains. In spite of the fact that either bushpig or warthog or both
have probably always been reasonably common around both sites and one or both show up
in fair numbers in comparable Later Stone Age sites, suid remains are rare or absent in the
Klasies and Die Kelders Middle Stone Age levels. Similarly, in spite of the fact that
eland was probably never the most common larger bovid in the southern Cape and is not
especially well represented at local Later Stone Age sites, eland is the most common
large bovid in the Middle Stone Age levels of both Klasies and Die Kelders. Pigs and
eland would be on nearly opposite ends of a continuum from ferocity to docility when attacked, and I have suggested elsewhere that this may account for the near absence of the
first and the commonness of the second in southern Cape Middle Stone Age levels.

      Small sample sizes make it difficult to determine if comparable faunal contrasts
might characterize the Middle Stone Age and Later Stone Age levels at relevant sites
outside the southern Cape, such as Bushman Rock Shelter or Border Cave. I have also
been unable to locate enough published quantitative information to determine if similar
contrasts might exist between like-aged culture-stratigraphic units outside of Africa.
However, I think it is pertinent that a survey of mostly nonquantitative information on
archeological faunas from northern Spain led Freeman (1973, p. 23) to suggest that wild
boar (*Sus scrofa*) remains were significantly more common in local Upper Paleolithic
levels than in preceding Mousterian ones. On present evidence, Mousterian and Middle
Stone Age artifact assemblages occupy very similar time intervals, and they are very
comparable typologically as well.

4. The Frequency of Carnivores in Archeological Sites

      At most archeological sites, fissiped carnivores are a relatively unimportant
part of the fauna. Table 13.8 presents summary frequencies of carnivores and ungulates for
a representative sample of southern Cape sites and shows that these sites conform to the

Table 13.8

Minimum Numbers of Fissiped Carnivores and of Ungulates in Various Southern Cape Archeological Sites and in the Nonarcheological Site of Swartklip 1

	Carnivores	Ungulates
Nonarcheological Site		
Swartklip 1	46 (22%)	159
Middle Stone Age Sites		
Klasies 1	38 (9%)	426
Die Kelders 1	28 (10%)	266
Later Stone Age Sites		
Nelson Bay Cave	46 (7%)	589
Byeneskranskop	31 (13%)	209
Die Kelders 1	33 (12%)	234
Buffelskloof	8 (10%)	76
Boomplaas	22 (13%)	148

Note: The frequency differences between Swartklip and each of the archeological sites can be demonstrated to be significant at less than the .001 level in most cases and less than the .02 level in all cases.

general rule. Relevant frequencies from Swartklip 1, also presented in table 13.8 suggest that one of the distinctive features of at least some nonarcheological accumulations is a relatively high proportion of carnivores.

The relatively low carnivore-to-ungulate ratios in archeological sites certainly reflect natural ratios to some extent and probably also a mutual avoidance relationship between people and at least the larger carnivores. Lions (*Panthera leo*) are extremely scarce in southern Cape sites, and hyenas (*Crocuta crocuta* and *Hyaena brunnea*) are only slightly less so. Leopards (*Panthera pardus*) occur more frequently, but there is the distinct possibility that their remains derive largely from animals who lived and died in the sites when people were absent. A nearly complete leopard skeleton in one of the Middle Stone Age horizons of Klasies Cave 1, showing no signs of injury, butchering, etc., lends weight to this argument. The leopards sporadically occupied Nelson Bay Cave is suggested by a correlation between the frequency of leopard and the frequency of one of its favored prey, baboon (*Papio ursinus*). At Paardeberg Cave, the excavators (H. J. Deacon, R. G. Klein, and a party of students) even found a leopard in residence when they arrived.

The specific effects that periodic carnivore visits might have on a basically hominid-derived bone accumulation are difficult to establish, but such visits might account for the frequent occurrence of fragmented and dispersed human bones in southern Cape and other archeological sites. Heretofore, the best explanation has been widespread and persistent cannibalism, even though this has been rarely observed ethnographically, especially among hunter-gatherers. I think a more likely explanation is that the isolated human bones represent the remains of bodies exhumed from shallow graves by scavengers, particularly hyenas. Hyenas have been observed in precisely this kind of activity in East Africa (Sutcliffe 1970).

5. The Impact of the Introduction of Domestic Ungulates on Wild Ungulates

Archeological investigations in various parts of the world have shown that the appearance of domesticated stock, particularly caprines, generally precipitates a decline in wild ungulate populations. The decline may be brought about by range alteration, the introduction of new diseases, or persecution by herdsmen of species that compete with their stock. Data from sites such as Nelson Bay Cave, Boomplaas, Byeneskranskop, and Die Kelders 1 indicate that the introduction of sheep may have adversely affected several larger ungulates in the southern Cape, especially the blue antelope (*Hippotragus leucophaeus*). This species, a southern Cape endemic, was common in presheep times, but is so far unrecorded in levels with sheep. It had become very rare by the time of European contact a few centuries ago and became extinct around 1800 A.D. (Mohr 1967; Klein 1974*b*). Although the evidence is less clear, it is possible the appearance of sheep had a comparable negative impact on another southern Cape endemic, the bontebok (*Damaliscus dorcas dorcas*), which was also very rare at time of European contact and only just escaped extinction.

6. Terminal Pleistocene Extinctions and the
    Ecology of Extinct Species

One of the most controversial issues in Early Man studies concerns the extent to which people were responsible for the wave of mammalian extinctions that swept much of the world at the end of the Pleistocene, some 12,000-10,000 years ago. The issue has been mostly joined in the New World where extinctions were especially dramatic, involving no less than thirty-one genera, and where there is reason to suppose the extinctions coincided fairly closely with the first appearance of people (Martin 1967, 1973, 1975). Africa was long assumed to have been immune from terminal Pleistocene extinctions, but it is now clear from excavations at Nelson Bay Cave, Byeneskranskop, Elandsbay, and other southern Cape sites that at least two extinct genera and five or six species in surviving genera disappeared locally around 12,000-10,000 B.P. (Klein 1974*c*). Most of these species were probably southern Cape endemics, but at least one of them (the "giant Cape horse" *Equus capensis*) and both the genera (the "giant buffalo" *Pelorovis* and the "giant hartebeest" *Megalotragus*) are known from early and mid-Upper Pleistocene sites to the north (Florisbad, Vlakkraal, Border Cave, Redcliff, etc.) where they are often accompanied by another extinct form which is not represented in the southern Cape, the hyperhypsodont springbok, *Antidorcas bondi*. Although it cannot be demonstrated at this moment, I think it is very likely that there were terminal Pleistocene extinctions in the southern African interior involving both Bond's springbok and the other now extinct forms which accompany it in Upper Pleistocene sites.

It seems logical that the rather considerable terminal Pleistocene to early Holocene environmental change which has been hypothesized for the southern Cape on climatological grounds (van Zinderen Bakker 1976) and which can be demonstrated on faunal evidence and on sedimentological/geomorphological observations made by Butzer (1973; Butzer and Helgren 1972; van Zinderen Bakker and Butzer 1973) played a role in bringing about extinctions. Specifically, a substantial decrease in grassland probably significantly reduced both the numbers and the distribution of extinct species, virtually all of which were primarily grazers. There is evidence from Klasies River Mouth that at least some of the extinct species also suffered a substantial decrease in numbers at the onset of the Last Interglacial, which means that the environmental change accompanying a glacial/interglacial transition is not a sufficient explanation for their final disappearance. I believe that the critical factor differentiating the Last Interglacial from the beginning of the present one was the presence of substantially more competent human hunters, as reflected in the contrast between the Klasies and Die Kelders Middle Stone Age faunas on the one hand and various Later Stone Age ones on the other.

For most of the mammalian species that became extinct in the southern Cape in the terminal Pleistocene, there is insufficient material to establish factors such as population trends or selective predation that might have a bearing on the causes of extinction. One partial exception is the giant buffalo as discussed previously. Giant buffalo remains are also common enough at Klasies River Mouth and Nelson Bay Cave to allow a search for statistical associations with extant species that might shed light on giant buffalo habitat preferences. Its relatively high-crowned tooth morphology, together with the feeding habitats of its closest living relatives (the various extant species of the Bovini), indicates clearly that the giant buffalo was primarily a grazer. A preference for relatively open habitats is strongly suggested by giant buffalo horn spans, which in some adult males must have exceeded three meters. However, both univariate and multivariate analyses of species frequency fluctuations through the multilevel sequences at Klasies River Mouth and Nelson Bay group the giant buffalo with eland and kudu and not with such unquestionable open-country denizens as the alcelaphines and springbok, or for that matter with the giant buffalo's closest living, local relative, the Cape buffalo. Statistical association with the eland is especially clear, and it seems possible that, like the eland (for example, as recently studied in East Africa by Hillman 1974), the giant buffalo preferred open country with important islands and galleries of tree and shrub growth to which it gravitated for shade, if not for food.

I also have some data bearing on the probable habitat preferences of Bond's springbok. As might be expected from its exceptionally hypsodont teeth, the frequency of this creature is correlated at both Border Cave and Redcliff with the frequencies of alcelaphine antelopes and other very hypsodont ungulates, indicating that it preferred relatively open, grassy, vegetational settings.

7. Artifact Types and Species Frequencies

The small backed tools that are common in the Holocene, Later Stone Age Wilton Culture are widely believed to have served as inserts in composite projectiles (arrows and spears). H. J. Deacon (1972, pp. 34-36; see also J. Deacon 1974, pp. 15-17) has pointed out that backed elements are relatively much more common in Wilton sites in northern Rhodesia and Zambia such as Gwisho (Gabel 1965; Fagan and van Noten 1971) than in comparable southern Cape sites such as the Wilton name site (J. Deacon 1972) or Melkoutboom (H. J. Deacon 1974). At the same time, the fauna of Gwisho is dominated by large grazing animals such as buffalo, wildebeest, lechwe, and zebra, while the fauna of southern Cape Wilton sites is dominated by small, nongregarious browsers such as grysbok, bushbuck, and duiker. The faunal contrast implies that in earlier Wilton times, just as now, the southern Cape was characterized by more closed vegetation than northern Rhodesia and Zambia, leading Deacon to suggest that the lower relative frequency of backed elements in southern Cape sites reflects the decreased utility of hunting with projectiles (vs. snares, traps, etc.) in more closed vegetational settings.

More than most of the other examples I have used here, Deacon's consideration of Wilton artifacts and fauna together makes explicit the complementarity of artifactual and faunal studies in archeology. In concluding here, I would like to point out that the complementarity extends deeper than correlations between the frequencies of artifact types and those of species or of body parts, and it involves greater dependence of the faunal analyst on the artifact specialist than I have specified so far. To begin with, I have often found that analyses of associated artifacts provide the best, sometimes the only rational basis for lumping species or body-part frequencies from adjacent stratigraphic

units to obtain figures that are large enough for statistical manipulation. More funda-
mentally, as is implicit in many of the examples I have used here, a knowledge of associated artifacts is obviously pertinent to sorting out the extent to which culture as
opposed to environment is responsible for the composition of a particular faunal assemblage.

Acknowledgments

I thank P. B. Beaumont, C. K. Brain, M. Brooker, C. K. Cooke, H. J. Deacon, J.
Deacon, Q. B. Hendey, R. R. Inskeep, H. Opperman, J. E. Parkington, R. Singer, F. R.
Schweitzer, and J. J. Wymer for making faunal materials they excavated available to me.
Among many people who have kindly discussed the results of the analyses with me, I thank
especially H. J. Deacon and Q. B. Hendey for their stimulation and their patience. All
the original analyses reported here were carried out in the Department of Palaeomammology
of the South African Museum, Cape Town. Financial support was provided by the National
Science Foundation.

THE SIGNIFICANCE OF BOVID REMAINS AS INDICATORS OF

ENVIRONMENT AND PREDATION PATTERNS

Elisabeth S. Vrba

Introduction

To what extent can fossil mammals be used to interpret past environments and food-chain interactions? Antelopes (Family Bovidae) are notable for their taxonomic diversity and their abundance, relative to other large mammal groups, in Plio-Pleistocene African fossil assemblages. They have a long history of environmental, particularly vegetational, specialization. Bovids have probably also been a major food source of large African carnivores for many millions of years. Such characteristics make this mammalian group particularly useful for paleoenvironmental reconstruction. Of course, success often depends heavily on adequate estimation of differences between taphocoenoses (death assemblages) and biocoenoses (life assemblages, or communities).

Bovidae as Indicators of Environment

The evolution of species is a function of environment. The morphology, and secondarily the taxonomy, of animals should thus contain information about their ancestral and present-day environments. Further, the larger and more mobile the animal, the more likely such information will be applicable to large areas, i.e., macroenvironmental. Of the large extant mammals, probably none have radiated so spectacularly during the past 20 my as the Bovidae, both in terms of range of habitat and in numbers of species and individuals. When compared to other related groups, the family as a whole appears to represent exploitation of vegetationally more open environments. In an investigation of the Lower Pliocene Hipparion fauna of China from many different localities, Schlosser (1903) and Kurtén (1952) found two broad faunal divisions: woodland and steppe. Bovidae were found (Kurtén (1952, fig. 2) to predominate heavily over other groups in the steppe fauna, and Cervidae in the woodland fauna. From one perspective we might say that within the group those that "remained behind" in habitats not far removed from the ancestral forests can be distinguished from those that "went forth" to colonize the more open and/or arid environments. In each faction a wide range of behavioral, size, and other morphological adaptations evolved to enable bovids to exploit available nutritional opportunities.

Schlosser (in Zittel 1918) split all bovids into two groups, the Boodontia and the Aegodontia, on the basis of skull and dentition characteristics. The boodonts probably came first in bovid evolution. The earliest described bovid genus, *Eotragus* Pilgrim, a boodont boselaphine from the Burdigalian and Vindobonian of Europe, also ap-

pears to be present in Miocene deposits from South-West Africa (Gentry 1970a). The
Boodontia exhibit the greater number of primitive characters, and most modern species that
inhabit the more wooded and/or moist habitats belong to this group. To facilitate discussion, a rough scheme of bovid evolution, with particular reference to Africa, is presented in figure 14.1. It will be obvious to anyone familiar with bovids in Africa, or
indeed Europe or Asia, that most of the successful exploiters of open and/or dry environments are aegodonts. These taxa have adapted successfully to seasonal conditions with the
migratory habit and such attendant behavioral adaptations as "follower" as opposed to
"hider" young (Lent 1974). Successful exploiters of open and/or arid habitats are also
present among the Boodontia, but are outnumbered in this group by more cover- and/or
water-dependent forms.

On the face of it, then, we would expect the Bovidae to be useful for paleoecological interpretation of mid-Miocene to later African fossil assemblages. Certain
aspects of how this expectation can be realized are investigated in this study. It includes two approaches: an assessment of modern taxonomic and morphological correlation
with African habitats, and extrapolation from the modern analogy to fossil assemblages,
in particular those of the Transvaal limestone caves, in the Sterkfontein Valley near
Krugersdorp, Republic of South Africa.

Extant Bovid Taxonomic Separation in African Habitats

Table 14.1 represents the percentages of bovid tribes in sixteen African game
reserves, parks, and other areas. Table 14.2 is a list of these game areas, including
sources of the censuses used for table 14.1.

A correspondence analysis (Benzécri 1973) was applied to these census data. As a
fuller discussion of the application of this procedure to the present data will appear
elsewhere (Browne and Vrba, in preparation), I will here confine myself to a few remarks.
The purpose of the analysis was to provide a description of the table of frequencies of
tribes in game reserves by representing both tribes and game reserves as points in
euclidean space. The dimension of the space to be employed is determined subjectively by
inspecting the eigenvalues, and by choosing only as many dimensions as are interpretable.
Principal axes in the space are interpreted in much the same way as factors in factor
analysis. The relative sizes of the eigenvalues indicate the relative contributions of
the axes to the explanation of the patterns of occurrence in the table.

The first axis was found to separate the game areas decisively into two groups:
Ngorongoro, Lake Turkana, Etosha, Kalahari, Serengeti, Nairobi all less than zero, and
all others greater than 0.5. I have interpreted this first axis, and thus the major
contribution to the explanation of the patterns in the original data matrix, as representing degree of vegetational cover (height and spacing of bushes and trees). On this basis
the game areas in table 14.1 were separated into types A and B.

Type A areas enjoy a generally higher rainfall than type B areas. Exceptions
seem to attain their closed versus open vegetational characteristics because of temperature, altitude, etc. Thus Quicama, with the lowest rainfall among less open parks, is
low-lying and subject to frequent mists (Huntley, personal communication); Serengeti and
Ngorongoro owe their open character partly to high altitude. An ideal approach would be
to relate tree/bush height and percentage over to altitude, annual temperature and rain
distribution, and soil factors. The present simple division, however, does point out
striking bovid tribal separation between more and less open habitats:

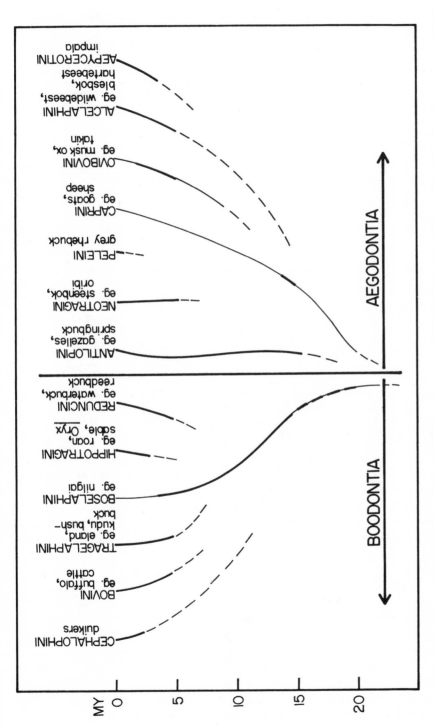

Figure 14.1. The evolution of bovid tribes, with particular reference to Africa south of the Sahara. A rough and tentative representation compiled from available literature (see publications of Gentry, Hendey, Pilgrim, Wells in reference list) and personal observation of such fossil assemblages as I have seen. Bold solid lines are drawn specifically on the basis of records from south of the Sahara (although such bold lines do not exclude presence in Europe and/or Asia). The bold dashes for early boselaphini indicate the probability that early African fossils like *Eotragus* Pilgrim from South-West Africa belongs here. Faint solid lines represent records in Europe and/or Asia. Broken faint lines are to my knowledge not supported by fossil occurrences anywhere in the world, but convey an impression of possible phylogenetic relationships. Classification after Simpson (1945), with modifications from Ansell (1971).

Table 14.1

Percentage Tribal Representation of Extant Bovidae in Some African Game Parks, Reserves, or Areas

Name of Area (Abbreviated from List in Table 14.2)	Bovid Tribes								
	Cephalopini	Bovini	Tragelaphini	Hippotragini	Reduncini	Neotragini	Antilopini	Alcelaphini	Aepycerotini
Type A: Areas with a Higher Proportion of Bush and Tree Cover									
Kruger	1	11	5	1	2	3		5	72
Manyara		66	1		2				31
Quicama	18	40	28	7	7				
Bicuar	18	4	17	18	6	13		18	6
Luando	14	2	12	39	30	3			
Cuelei	12		17	5	37	23		6	
Mkuzi	<0.5		5		1	<0.5		12	82
Hluhluwe	<0.5	15	21		7			24	33
Kafue	4	13	6	5	30	13		28	<0.5
Wankie	5	33	14	7	3	10		7	21
Type B: Areas with a Lower Proportion of Bush and Tree Cover									
Kalahari	1		9	23		2	35	30	
Nairobi	<0.5		3		3	<0.5	28	45	21
Lake Turkana				30			24	46	
Serengeti		7	1	1	1		24	58	8
Ngorongoro		<0.5	2		1		26	71	
Etosha	1		10	19		3	48	19	

Note: In the case of Serengeti I have added migratory, woodlands resident and plains resident bovid species of Schaller's (1972) table 32, and arbitrarily split his category "Other," containing 10,000 individuals, among roan plus *Oryx* (5,000), reedbuck (2,500), and bushbuck (2,500). Maximum percentages per area are underscored.

1. Antilopini are totally absent in all Type A areas here included. In Type B they are either the most abundant or second most abundant tribe in five out of six areas, while being third in the remaining area.

2. Peak abundance in Type A is always attained by a tribe other than Alcelaphini or Antilopini, except in Bicuar, where peak abundance is shared by Cephalophini, Hippotragini, and Alcelaphini.

3. Alcelaphines are the most abundant in four, and second most abundant in the remaining Type B area. The only alcelaphines recorded in Type A areas are *Connochaetes taurinus*, the most environmentally versatile alcelaphine species, and *Alcelaphus lichtensteini*. *A. buselaphus* and *Damaliscus* are absent from the Type A sample.

4. The only tribe which in its relative abundance rivals alcelaphines and antilopines in Type B habitats is the tribe Hippotragini. In fact, all Type B hippotragine records belong to the genus *Oryx*, apart from a small number of roan antelopes in the Serengeti. All Type A hippotragine totals include only the genus *Hippotragus*.

On the basis of the literature supporting table 14.1, and information from Pienaar (1974), Estes (1974), and Dorst and Dandelot (1970), the very much simplified scheme in table 14.3 was drawn up. This table again emphasizes the gravitation of alcelaphines and antilopines towards open and/or arid habitats. It also points out genera in other tribes which have successfully exploited open and arid environments:

1. *Oryx* and *Addax*: Of the genera belonging to tribes other than Alcelaphini and Antilopini in the lower right quadrant of table 14.3, only *Oryx* and *Addax* are specialized on open-arid environments to the near exclusion of other habitats. In table 14.1 the percentages of *Oryx* for each of the Kalahari and Lake Turkana areas are among the top three percentages. The question arises whether Hippotragini should be included among open/arid indicator tribes. This is unlikely to be successful for the following reasons. Hippotragini are more drastically split in terms of environmental indications than any other extant African bovid tribe. This is emphasized by table 14.1 where, in addition to the high representations in Type B areas, hippotragines achieve peak percentages at Bicuar and Luando of Type A. The tribe as a whole is known in the fossil record only since Plio-Pleistocene times (fig. 14.1). It is likely that *Oryx* and *Addax* lineages evolved from bush-loving hippotragines, but this is by no means clear from the fossil finds. In the absence of diagnostic horn-core and skull material, it may be difficult to distinguish which of the two environmental specializations is represented in early hippotragine material. However, in cases where *Oryx* and *Addax* fossils can be securely identified they should provide excellent indications of paleoenvironment.

2. *Taurotragus, Raphicerus,* and *Sylvicapra*: These genera are apparently exploiting a wide variety of habitats, and are therefore by themselves not diagnostic of rainfall and cover conditions.

Total (*abcd*) alcelaphine plus antilopine percentages for all weight classes in table 14.4 show no overlap between Type A and B areas, but are always less than 30% in the former, and more than 60% in the latter. This is quite a remarkable split which may not appear once additional game areas are included.

The present data support the hypothesis that alcelaphine/antilopine percentages, obtained by random sampling over all bovids present in a particular modern context, provide environmental information.

Correlation between Bovid Morphology and Environment

The fundamental variables, measurements of which should correlate environmentally, are the morphological, metabolic, and behavioral characteristics that result in taxonomic differentiation. Some morphological features which characterize aegodonts as a whole (fig. 14.1), particularly the progressive open-country forms, are (1) hypsodont teeth, (2) shorter premolar rows, (3) limb bones that are longer with respect to volume and show various other cursorial features, and (4) braincases that are "bent down" or angled with respect to the facial axis.

These developments are present among members of the Boodontia as well, as indeed they are in other groups such as equids, rhinocerotids, suids, etc. Among African bovid tribes, however, the combination of these features is nowhere as markedly and consistently expressed as in the alcelaphini and antilopini.

Hypsodonty, generally accepted as being an adaptation to a tough, fibrous diet, would be a useful environmental indicator if one could measure it accurately. Several East and South African fossil assemblages consist predominantly of dentitions. However, any meaningful comparison of hyposodonty can only be conducted among specimens of similar age/wear stage. Further, tooth crown height should be related to occlusal surface area (ideally the enamel area) and to the overall size of the animal. A study on modern

Table 14.2

Data on Game Reserves and Parks from Which Censuses Were Obtained

Name and Country	Area (km²)	Source of Census	Mean Annual Rainfall (mm)	Vegetation
Kruger National Park, South Africa	19,084	G. de Graaff (National Parks Board of Trustees), personal comm., 1975	400-600 in central and northern areas; 600-800 in south (Clark 1967)	Lowveld, Sour, Bushveld, Arid Lowveld, Mopani Veld, Mixed Bushveld in ratios 5:30:25:39:1 (Edwards 1974)
Manyara Park, Tanzania	91	Schaller 1972		Evergreen forests in north; *Acacia* woodland in center; riverine forest and scattered trees in south; marshes and alkaline flats along Lake (Makacha and Schaller 1969)
Quicama	9,960	All information on Angolan areas from B.J. Huntley (C.S.I.R., Pretoria), personal comm. 1976	350-600	Mangrove, floodplain, *Setaria welwitschia* grasslands, dry thicket and gallery forest ecosystems
Bicuar National Park, Angola	7,900		750	*Baikaiea, Brachystegia,* and *Julbernardia* woodland, grassland, dry thicket, and riverine ecosystems
Luando National Park, Angola	8,280		1200	Woodland, riverine forest, swamp forest, and floodplain grassland ecosystems
Cuelei Regional Nature Park, Angola	4,500		800	*Brachystegia* woodland, dry thicket, and riverine ecosystems
Mkuzi Game Reserve and Nxwala State Lands, South Africa	323 (Brooks 1975)	Obtained from J. Hanks (Natal Parks Board) and based on 1972 census by P. M. Hitchins (Natal Parks Board)	600-800? (Clark 1967)	Lowveld (Edwards 1974), foothills: open woodland; the rest: closed and open woodland (*Acacia*) mosaic (Brooks 1975)
Hluhluwe/Corridor/Umfolozi Game Reserve Complex, South Africa	960 (Bourquin et al. 1971)	As for Mkuzi	730 (Umfolozi) 985 (Hluhluwe) (from Bourquin et al. 1971)	Zululand thornveld (Edwards 1974) forest, thicket, woodland, tree savannah (Bourquin et al. 1971)

Location		Reference		Description
Kafue National Park (Ngoma area only), Zambia	91 (Dowsett 1966)	Dowsett 1966		*Baikaiea plurijuga* forest, thickets, *Terminalia/Burkea* woodland, grassland (Dowsett 1966)
Wankie National Park, Rhodesia		All information from J. Rushworth and B.R. Williamson (pers. comm.), Research Unit, Wankie National Park	640	Two-thirds of park: *Baikaiea* woodland and *Terminalia/Acacia* scrub; rest: mopane woodland, *Combretum/Terminalia* open woodland, some *Brachystegia* woodland, grassland less than 5% of park area
Kalahari Gemsbok National Park (South Africa), Kalahari Gemsbok Park (Botswana), and Mabua Reserve (Botswana); all continuous	36,692 (G.de Graaff, J. du P. Bothma, and E. Moolman, pers. comm.)	G. de Graaff, J. du P. Bothma, and E. Moolman, pers. comm.; maximum figures for each species recorded from March 1973 to April 1974 were used in this study	Most of area has 100-200; extreme northeast has 200-400 (Clark 1967)	Main part consists of arid shrub savannah: sparse vegetation, rolling dunes, widely scattered trees; in northeast there is Southern Kalahari bush savannah: rolling sandy country with flatter dunes, wide plains, depressions and pans, growth of tall trees, and shrubs on rises (Weare and Yalada 1971)
Nairobi National Park, Kenya	122 (Foster and Kearney 1967)	Foster and Kearney 1967 (1966 census)	506 (Stewart and Zaphiro 1963)	Rolling plains, wooded valleys, and a small forested area (Foster and Kearney 1967), scattered tree-grassland (*Acacia themeda*) (Stewart and Zaphiro 1963)
Lake Turkana, an area east of the lake between Allia Bay and the northeast corner of the lake, Kenya	2,050 (Stewart 1963)	Stewart 1963	200-400 ? (Clark 1967)	Census done on narrow grassy strip beyond which the terrain becomes abruptly one of harsh arid rock ridges or sand bearing desert scrub (Stewart 1963)
Serengeti Ecological Unit (as in Schaller 1972) Tanzania	25,500 (Schaller 1972)	Schaller 1972	772 (at Banagi) (Schaller 1972)	Some 5200 km^2 of grass plains with trees absent or limited to banks of streams; otherwise wooded grasslands with widely scattered trees (Schaller 1972)

Table 14.2—Continued

Name and Country	Area (km^2)	Source of Census	Mean Annual Rainfall (mm)	Vegetation
Ngorongoro Crater, Tanzania	360 (Kruuk 1972)	Kruuk 1972	750 (Kruuk 1972)	Mainly flat, treeless grasslands (Kruuk 1972)
Etosha National Park, South West Africa		All information from E. Joubert (personal communication), Nature Conservation and Tourism Windhoek	350	An arid tree and shrub savannah, with *Colophspermum mopane* one of the dominant trees

Table 14.3

Presence of Some Bovid Genera in Four Basic Types of Habitat

	Forest-Thicket-Woodland		More Open	
Moister	Cephalophus	Syncerus	Ourebia	Alcelaphus +
	Neotragus	Aepyceros ?	Redunca	Damaliscus +
	Boocercus	Taurotragus *	Kobus	Connochaetes +
	Tragelaphus	Sylvicapra *	Hippotragus	Taurotragus *
	Hippotragus	Raphicerus *	Syncerus	Sylvicapra *
	Kobus		Gazella +	Raphicerus
More Arid	Madoqua	Oryx	Oryx	Damaliscus +
	Aepyceros	Connochaetes ? +	Addax	Connochaetes +
	Tragelaphus	Alcelaphus ? +	Gazella +	Taurotragus *
	Syncerus	Taurotragus *	Antidorcas +	Sylvicapra *
	Litocranius +	Sylvicapra *	Alcelaphus +	Raphicerus *
	Gazella +	Raphicerus *		
	Antidorcas +			

Note: Errors have undoubtedly been introduced by this drastic reduction of the often-used, numerous forest-desert and moist-arid classificatory divisions. This table is meant to provide no more than a quick impression of habitat specializations of alcelaphini and antilopini. Genera belonging to these tribes are noted by pluses; asterisks mark genera which apparently occur in each of the four divisions. (Adjacent or successive placement of genera does not necessarily denote taxonomic relationship.)

springbuck (Vrba 1973) showed that, within a random sample of individuals of similar age, the hypsodonty index

$$z = \frac{100 \text{ x tooth crown height}}{\text{occlusal length x breadth}}$$

declines as skull length increases, and is probably allometrically related to body size.

Premolar length, as a ratio of molar length, is easy to determine. A probable explanation of the shortened premolar rows of many grazers (and/or browsers of tough vegetation) is that the greatest bite force is developed nearest the fulcrum of the jaw, which can be described as a second-class lever (Vrba 1970)--i.e., between teeth which are situated most distally with respect to origins and insertions of masseter and pterygoideus muscles. Selection in grazers would favor expansion and increasing hypsodonty of posterior teeth, while the less-used anterior teeth atrophy. Lending support to this suggestion is the fact that premolar reduction is generally accompanied not only by buccolingual molar expansion, but also by a distal (posterior) extension of metastyle and metastylid of M^3 and M_3, respectively. Thus, there is a suite of dentition characters, to which can be added diastema length and height and angulation of the ascending mandibular ramus, which should give excellent indications of feeding behavior and environment.

Gentry (1970a) has compiled a useful and comprehensive list of cursorial features of bovid limb bones. His "Feature 4," the angulation of the braincase with respect to the long axis of the face, could be correlated with specialized head support necessary either for cursorial locomotion or for intraspecific horn fighting. Schaffer and Reed (1972) discuss how in Caprini an emphasis on head-to-head butting or ramming, as a major form of intraspecific competition, is associated with alteration in the shape and relative propor-

Table 14.4

Percentages of Extant Alcelaphines Plus Antilopines from Some African Areas within Bovid Weight Classes, and within Combinations of Weight Classes

Name of Area	Weight Classes									
	a	b	c	d	ab	bc	cd	abc	bcd	abcd

Type A: Areas with a Higher Proportion of Bush and Tree Cover (Figures in *ab, bc, cd, abc, bcd, abcd* which Exceed 40% are Underscored).

Name of Area	a	b	c	d	ab	bc	cd	abc	bcd	abcd
Kruger	0	0	42	0	0	6	21	5	5	<.05
Manyara	0	0	0	0	0	0	0	0	0	0
Quicama	0	0	0	0	0	0	0	0	0	0
Bicuar	0	0	40	0	0	33	31	21	27	19
Luando	0	0	0	0	0	0	0	0	0	0
Cuelei	0	0	17	0	0	10	15	6	9	6
Mkuzi	0	0	87	0	0	12	87	12	12	12
Hluhluwe	0	0	66	0	0	28	47	28	24	24
Kafue	0	0	59	0	0	44	46	33	35	28
Wankie	0	0	27	0	0	15	11	11	8	7

Type B: Areas with a Lower Proportion of Bush and Tree Cover (Figures in *ab--abcd* which are Less than 60% are Underscored).

Name of Area	a	b	c	d	ab	bc	cd	abc	bcd	abcd
Kalahari	0	100	56	0	91	74	48	71	66	64
Nairobi	95	43	94	0	56	71	68	74	70	73
Lake Rudolf	?	100	60	0	100	70	60	70	70	70
Serengeti	100	13	98	0	73	86	88	89	78	83
Ngorongoro	100	96	99	0	99	99	87	99	96	97
Etosha	0	100	40	0	92	71	39	68	70	67

Note: Weight estimates for individual species were obtained by calculating median values from Brain's (1974) weight ranges, and in cases of species not included in this reference, from Dorst and Dandelot's (1970) ranges. Cube roots of weight estimates were calculated.

Weight class *a*:   (1-3) $\sqrt[3]{kg}$ Weight class *b*:   (3-5) $\sqrt[3]{kg}$
Weight class *c*:   (5-7) $\sqrt[3]{kg}$ Weight class *d*:   (7-9) $\sqrt[3]{kg}$

tions of the skull. This is a feature that is more pronounced among aegodonts than boodonts, particularly in some alcelaphines.

Unfortunately, an objective assessment of how these characters relate to environment is at present impossible. None of the four features has, to my knowledge, been measured on adequate samples of modern bovid species. They have, however, been used very successfully by paleontologists: Gentry (1970a) pointed out limb bone and dentition differences between caprine (aegodont) and boselaphine (boodont) elements at Fort Ternan, showing that at this early time in African bovid evolution the split in environmental adaptations was clear. Kurtén (1952) showed that the woodland faunal elements of Chinese Hipparion levels include more brachyodont dentitions among the bovidae, and among Giraffidae and Rhinocerotidae as well, while species of the same families in the steppe fauna are generally more hypsodont. Simpson (1944) assessed comparative rates of evolution in horse lineages by hypsodonty measurements.

Environmental Indications in Fossil Assemblages: General

On the basis of the data in table 14.1 it appears that the calculation of bovid taxonomic representation in fossil samples may provide useful environmental information. The use of the alcelaphine/antilopine criterion should be restricted to late Pliocene to Pleistocene assemblages (fig. 14.1). In earlier fossil samples other contrasts such as aegodont versus boodont, or caprine plus antilopine versus boselaphine percentages can probably be usefully employed. Morphological contrasts should be used wherever possible.

The necessary prerequisite for the use of either a taxonomic or morphological criterion is a fossil sample which is unbiased in its proportions of environmental indicators of the original community. Information essential for successful assessment of paleoenvironment on the basis of fossil assemblages includes the following.

1. The time sampled. Assemblages accumulated over very short periods may result in distorted environmental indications. Fossil samples representing single events in time such as catastrophic death through drought, flood, epidemic, etc., are of this kind. The same would be true of seasonally accumulated assemblages, e.g., the debris of seasonal hominid occupation sites. Ideally, the time sampled should be long enough to ensure that catastrophic inclusions and seasonal patterns cannot dominate the "normal" long-term pattern. On the other hand, a long-term accumulation which straddles a climatic/environmental change not readily recognizable in the stratigraphy will lead to mixed environmental information. This is perhaps more likely to occur in cave fills than in other sedimentary contexts.

2. The area sampled. While macroenvironmental indications are more informative (for the purposes of this study) than microenvironmental ones, the area sampled should not be so large that distinctly different climate/habitat zones are included. The autochthonous or allochthonous character of an assemblage should be known before attempting paleoenvironmental reconstruction.

3. The direction of sampling bias. Did all the transporting/accumulating and other taphonomic biases sample randomly from proportions of environmental indicator animals in the living community? This would result in an undistorted fossil sample drawn in true proportion to taxonomic abundance in the original community. Alternatively, if the bias results in disproportionate inclusion of one or another indicator group, this would lead to a sample which cannot be readily interpreted in terms of paleoenvironments.

4. Paleoenvironmental comparability between samples. Even if factors 1-3 should all be optimal, and even if the morphologic and/or taxonomic criteria employed are valid, we also need to know about possible time differences between samples that are to be compared. It is obviously easiest to sort out paleoecological differences when the fossil samples are more or less contemporaneous. In cases where there is no good reason to assume contemporaneity, we are faced with a major problem: Is the observed degree of difference between samples the result of time, environment, or both? In any one area, ideally, we might expect to record progressively more hypsodont or cursorial features, and possibly also taxonomic changes, over a time span during which the environment has not changed. In this case, low faunal similarity would be due only to time differences. However, low faunal similarity coefficients can occur between even closely spaced faunas living at the same time due to ecological differences (e.g., modern Wankie and Kalahari censuses in table 14.1). Attempting to correlate such similarity coefficients only with either time or environment is thus potentially fallacious.

	Time Influence Absent	Time Influence Present
Environmental Influence Absent	similarity maximum	potential similarity due to equivalent environments
Environmental Influence Present	potential similarity due to equivalence in time	similarity minimum

Considering the quadrants of the above diagram as endpoints in a spectrum of time-environment influence on faunas, one has to assess where any particular fossil sample may lie within this system before interpreting faunal similarities or differences in terms of their paleoecological or evolutionary significance.

Environmental Indications in Fossil Assemblages: The Particular Case of Sterkfontein Valley (Krugersdorp) Cave Accumulations

The Transvaal australopithecine sites, Sterkfontein, Swartkrans, and Kromdraai, are situated in the Sterkfontein Valley not more than a kilometer apart. For background on the area and the cave fills the reader is referred to Brain (1958). Within each cave, breccias representing more than one time level (site units) have been identified. The assemblages I shall discuss here are from the following sites.

STS - Sterkfontein Type Locality; Member 4 (Partridge 1978)

SKa - Swartkrans Member 1 (Brain 1976), formerly called the "pink" breccia (Brain 1958)

KA - Kromdraai Faunal Site

SE - West Pit of Sterkfontein Extension Locality; Member 5 (Partridge 1978)

SKb - Swartkrans Member 2 (Brain 1976), including what was formerly known as "brown" breccia (Brain 1958), as well as fills from channels forming at a relatively late stage through both "pink" and "brown" breccias (Brain, Vrba, and Robinson 1974).

An estimate of the time relationships of these site units, based on bovid fossils (Vrba 1974, fig. 1), places them chronologically in the sequence (with STS earliest) given above and elsewhere in this paper.

I want to discuss to what extent the four factors mentioned above are likely to affect paleoenvironmental reconstruction on the basis of the bovid assemblages from these cave breccias.

1. The time sampled. Reliable estimates of the duration of accumulation are not available for any of the breccias or site units here discussed. It is fairly certain, however, that the time sampled in each case is long enough to reduce to insignificance the effects of possible catastrophes or seasonal patterns in the final sample. I hope to discuss elsewhere the possible existence of discrete environmental and temporal patterns within the SKb bovid sample. The time factor is probably optimal for environmental reconstruction in most of the site units.

2. The area sampled. As bovids do not live in caves, the vast majority of their remains must have been carried there by other animals. These could have been either primary predators (e.g., carnivore or hominid), scavengers (e.g., carnivore or hominid), or collectors (e.g., porcupine, hominid). (Refer to table 14.5.) Primary predators other than leopards do not generally move prey for any appreciable distance from the place of its death. Where such records exist, distances never exceed a few hundred meters, including records for leopards (Prof. F. C. Eloff, personal communication; Schaller 1972; Pienaar 1969; Turnbull-Kemp 1967). Records for hyenas are occasionally in the vicinity of a few kilometers (e.g., Sutcliffe 1970). The Dobe area bushmen of Botswana occasionally go on hunting trips up to 30 km from their home bases (Lee 1965). For the vast majority of fossils from any of the cave breccias, an area of live origin not exceeding a 10 km radius around the Sterkfontein Valley can probably be safely assumed. From the point of view of environmental reconstruction, accumulations of bones in caves can be regarded as optimal.

Table 14.5

Classification of Major Transport Mechanisms Responsible for the Accumulation of Fossil Assemblages, with Particular Reference to the Assemblages of the Sterkfontein Valley

TRANSPORT MECHANISMS				STERKFONTEIN VALLEY ASSEMBLAGES
Mechanical transport	water, mudflow, wind, etc.			
Animal transport	autopod (remains "arrive on their own feet")			
	allopod (remains "arrive on other feet")	primary	Carnivora	STS SKa KA ?
			Hominidae	SKb ?
		secondary	scavengers — Carnivora	
			scavengers — Hominidae	SE
			collectors — Hominidae	
			collectors — Other	

3. The direction of sampling bias.   It seems reasonable to suppose that bone collections made by hominids or other animals would reflect the environments represented in the death assemblages from which they are drawn. Studies of modern carnivore predation have shown that the most important factors influencing prey selection appear to be size and avail- ability (Schaller 1972; Kruuk 1972; Pienaar 1969). Primary predation should produce re- mains which are markedly sorted with respect to size, the size of the predator being the determining factor. Predator preferences for particular prey species are often recorded in the literature. However, they seem to vary for one predator species from area to area and ultimately reflect availability in that area of prey species of the particular size range preferred by that predator. Thus in Manyara Park the most frequently occurring animal in lion food is the buffalo (62%); in the Masai and Seronera areas of Serengeti it is Thomson's gazelle (50%) (Schaller 1972). In the Kruger Park lions concentrate mainly on impala in the southern district, and on the zebra in the northern mopani-veld, although the most abundant prey species overall in the park is wildebeest (Pienaar 1969). A close look at the prey community in these various areas suggests that prey size and availability remain the most important factors. If the dominant processes in a bone accumulation can be specified as predation, then, given the long time periods represented in the cave site units, the bone samples will very likely record prey size and availability. Availability is precisely what we need to know in order to interpret paleoenvironment; prey size, however, is not.

Between the death of animals and the interment of their remains in the cave breccias, numerous taphonomic factors intervene which in most cases constitute further size selection. Thus damage through scavenging, creatures walking over bones, weathering, etc., probably all produce bias against small and/or juvenile forms in the assemblages we find.

4. Paleoenvironmental comparability between samples.   Large assemblages predating STS are known in South Africa. From both Langebaanweg and Makapansgat Grey Breccia, alcela- phines and antilopines are recorded (Gentry 1970b; Hendey 1969; Wells and Cooke 1956), suggesting that the taxonomic ratio test may be fruitfully applied in the Sterkfontein Valley context. The problem of whether we can detect any effects of environmental change in the assemblages which can be distinguished from those due to time only, using this criterion, is a knotty one and will be discussed below.

We can conclude that, with respect to factors 1 and 2, the Transvaal cave as- semblages offer relatively good opportunities for broad-scale paleoenvironmental recon- struction. In these respects each fossil sample is, as expressed by Behrensmeyer (1975), "a fossil assemblage composed of bones from a fragmented thanatocoenose that has not been transported" (at least not substantially) and "should preserve the best evidence for the paleoecology of the fauna." The major problem under the heading of point 3 seems to be one of size selection, and I now wish to discuss how this interferes with environ- mental reconstruction in the Sterkfontein Valley context.

Does Size Selection Produce Environmentally
Biased Samples?

I shall test this against the alcelaphine-plus-antilopine criterion, and parti- cularly from the point of view of predator size selection.

Figure 14.2 shows the distribution of extant southern African species throughout the bovid weight range. Alcelaphines and antilopines have been separated from other tribes to show how their size and weight distributions compare. They are fairly evenly distri-

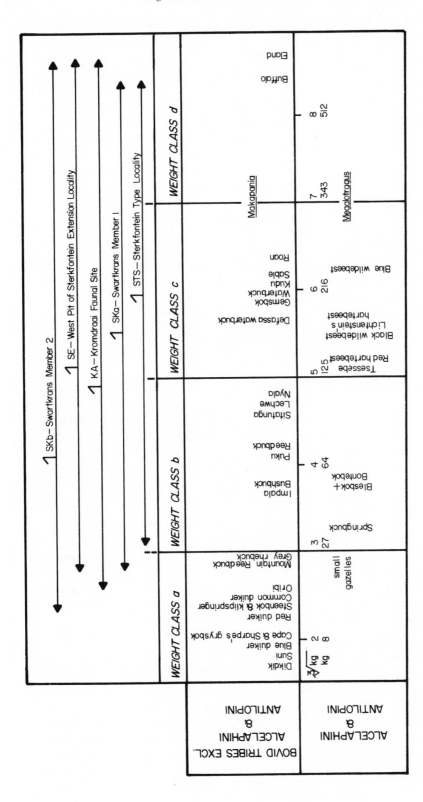

Figure 14.2. Weight distribution in extant southern African bovid species. Three extinct forms, present in South African fossil assemblages, which extend the given range of mean weights, are included in horizontal print. Estimates are based on medians obtained from adult weight ranges in Brain (1974) and Dorst and Dandelot (1970). Estimated mean weights and ranges of some Krugersdorp fossil bovid assemblages are included.

buted throughout weight class *b* and *c*, but are absent from modern southern African weight
classes *a* and *d*.   Gazelles of weight class *a* exist today in other parts of Africa,
and are known from several South African fossil assemblages. The large extinct alcelaphine
genus *Megalotragus*, known from East and/or South African fossil assemblages from Plio-
Pleistocene until late Pleistocene times (Klein 1974), would have extended the available
alcelaphine-antilopine size range into weight class *d*. From figure 14.2 one might expect
size-selected samples with means in weight classes *b* and *c* to truly reflect abundances
in the original community. Table 14.4 tests this expectation more stringently. Total
(*abcd*) alcelaphine-plus-antilopine percentages in the various game areas are contrasted
with the same percentages calculated for individual and combined weight classes *a–d*. It
is clear from table 14.4 that a selection mechanism that restricts itself to a single
weight class, especially weight classes *a* and *d*, may produce a sample to which the
alcelaphine/antilopine criterion cannot be applied. When a combination of adjacent weight
classes is sampled, however, only two Type A percentages fall above 40%, while only three
Type B percentages fall below 60%.

　　　　How do actual predation samples compare in terms of weight distribution with
that present in the living community? Table 14.6 gives relevant figures from the Kruger

Table 14.6. *Percentages of Bovid Prey Availability and Predator Sampling in Weight
Classes a–d in the Kruger National Park (Calculated from Pienaar 1969)*

	Weight Class			
	a	*b*	*c*	*d*
Live bovid population (census Pienaar 1969)	1	71	17	11
Lion	<0.5	26	61	13
Leopard	4	87	9	<0.5
Wild dog	2	90	8	<0.5
Cheetah	3	77	20	<0.5
Spotted hyena	1	64	33	2
Brown hyena	1	25	72	2
Total prey sampling percentage	2	56	36	6

National Park. Although individual predation patterns overemphasize either weight class
b or *c*, the combined weight distribution of all kills deviates from that in the live
population by less than 20%. Table 14.6 shows two cases in which alcelaphine-plus-
antilopine percentages are essentially similar in carnivore food samples and in the live
communities from which these are derived.

The Environment of the Sterkfontein Valley during
STS, SKa, KA, SE, and SKb Times

　　　　Figure 14.3 shows the distribution of bovids in weight classes *a–d* at the relevant
site units. In each case the alcelaphine-plus-antilopine content is indicated. Figure
14.4 represents the overall alcelaphine-plus-antilopine percentages as well as hominid and
cultural associations, as currently interpreted. SKa, KA, and SE percentages are over 80%.
That of the "brown" breccia component of SKb would be over 80% as well, if one could
remove a later, Middle Stone Age, bush-loving bovid component. In figure 14.3, highest
and second-highest numbers of bovids belong to either weight class *b* or *c* in each case.

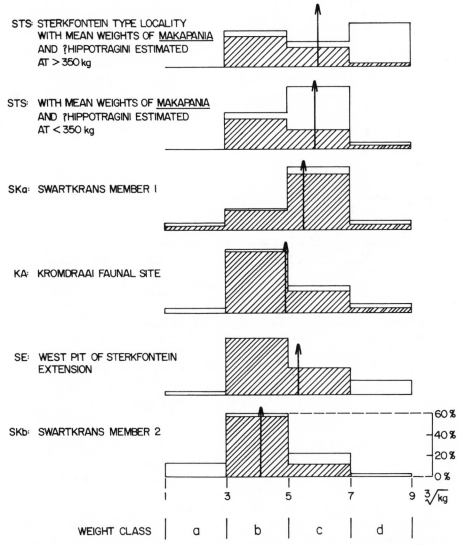

STS: STERKFONTEIN TYPE LOCALITY
WITH MEAN WEIGHTS OF <u>MAKAPANIA</u>
AND ?HIPPOTRAGINI ESTIMATED
AT > 350 kg

STS: WITH MEAN WEIGHTS OF <u>MAKAPANIA</u>
AND ?HIPPOTRAGINI ESTIMATED
AT < 350 kg

SKa: SWARTKRANS MEMBER I

KA: KROMDRAAI FAUNAL SITE

SE: WEST PIT OF STERKFONTEIN
EXTENSION

SKb: SWARTKRANS MEMBER 2

60%
40%
20%
0%

1   3   5   7   9 $\sqrt[3]{kg}$

WEIGHT CLASS | a | b | c | d |

Figure 14.3. Histograms and means of estimated bovid weights in some Krugersdorp
(Sterkfontein Valley) fossil assemblages; cross-hatching indicates alcelaphine-plus-
antilopine component.

There are six or seven alcelaphine-plus-antilopine species present, constituting 40% or
more of the total number of species identified. The only case in Type A areas (table
14.4) where sampling in weight classes *b* or *c* only, or *b* and *c* combined, resulted in
alcelaphine-plus-antilopine percentages in excess of 80% (indeed, over 50%) is that
of Hluhluwe. As mentioned before, there is only one alcelaphine species in this area,
and this species attains a higher abundance in weight class *c* than other species. I
conclude on the basis of this evidence that SKa, KA, SE (and probably the "brown"
breccia component of SKb) represent phases of very little bush cover in the Sterkfontein
Valley, probably comparable to the situation there today, and to that in Type B (table
14.1) areas.

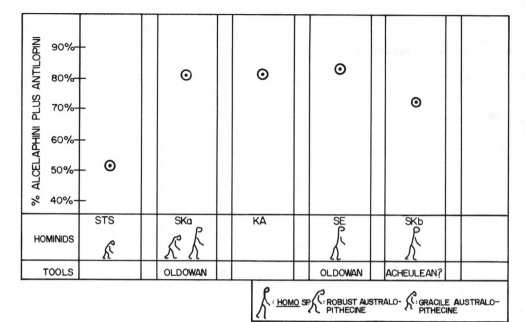

Figure 14.4. The percentage constituted by the added minimum numbers of alcelaphine and antilopine individuals of the total minimum number of bovid individuals, per site unit. Solid line hominid figures in the cases of STS, SKa, and SKb express widely accepted taxonomic evaluations; the SE record is based on the recent recovery of a skull that probably belongs to the genus *Homo* (Hughes and Tobias 1976) from this breccia; cultural associations of SKa and SE as in Leakey (1970), of SKb according to Brain (personal communication). (Adapted from Vrba 1976, fig. 19.)

The STS overall percentage of 51% contrasts markedly with the others in figure 14.4. The difference in percentages between STS and SKa is statistically highly significant (Vrba 1975). If weight class *a* bovids were as well represented at STS as they are at other sites, this would further depress the percentage. Nonetheless, about half of STS species belong to either alcelaphines or antilopines. This fact, together with an overall 51% obtained by sampling either combined weight class *bcd* fairly evenly, or mainly *bc* (figure 14.3), would seem to indicate that there was decisively less bush cover at STS times than is present in any Type A areas of tables 14.1–14.4. The Sterkfontein Valley environment apparently belonged in the vegetationally open part of the African habitat spectrum during all known fossil deposition phases. Seen from this broad perspective, these bovid data seem to corroborate the view (Butzer 1971) that each of the breccias in question basically represents colluvial sediments compatible only with incomplete vegetation cover. Was there, within this overall environmental characterization, a difference between STS and SKa habitats? Considerable interest centers on the answer to this question, as the STS and SKa hominids are different.

The two taxa which are mainly responsible for the low alcelaphine-plus-antilopine percentage at STS are *Makapania* cf. *broomi* and what I have called ?Hippotragini (Vrba 1975, 1976). *Makapania* is an extinct genus and the other specimens probably also belong to an extinct genus (or perhaps two extinct genera). It is very difficult to predict weight and habitat preference of these taxa from the scrappy dental remains available at STS. On tooth size they are larger than the blue wildebeest but much smaller than buffalo or eland. I estimated their weights to lie between 300 and 400 kg. The median,

350 kg, is almost exactly situated on the boundary between weight classes *c* and *d* (343 kg). This is why STS is shown by two alternative histograms in figure 14.3. The carnivore predators present in the STS assemblage include *Megantereon*, a true saber-toothed cat; *Dinofelis*, a false saber-toothed cat; and a leopard (Ewer 1955, 1956). These did not differ substantially in size from the array of modern predators in table 14.7. Table 14.7 shows that the largest predator, the lion, did not sample weight class *d* out of

Table 14.7

Alcelaphine-plus-Antilopine Percentages in Carnivore Kill (or Kill plus Scavenge) Samples, as Contrasted with Alcelaphine-plus-Antilopine Percentages of Live Bovid Populations in the Serengeti and in the Kruger National Park[+]*

	Percent of Alcelaphines-plus-Antilopines
Serengeti (Census from Serengeti Ecological Unit; Kill plus Scavenge Data from Serengeti Park)	
Live bovid population (as in table 14.1)	82
Lion (kill plus scavenge)	90
Cheetah (kill)	98
Leopard (kill)	87
Wild dog (kill)	97
Percentage in added (kill plus scavenge) samples	92
Kruger National Park	
Live bovid population[+]	10
Lion (kill)	31
Cheetah (kill)	6
Leopard (kill)	2
Wild dog (kill)	<0.5
Spotted Hyena (kill)	12
Brown hyena (kill)	4
Percentage in added kill samples	16

*Calculated from Schaller (1972).

+Calculated from Pienaar (1969).

proportion to its actual live presence in the Kruger National Park. Similarly, lion predation figures from Manyara in Schaller (1972) indicate that 62% of animals consumed were buffalo, which constitute 66% of the live prey community. From all this it appears likely that the high percentage of the larger-than-wildebeest *Makapania* and ?Hippotragini in the fossil sample may reflect a comparable high abundance in the living community at STS times (provided the supposition that this is predominantly a carnivore assemblage is correct; see below). It is worthy of note that bovids larger than the blue wildebeest never exceed 10% of all bovids in any of the Type B open areas which form the bases for tables 14.1 and 14.4. In Type A bush environments, however, they generally exceed 10% (except at Cuelei, Luando, and Mkuzi) and sometimes reach very high proportional represent-ation (e.g., 66% at Manyara, 53% at Quicama). This is not a very conclusive indication of STS habitat. I hope to examine the limb bones of these two STS species for cursorial characters. The dentitions are certainly less hypsodont than those of contemporaneous alcelaphines, while the premolar rows are longer. Unfortunately, this cannot be quantified as the relevant specimens are in poor condition.

A 2x2x4 χ^2 complex contingency table analysis was performed on the STS and SKa bovid data, to determine which of the factors S = site (SKa, STS), W = weight (four weight classes), and T = tribe (alcelaphines plus antilopines, other bovid tribes) are dependent. To obviate the difficulty of estimating weights for the two crucial large STS species, the weight classification for the test groups together individuals of (2-4) $3\sqrt{kg}$, (4-6) $3\sqrt{kg}$, (6-8) $3\sqrt{kg}$, and (8-10) $3\sqrt{kg}$. It is certain that *Makapania* and ?Hippotragini fall comfortably within the (6-8) $3\sqrt{kg}$ group. The results are given in table 14.8.

Table 14.8

Results of a 2x2x4 χ^2 Complex Contingency Table Analysis of STS and SKa Bovid Data

Component	Degrees of Freedom	χ^2 (Parameters Estimated)
S-W independence	3	2.64 n.s.
S-T independence	1	12.23*
T-W independence	3	28.11*
S-T-W interaction	3	28.43*
Totals (*S-T-W* independence)	10	71.41*

*Highly significant

In this test, weight was not found to differ significantly between STS and SKa. The two bovid assemblages, however, differed significantly with respect to alcelaphine-plus-antilopine content. Further, weight and tribes were found to be significantly dependent. A 2x2 χ^2 contingency table 2 test (corrected for continuity) was then performed to determine independence between sites and tribes, including only weight classes a and b. The resulting difference between STS and SKa was not significant ($\chi^2_1 = 1.94$; $\chi^2_{1;0.95} = 3.841$). An analogous test using only weight classes c and d indicated that SKa has a significantly larger alcelaphine-plus-antilopine component ($\chi^2_1 = 1081$).

The bovid evidence available on paleoenvironmental differences between STS and SKa is inconclusive. The most that can be said is that the several available indications all seem to favor a greater vegetation cover during STS deposition times. Further indications in this direction include the following: large alcelaphines (of about *Connochaetes*, i.e., blue wildebeest, size) form a markedly low proportion of the STS assemblage (1 out of 43 bovid individuals, or 2%) when compared with that in the SKa assemblage (19 out of 80 individuals, or 24%). Yet the weight estimate for this size group is very close to the STS weight mean estimate (fig. 14.2). *Connochaetes* has been reported as one of the four most abundantly represented bovid species from the Makapansgat Grey Breccia (Wells and Cooke 1956), which is generally taken as preceding STS closely in time. The modern alcelaphine species of this size, the blue wildebeest, is characteristically very abundant in all the Type B open country areas, but is uncommon, if present at all, in more bush- and tree-covered Type A game reserves. If we accept that predation patterns underlie the STS assemblage, and assume a long time span for accumulation of the assemblage, size and availability of bovids are more likely to be recorded than specific predator preferences out of the context of prey size and availability. Therefore, I regard the marginal presence at STS of *Connochaetes*-sized alcelaphines as ecologically significant.

The time gap between STS and SKa accumulations is not known. Most current estimates are less than 1 my. Yet a major change in bovid tribal representation took place over this time period. The STS *Makapania* cf. *broomi* probably represents the last record of the tribe Ovibovini in Africa. The peak abundance of ?Hippotragini at STS is reduced to a marginal representation (one specimen) in the SKa assemblage. Although a proportion of this representational discrepancy may be due to the mean weight difference between STS and SKa bovid assemblages (fig. 14.2), the conclusion is inescapable that a replacement of *Makapania* and ?Hippotragini occurred, by SKa forms like *Connochaetes*, in the Sterkfontein Valley environment. Such replacement is suggestive of environmental change. The whole faunal change between STS and earliest Swartkrans deposition may on closer scrutiny deserve the term "community succession" (see Olson, this volume).

It seems probable that the STS gracile australopithecines lived in an environment that was essentially open (somewhere between the modern Type A and B habitats of tables 14.1 and 14.4), but with a proportionally greater bush and tree cover than succeeding robust australopithecine phases in the Sterkfontein area. To what extent such a vegetational change could be due to rainfall and/or temperature changes will not be discussed here.

Bovidae as Indicators of Predation Patterns

The minimum numbers of bovid individuals, on which conclusions of both environmental and predation patterns were reached, are based on cranial fossil material only. Klein (1975, table 4) has pointed out that in several fossil assemblages the proportions of postcranial to cranial elements differ between large and small bovid species. However, Brain (personal communication) has informed me that, certainly in the vast majority of bovid species, and maybe all, from STS, SKa, SKb, KA, and SE, postcranial elements are less frequently preserved than cranial ones. This means that the ultimate minimum numbers of individuals, based on future tallies of cranial plus postcranial material, will still be based on the more frequently occurring cranial remains. The paleoenvironmental conclusions reached above and elsewhere (Vrba 1974, 1975, 1976) are unlikely to be affected once the ratios of postcranial to cranial elements are accurately calculated. Such information is, however, essential to the prediction of past accumulation patterns and agents. The comments and questions that follow are based on bovid cranial material only and await the analyses on all faunal and skeletal elements of Brain (in preparation).

In previous writings (Vrba 1975, 1976) I set up the model shown in table 14.9 for interpreting adult-juvenile and weight data of fossil bovids in terms of accumulation patterns.

In figure 14.5 the relevant data for STS, SKa, KA, SE, and SKb are given. Some data on adult-juvenile proportions in modern bovid populations may be compared with figure 14.5a. In all cases the correspondence of "juveniles," as used here, and "immatures" or "subadults," as cited, was checked. Herbert (1972) found that, in a population of waterbuck, *Kobus ellipsiprymnus*, in the Sabi-Sand Game Reserve, about 33% of the population was composed of immatures or young animals (59% adult and 8% unclassified). In age counts of *Gazella thomsoni* from the Serengeti National Park, Hvidberg-Hansen and de Vos (1971) found the average percentage of subadults to be 25% (i.e., adults make up 75%). Dasmann and Mossman (1962) did age counts on some ungulate populations in Southern Rhodesia and observed the following (p. 268): "The pattern observed for kudu, waterbuck, impala, duiker and steenbuck was that well-situated populations consist of 40-50 percent immature animals (under 2 years of age), and 22-40 percent young-of-the-year. Among

Table 14.9

Model for Interpreting History of Bone Accumulations Using Adult-Juvenile and Weight Data

Stage 1

Bovids are killed by predation in the vicinity of the cave, or die a natural death in or near the cave.

Stage 2

Remains of dead animals are brought into the cave to eventually constitute:

PRIMARY ASSEMBLAGES, those brought into the cave, which serves as a lair or shelter, by the primary predators (hominid hunters are included as a subset of the set of predators throughout).	SECONDARY ASSEMBLAGES, those brought into the cave by scavengers or collectors.
Unlikely to contain a low percentage of juveniles.	Likely to contain a low percentage of juveniles.
Most individuals fall into a restricted body weight range (only where one predator or predators preying on prey of similar size predominate(s)).	The body weight distribution may have a high variance.

zebra, giraffe, bushbuck, wildebeest, sable and reedbuck, the percentage of immature animals was less, varying from 25 to 40 percent." Juveniles were found to constitute 20.4% of a black-faced impala (*Aepyceros petersi*) population from the Kaokoveld in South-West Africa, and 21.2% of the total impala population (*A. melampus*) in the Mkuzi Game Reserve, Natal (Stewart and Stewart 1966, quoted in Joubert 1971). Juveniles of smaller species are often eaten totally, or almost totally, by predators (Pienaar 1969). In view of this and other taphonomic factors which militate against the preservation of juveniles in fossil assemblages, the percentages in figure 14.5a for STS, KA particularly, and also those for SKa and SKb, can be regarded as high when compared with average annual juvenile percentages in live populations. A very likely source of such high juvenile representation is primary predation.

I have suggested (Vrba 1975, 1976) that the low juvenile percentage at SE might reflect a predominantly scavenged origin of that bovid assemblage. I have found very little in the literature to test my hypothesis that scavenged assemblages should generally contain lower percentages of juveniles than primarily predated ones. Kruuk presents hyena scavenging and killing totals of adult and juvenile wildebeest, zebra, and gazelle in Serengeti and Ngorongoro (Kruuk 1972, table 22). Most of these data (excepting zebras at Serengeti) indicate that fewer juvenile than adults are scavenged by hyenas, while the reverse is apparent in the killing totals.

While it is clear from modern predation studies such as used in table 14.6 that

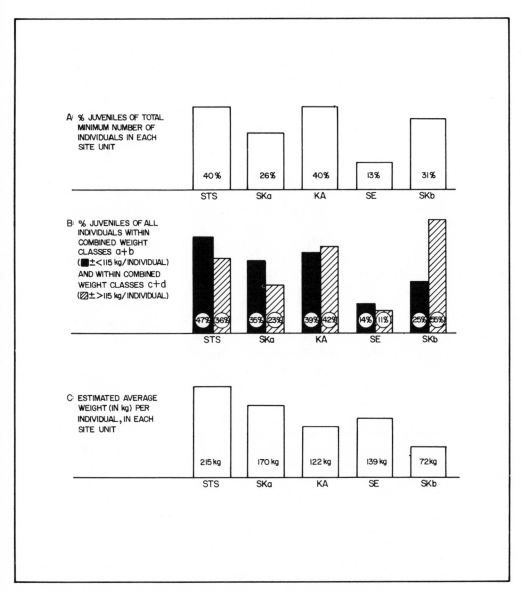

Figure 14.5. Changes in average body weight and in juvenile proportions in Krugersdorp Valley assemblages; weight estimates as in fig. 14.2. (Adapted with slight alterations from Vrba 1976, fig. 18.)

kill samples show high peaks in particular bovid weight categories (i.e., a low weight variance), I am no longer as confident as I was (Vrba 1975) that high peaks accompanied by low variance in fossil bovid weight histograms necessarily indicate primary predation. In table 14.6 the live population of the Kruger Park shows a high peak in weight class *b*. In other censuses of game areas available for this study, similar peaked weight distributions of the live bovid communities are evident. A predation peak only becomes definitive when contrasted with the weight distribution of the underlying live population, which is unavailable in the case of fossil assemblages. Further modern contrasts between kill samples and living communities are required to establish the weight characteristics that could be used to separate kill samples from death assemblages in which predation has had no part.

Nonetheless, the histograms of figure 14.3 (except the first alternative for STS) do correspond well to weight distributions observed in modern kill samples of different predators. Taking these together with the juvenile percentage data, it does seem likely that the fossil bovid samples, except SE, may represent primary predation assemblages, with possible scavenging and/or collecting influences relegated to minor importance. The phenomenon that juvenile percentages do not increase in higher weight classes in STS, SKa, and KA (fig. 14.5b) is not characteristic of single predatory patterns. Tables 19-28 of Pienaar's (1969) Kruger National Park study, for instance, indicate a steady (not invariable, but general) increase in juveniles as the predators sample increasing prey sizes, from impala to wildebeest and zebra. The first three histograms of figure 14.5b may reflect the overlap of larger and smaller prey preference patterns, e.g., saber-tooth and leopard. It is interesting that average bovid weights steadily decline with time, from STS to SKa to KA to SKb. There is no reason to believe that average weights of the corresponding live populations were doing the same. This decline may reflect the decreasing role of false and true saber-toothed cats (which indeed decrease correspondingly among carnivore fossils) in favor of carnivores of smaller prey preference, like leopards and hyenas, or hominids.

The contrast between the SE and STS bovid assemblages is great, both in terms of juvenile proportions in figure 14.5a and weight distribution in figure 14.3 (the latter referring only to the location of the weight peak). It seems highly likely that the bovid fossils at SE resulted from food remains of the hominids that left hundreds of stone tools interspersed with the fossils. If the criterion of high versus low juvenile percentages is accepted, then the Sterkfontein assemblages, STS and SE, reflect an earlier primary predator occupancy and a later scavenger occupancy of the same cave. Was *Australopithecus* the STS cave occupant, and could the assemblage largely represent his food remains? Such a hypothesis must suppose that the earliest, smallest hominid found in the Sterkfontein Valley, apparently without stone tools, hunted the assemblage of largest average body weight, while a later cave occupant, with stone tools, habitually scavenged. In the light of the present interpretation of the bovid data it would seem far more likely that STS *Australopithecus* was not the "mighty hunter," as has been claimed for the temporally and taxonomically closely related Makapansgat Grey Breccia apeman (Dart 1957), but in fact among "the hunted." The present interpretation suggests that SE represents the first major hominid occupation of caves in the Sterkfontein Valley. It is not clear to what extent subsequent assemblages like SKb reflect hominid accumulating influences. The SKb *Homo* must have occupied the cave for at least some time, as attested to by the Swartkrans stone tools, but the patterns in figure 14.5a,b, and c are equally compatible with those expected of predators of smaller prey preference, like leopards and hyenas. These tentative suggestions await the application to the overall assemblages of additional criteria such as Klein's (1975) observation that carnivore-ungulate ratios are higher in carnivore than hominid-accumulated assemblages. The analyses of C. K. Brain on the Sterkfontein Valley assemblages, currently in progress, will provide a more firmly based picture than was here obtained on Bovidae alone.

Acknowledgments

I am very grateful to the following people and institutions for allowing me access to unpublished game censuses: G. de Graaff of the National Parks Board of Trustees,

Pretoria, for the 1975 census of Kruger National Park; G. de Graaff, J. du P. Bothma of
the University of Pretoria, and E. Moolman of the National Parks Board of Trustees,
Pretoria, for the 1973-74 census of the Kalahari Gemsbok National Park and adjoining
areas; B. J. Huntley of the Council for Scientific and Industrial Research, Pretoria, for
censuses of all the Angolan game areas included in this work; J. Hanks of the Natal Parks
Board, Pietermaritzburg, for censuses of Natal game areas; E. Joubert of Nature Conservation and Tourism, Windhoek, South-West Africa, for the census of Etosha National Park;
J. Rushworth and B. R. Williamson of the Research Unit, Wankie National Park, Rhodesia,
for the census of Wankie National Park.

I am very grateful to M. W. Browne, of the University of South Africa, Pretoria,
for suggesting the use of correspondence analysis on the bovid census data, and for
supervising the analyses at the said institution. I also want to thank him for explanation and discussion, particularly on the interpretation of results.

I acknowledge with pleasure having most helpful discussions with, and receiving
information from, the following: mainly C. K. Brain, also F. C. Eloff, G. de Graaf, B. J.
Huntley, and F. E. Steffens.

I am grateful to those that have allowed me to look at fossil bovid collections
in their care: P. V. Tobias, A. R. Hughes, J. Kitching, Q. B. Hendey, R. E. Leakey, J. M.
Harris, and H. Obernolzer.

Judith A. H. Van Couvering

Introduction

Definitions of the term *community* vary from that of a mere aggregation of plants and animals to that in which the community is considered to be almost a superorganism (see Shelford 1963; Ricklefs 1973; Valentine 1973; Kauffman and Scott 1976). I will use the term to mean an *aggregation of plants and animals with distinct dynamic interrelationships*. These distinct dynamic interrelationships may be defined, for the purpose of studying paleocommunities, in terms of *community structure*. This consists of the relative distributions of species in body size, feeding type, and locomotor type as well as species richness, species diversity, and trophic structure. Olson (this volume) has pointed out a distinction between *community* and *community type* which will be followed here. A true community is one in which the plants and animals live together in a particular area as, for example, the Budongo rain forest community. A community type is, however, an abstraction, such as a "lowland rain forest community." Olson also points out that there is a difference between *community evolution* and *community succession,* and suggests that the former should "encompass only those changes that occur within integrated complexes of organisms that maintain direct continuity through time by persistence of their basic ecological structure." Community succession is, according to Olson, the process whereby one community succeeds another either by evolution in situ, immigration from elsewhere, or a combination of both, resulting in a change in trophic structure and population dynamics as well as a change in biota. The term community succession is also used in ecology to mean "replacement of populations in a habitat through a regular progression to a stable state" (Ricklefs 1973). Both types of succession are identifiable in the fossil record (Johnson 1972), and it should be made clear which type is being discussed. Referring either to *ecological succession* or to *evolutionary succession* might eliminate this confusion.

Community evolution and evolutionary succession could be studied in some detail within certain sequences in East Africa such as the Kisingiri Early Miocene sequence (Andrews and Van Couvering 1975), the Tugen Hills Mid-Late Miocene sequence (Bishop and Pickford 1975), and the Plio-Pleistocene sequences of the Omo River basin and the East Lake Turkana area (Coppens et al. 1976). Detailed taphonomic and stratigraphic work is, however, essential to an analysis of this sort. Stratigraphic work has been at least partially completed in all of these areas, and the taphonomy is in progress, or complete, in both the Kisingiri and Omo River sequences.

The approach in this paper has been to study the evolution and succession of community types based on data derived in part from contiguous and in part from noncontiguous communities.

The term *diversity* refers, in this paper, to species richness (number of species) combined with species equability (evenness of distribution of individuals among these species) and, in general, to within-habitat, or alpha, diversity. This may include a "habitat" which encompasses an entire drainage basin, or at least proximal and distal stream communities. The modern communities analyzed have been chosen to correspond in size to those areas sampled in fossil communities.

Methods

I have used two new techniques in analyzing the fossil assemblages: *habitat spectra* and *ecological diversity spectra*. In constructing habitat spectra each species is assigned a habitat range based on the present range of its nearest living relative, its morphology, and well-documented associations of the species in the world fossil record. Weightings of 1 through 6 were given to each species (6 if the species itself is living today, 5 if the genus is living but not the species, and so on; 4 for tribe, 3 for subfamily, 2 for family, and 1 for any higher category). This method was modified with the help of Nikos Solounias from one suggested by Alan Walker (personal communication, 1974). Habitat spectra, or histograms, were constructed for a number of selected localities (see fig. 15.1) by figuring the weighted percentage of taxa adapted to each habitat zone (sum of species in a given habitat column x respective weighting factors/sum of total species in a fauna x respective weighting factor). Several localities for each time period were studied. Using only presence and absence data, we ignored the details of particular sites and instead combined the data from the sites within each major locality. This establishes a general picture of the environmental spectrum in each of the drainage basins studied. The consistency of the habitat spectra drawn for particular time periods suggests that the pattern that emerges is real. Those localities shown in figure 15.1 are typical of their respective time periods. Habitat spectra have been constructed even for those collections in which there is gross taphonomic bias. Although in most cases the overall shape of the histogram is changed very little from those of relatively unbiased collections, knowledge of this bias is necessary in order to properly interpret a spectrum.

The advantages to this method are as follows: (1) The *whole* mammalian fauna sampled (or in an ideal situation, the entire sampled biota) is analyzed. This reduces the chance of error inherent in predictions of past habitats based on selected key species. (2) Each species is given a *habitat range* rather than assuming that it was confined to a particular habitat, either in time or in space. This conforms with the realities of habitat usage by modern mammals and gives proper significance to animals that are truly restricted to a narrow habitat. (3) Greater relative weight is given to those species which are more closely related to modern species, thus compensating for greater error in assigning habitat preference to more distantly related species. The basic assumptions are (*a*) that the more closely a fossil species is related to a living species, the more similar they are in habitat preference; and (*b*) that morphology reflects habitat. These assumptions may be wrong or misinterpreted in some cases. In addition, predictions of the habitats of fossil animals which were members of diverse groups in the past and which are today represented only by relics are not reliable (e.g., were the four species of *Dorcatherium* from the Early Miocene of East Africa all waterside forest

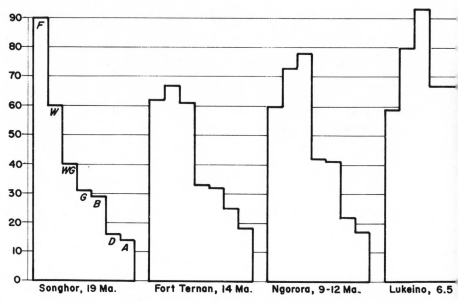

Figure 15.1. Habitat spectra for selected localities of the East African Neogene. The selected localities shown here are *typical* of the spectra of their particular time period. See text for discussion of compilation method. F: equatorial rain forest; W: woodland; WG: wooded grassland; G: grassland; B: bushland; D: desert; A: aquatic margin.

animals like the single modern species of African tragulid, *Hyemoschus aquaticus?*). A pooled analysis of the whole fauna helps to overcome these problems.

Ecological diversity histograms were constructed by Peter Andrews and his colleagues J. M. Lord and E. N. Evans for twenty-two modern and five fossil localities following the method of Fleming (1973). Typical cases are shown in figure 15.2. Ecological diversity histograms can only be used to demonstrate community structure for fossil assemblages that seem to reflect the living assemblage without too much bias—for instance Rusinga, Fort Ternan, and many of the Plio-Pleistocene localities. They should be used for studying single communities derived from areas of comparable size.

Discerning and explaining the differences between fossil communities and modern communities are vexing problems in paleoecology. Uniformitarianism, if used too strictly, can blind us to the differences between the past and the present. If a modern community type is taken as an exact standard for comparison with fossil communities, the differences between the two are often explained away as the result of sampling or other taphonomic bias, or an otherwise faulty record; this is referred to by Olson as "me-too-ecology" (personal communication, 1976). In this manner, through circular reasoning, we may conclude that the paleocommunity that we are studying was exactly like the modern community type with which we have compared it. This failure to discern differences between the past and the present has had important consequences. Olson has pointed out (personal communication, 1976) that it was this type of reasoning that at first blinded him to the idea that the food chain in the Early Permian was basically different from that of the modern world (see Olson 1961). In another case, the use of modern reptiles as strict metabolic models for dinosaurs blinded us to the possibility that these animals could have been functionally warm-blooded (see Bakker 1975). We must realize that, not only may the animals and the plants of the modern and fossil biota differ, but also the pos-

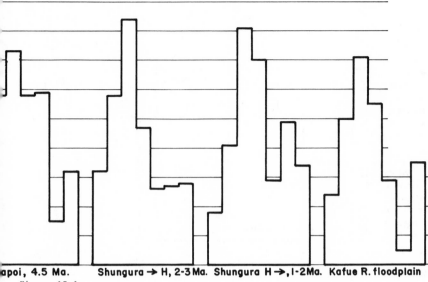

apoi, 4.5 Ma.   Shungura → H, 2-3 Ma. Shungura H →, I-2Ma. Kafue R. floodplain

Figure 15.1.—*Continued*

sibility must be considered that diversity, food chains, population dynamics, and other
basic biological interrelationships may have been different in the past as well. We
should look for these differences, as well as for similarities, as one of our first orders
of business. The techniques used here, although they employ only presence or absence
data, are especially valuable in discerning differences, and in showing similarities,
between modern and fossil communities and among the fossil communities themselves. For
example, (1) the *habitat spectra* show that woodland is of greater importance in the Middle
Miocene fossil record of East Africa than it is in any modern African community, and (2)
the *ecological diversity spectra* indicate that in the Early Miocene forest community
there is an excess of large, ground-dwelling browsers in comparison with modern equatorial
lowland rain forest. The interpretation of this difference is a difficult problem, as
can be seen in the discussions which follow.

Rain Forest and Savanna-Mosaic Community Types

Modern Equatorial Rain Forest

There are three major types of equatorial rain forests: (1) lowland evergreen;
(2) lowland deciduous; and (3) montane. Because it is difficult to separate these
habitats in the fossil record using the evidence currently available, they are here
lumped for convenience under the category "equatorial rain forest." Equatorial rain
forests occur today in Africa both at high altitudes and in areas where the low-pressure
equatorial trough conditions dominate, so that the climate is characterized by high and
year-round rainfall. Although slight monsoonal variations in rainfall can be measured
in many of the regions that are dominated by equatorial trough conditions, these differences are not great enough to affect the overall dynamics and structure of the rain
forest community. In each of the three major types of equatorial rain forest, trees
develop fruit and new growth throughout the year, leaves fall from the trees throughout the
year, the litter decomposes throughout the year, and the animals breed (and die)

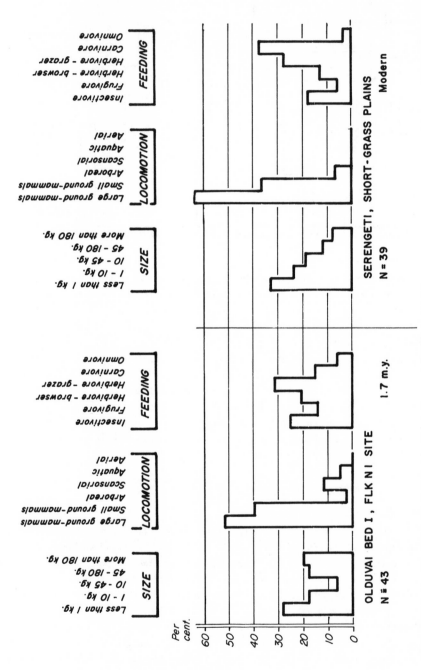

Figure 15.2. Ecological diversity spectra of selected fossil and modern mammalian communities. The histograms showing modern community structure are *typical* for those communities based on histograms drawn for 22 communities. *N* refers to number of mammalian species in the community. Compiled according to the method of Fleming (1973) by Peter Andrews (except Rusinga-Hiwegi which was compiled by the author). See text for discussion.

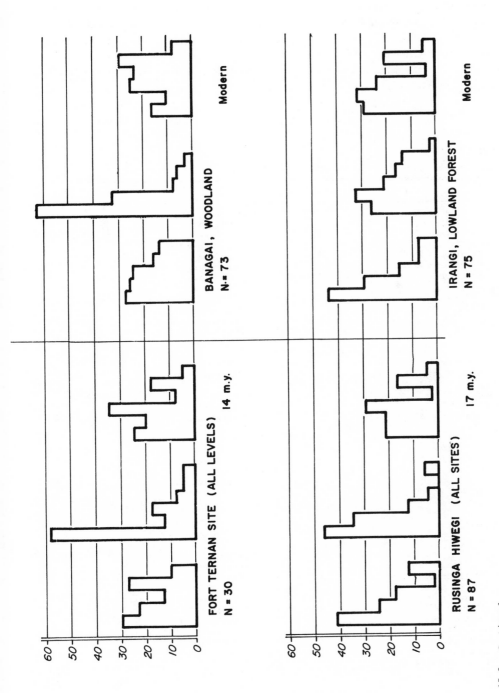

Figure 15.2.—Continued

throughout the year. The lowland deciduous rain forests are, however, more seasonal and are more widely distributed than the evergreen forests.

Equatorial forest physiognomy can be defined as "a closed stand of one or more storeys, with an interlaced upper canopy, rising 7.5—40 m or more in height. The ground cover is dominated by herbs and shrubs, and lianes and epiphytes are characteristic" (Pratt et al. 1966). The canopy structure allows little light to penetrate to lower levels. The lianas, epiphytes, saprophytes, and parasites all provide microhabitats for animals, as do the herbs and shrubs of the forest floor. The forest is broken in places by grassy glades, perhaps an important feature in the evolution of more open biotopes.

A list of genera present at two African evergreen rain forest localities is given in table 15.1. Figure 15.2 shows the trophic, size, and locomotor structure of two similar

Table 15.1

Mammalian Genera of Recent and Fossil Equatorial Rain Forest Communities

	Medje	Avakubi		Rus	Son
Lipotyphyla			Lipotyphla		
Potamogalidae			Tenrecidae		
Potamogale	1	1	†*Protenrec*	1	1
Soricidae			†*Erythrozootes*	-	1
Crocidura	8	5	Soricidae		
Sylvisorex	2	-	*Crocidura*	1	-
Scutisorex	1	-	Erinaceidae		
			†*?Galerix*	1	1
			†*Gymnurechinus*	2	1
			†*Amphechinus*	1	1
			†*Lanthanotherium*	1	1
			Chrysochloridae		
			†*Prochrysochloris*	-	1
Macroscelidea			Macroscelidea		
Rhynchocyon	1	1	**Rhynchocyon*	2	2
Primates			Primates		
Lorisidae			Lorisidae		
Perodicticus	1	1	**†Progalago*	2	2
Galago	1	1	**†Komba*	2	2
Cercopithecidae			**†Mioeuoticus*	1	-
Cercocebus	1	1	Hylobatidae		
Cerocopithecus	3	4	†*"Dendropithecus"*	1	1
Colobus	-	4	Pongidae		
Papio	-	1	†*Limnopithecus*	1	1
Pongidae			†*Dryopithecus*	4	4
Pan	1	1			
Lagomorpha			Lagomorpha		
			†*Kenyalagomys*	2	-
Rodentia			Rodentia		
Sciuridae			Sciuridae		
"Aethosciurus"	1	-	†*Vulcanisciurus*	1	1
Heliosciurus	1	1	Anomaluridae		
Funisciurus	2	2	**†Paranomalurus*	2	3
"Tamiscus"	2	2	**†Zenkerella*	-	1
Protoxerus	1	1	Cricetodontidae		
Anomaluridae			†*Afrocricetodon*	1	1
Anomalurus	3	2	†*Protarsomys*	1	1
Idiurus	1	1	†*Notocricetodon*	1	2
Cricetodontidae			Phiomyidae		
Dendromus	1	-	†*Phiomys*	1	1
Deomys	1	-	†*Paraphiomys*	2	2
Cricetomys	1	-	†*Epiphiomys*	1	1

Table 15.1—Continued

	Medje	Avakubi		Rus	Son
Cricetidae			**Phiomyidae (continued)**		
Lophuromys	3	1	†*Diamantomys*	1	1
Malacomys	1	1	†*Kenyamys*	1	1
Colomys	1	-	†*Simonomys*	1	1
Lemniscomys	1	1	†*Myophiomys*	1	1
Hybomys	1	1	†*Elmerimys*	1	1
Stochomys	1	-			
Muridae					
Dasymys	1	-			
Oenomys	-	1			
Mastomys	1	1			
Praomys	1	1			
Hylomyscus	1	1			
Thamnomys	1	1			
"Leggada"	1	-			
Chocromys	1	-			
Gliridae					
Claviglis	1	1			
Hystricidae					
Atherurus	1	-			
Creodonta			**Creodonta**		
			†*Kelba*	1	1
			†*Teratodor*	-	2
			†*Anasinopa*	2	-
			†*Dissopsalis*	1	1
			†*Metapterodon*	2	-
			†*Pterodon*	1	-
			†*Leakitherium*	1	-
			†*Hyaenodon*	3	2
			†*Megistotherium*	1	-
Carnivora			**Carnivora**		
Viverridae			**Viverridae**		
Civettictis	1	1	†*Kichechia*	1	1
Genetta	3	1	**Felidae**		
Poiana	1	-	†*Metailurus*	1	1
Nandinia	1	-	**Amphicyonidae**		
Herpestes	1	-	†*Hecubides*	2	1
Xenogale	1	1			
Atilax	1	-			
Felidae					
Panthera	1	-			
Profelis	1	1			
Mustelidae					
Lutra	-	1			
Aonyx	-	1			
Pholidota			**Pholidota**		
Manis	1	1			
Tubulidentata			**Tubulidentata**		
			†*Myorycteropus*	1	-
Hyracoidea			**Hyracoidea**		
Dendrohyrax	1	1	†*Merochyrax*	1	-
			†*Pachyhyrax*	2	1
Proboscidea			**Proboscidea**		
Loxodonta	(1)	(1)	†*Prodeinotherium*	1	1
			†*Platybelodon*	1	1
Perissodactyla			**Perissodactyla**		
Rhinocerotidae			**Rhinocerotidae**		
Diceros	(1)	(1)	†*Aceratherium*	1	-
			†*Chilotheridium*	1	1
			*Dicerorhinus	1	1

Table 15.1—Continued

	Medje	Avakubi		Rus	Son
			Chalicotheriiidae		
			†*Chalicotherium*	1	1
Artiodactyla			Artiodactyla		
Suidae			Suidae		
Hylochoerus	(1)	(1)	†*Hyotherium*	2	2
Potamochoerus	(1)	(1)	†*Bunolistriodon*	1	1
Hippopotamidae			†*Listriodon*	1	–
Choeropsis	(1)	(1)	†*Xenochoerus*	1	1
Tragulidae			Anthracotheriidae		
Hyemoschus	(1)	(1)	†*Masritherium*	1	–
Giraffidae			†*Brachyodus*	2	–
Okapia	(1)	(1)	Tragulidae		
Bovidae	(11)	(11)	*†*Dorcatherium*	3	1
			Gelocidae		
			†*Gelocus*	1	1
			Palaeomerycidae		
			†*Propalaeoryx*	1	1
			Giraffidae		
			†*Canthumeryx*	1	–
			Bovidae	1	1

Note: Orders have been aligned to make comparisons easier. Numbers in columns refer to numbers of species present; dash means genus absent. *Medje*: modern forest site between Ituri and Uele Rivers, Zaire; species list from American Museum Congo Expedition (AMCE) reports. *Avakubi*: modern forest site on Ituri River, Zaire; species list from AMCE reports. *Rus*: Early Miocene locality (Hiwegi Fm. only), Rusinga Island, Kenya; species list from Van Couvering and Van Couvering (1976). *Son*: Early Miocene locality, Songhor, Kenya; species list from Van Couvering and Van Couvering (1976). Species lists from fossil localities include only those species thought to have lived in forest.

†Genus extinct.   *Forest indicators

communities. Important structural features peculiar to modern evergreen rain forest communities are that (1) about 50% of the mammalian species are less than 1 kg in size and about 75% are less than 10 kg; (2) only 25% of the mammalian species are large ground dwellers, while around 50% are arboreal, scansorial, aerial, or aquatic; (3) insectivore and frugivore species predominate (about 55%) and grazing herbivores are unimportant (less than 5%). The structure of a modern evergreen rain forest community is very different from that of a savanna-mosaic community (fig. 15.2) even when only the wooded habitat in that community is investigated. Important features of the population dynamics and behavior of forest animals are listed in table 15.2.

Early Miocene Equatorial Rain Forest

The equatorial rain forest appears to have extended as far east as the modern Eastern Rift and perhaps all the way to the Indian Ocean in the Early Miocene (Andrews and Van Couvering 1975). Although there are few fossil floras of this age known from Africa (Chaney 1933; Chesters 1957; Hamilton, in Walker 1969), the available evidence suggests that many of the plant types which make up the African equatorial rain forest today were already present in the Early Miocene, which is as far back as the record goes in the Cenozoic of this area.

Analysis of the Early Miocene environments at Rusinga (Andrews and Van Couvering 1975) suggests that the Rusinga mammalian fauna differs from the fauna of a typical modern equatorial rain forest in that it has relatively more species of large ground mammals (50% rather than 25-30% of the total species) and browsers (30% rather than 20%),

Table 15.2

Structure and Dynamics of Equatorial Rain Forest and Savanna-Mosaic Communities

Forest	†Savanna-Mosaic
Climate	
Equatorial trough conditions	Monsoon conditions
Year-round rainfall	Seasonal rainfall
No fire	Fire during dry season
Water unlimited	Water limited seasonally
Exchangeable ions unlimited	Exchangeable ions limited seasonally
Plants	
Constant species numbers	Variable species numbers
Year-round growth	Seasonal growth
Year-round reproduction	Seasonal reproduction
Constant decay	Seasonal decay
Mammals	
Species rich	Species rich
High species diversity	*Low species diversity
Constant species numbers	Species numbers variable seasonally
Low population numbers	High population numbers
Low turnover rate	?Mixed turnover rate
K selection dominant	Dominant type of selection unknown
Small size dominant	Large or medium size dominant
Single individuals or small family groups	Large groups dominant
Pair bonding dominant	Polygyny dominant
Sedentary habits dominant	Migratory and dispersal habits dominant
Insectivores, omnivores, small herbivores and carnivores dominate	Ungulates, small herbivores and carnivores dominate
Small mammals (< 10 kg) dominate	No dominant size group
Arboreal, scansorial, aerial dominate	Large ground mammals dominate

Note: These features apply to the majority but not all of the component taxa. See discussion in text.

†Woodland component of the mosaic is similar to forest, but with seasonal plant growth and breeding.

*In many communities a few species dominate in numbers and biomass (see text). Primarily from Bourlière (1973) and Bourlière and Hadley (1970).

although the overall size spectrum is very similar (fig. 15.2). It is the presence of relatively more species of large ground-dwelling browsers that creates this difference. Whether the excessive numbers of mammals in this guild (a guild being "a group of species that exploit the same class of environmental resources in a similar way" [Root 1967, p. 335]) are due to (1) the presence in the fossil deposits of exotic, non-rain forest inhabitants from another nearby community; (2) a modern fauna "deplete" in this guild; (3) a slight difference in the plant community; or (4) the admixture of successional habitats, is not easy to answer. Each of these explanations has supportive circumstantial evidence, and none of them can be shown to be completely erroneous.

In reference to the first point, we know that when the Rusinga Hiwegi fauna is pooled there is some mixture of animals with different habitat preferences, but this does not eliminate the problem. Detailed excavations at small sites have produced both definite forest-dwelling mammals and those possibly from more open habitats buried together in flood basin or backwater deposits (e.g., sites KG and KF; Andrews and Van Couvering 1975, fig. 3).

The second point is also complex. A cursory review of the Early Tertiary record of North America suggests that there are more large, ground-dwelling browsers in those forest faunas than in modern-day forest communities, indicating that the modern-day forest is truly "deplete" in this guild. In order to demonstrate this, however, the entire record must be investigated in detail.

The possibility of a mixed forest-woodland is suggested by the fact that the differences in ecological diversity structure between the Rusinga Hiwegi fauna and modern equatorial rain forest communities are in the direction of woodland. This is supported to some degree by the floral and molluskan faunal evidence. The majority of the fossil flora from Rusinga (Chesters 1957) is made up of equatorial rain forest or potentially equatorial rain forest genera, but it also contains genera which today are restricted to woodland, bushland and even wooded grassland (see Andrews and Van Couvering 1975, table I). The ecological ranges of the fossil species are, however, uncertain. The majority of the gastropod fauna is also made up of genera which live in equatorial rain forest habitats today, but there are a few genera which are restricted to woodland or bushland (ibid., table II). Again, the ecological ranges of the fossil species are not certain.

The fourth possibility is that this proposed habitat mixture was due to successional events rather than a climax condition. This is suggested by the occurrence of between 100 and 1000 separate deposits of ash from successive volcanic eruptions in each of the major clastic formations on Rusinga (John Van Couvering, personal communications). This destruction of habitats with consequent repopulation and succession may have been of major importance in determining available habitats.

I have discussed this problem in some detail in order to point out the difficulties in interpretation once differences have been discerned. The mere discovery of differences (or similarities) is, nevertheless, a major step forward. The question of *cause* is, however, important to the interpretation of population dynamics in the Early Miocene rain forest community chronofauna. We need to know whether the climate was more seasonal than it is in a typical rain forest community type today. For the discussion which follows I am assuming that, even if the climate was more seasonal at the longitude of the East African sites, it was not more seasonal in the central African forest block.

Evolution of the Equatorial Rain Forest Community

There are two major points which I wish to discuss in this section: (1) the nature of the extreme faunal turnover in the mammalian part of the community; and (2) the cause of this taxonomic turnover. Although the dynamics and structure of the Early Miocene equatorial rain forest community and that of the same community type today are probably similar at least in the central forest (see above), the fauna itself has changed dramatically (table 15.1 and Van Couvering and Van Couvering 1976). The evolution of the equatorial rain forest community from the Early Miocene to the present cannot be studied directly because no true forest fauna is certainly known from the fossil record after 17 my (but see Lothagam, etc., below). Thus, we must indirectly study the changes which have taken place in the forest by examining the open-habitat faunas, for which we do have some record.

Table 15.3 indicates that replacement was gradual, with more marked episodes in the Middle Miocene and the Pleistocene. However, it should be noted first of all that the data on first and last appearance have been assembled from primarily nonforest faunas (table 15.3); and, secondly, that the forest is an ancient community which has, in some cases, served as a refuge for ancient lineages of forest-adapted animals. It is thus

Table 15.3

Geological Time Range of Genera in the East African Neogene

	Kis	Tin	Ngo	Lot-Luk	Kan-Che	Pli-Plei	Rec
*Tenrecidae	x	...	...	...	...	...	x
†*Protenrec*	x	...	...	...	...	...	...
†*Geogale*	x	...	...	...	...	...	...
†*Erythrozootes*	x	...	...	...	...	...	...
Chrysochloridae	x	...	...	...	...	...	x
†*Prochrysochloris*	x	...	...	...	...	...	...
Erinaceidae	x	x	...	...	...	...	x
†*Lanthanotherium*	x	...	...	...	...	...	...
†*Galerix*	x	...	...	...	...	...	...
†*Gymnurechinus*	x	...	...	...	...	...	...
†*Amphechinus*	x	...	...	...	...	...	...
Soricidae	x	...	...	...	...	x	x
Crocidura	x	...	...	...	...	...	x
Suncus	...	...	...	...	...	x	x
Macroscelididae	x	x	x	...	...	...	x
†*Myohyrax*	x	...	...	...	...	...	...
Rhynchocyon	x	...	...	...	...	...	x
Pteropodidae	x	...	...	...	...	x	x
‡*Propotto*	x	...	...	...	...	...	...
Eidolon	...	...	...	...	...	x	x
Megadermatidae	x	...	...	...	...	...	x
Hipposideridae	x	...	...	...	...	...	x
Hipposideros	x	...	...	...	...	...	x
Emballonuridae	x	...	...	...	...	x	x
Taphozous	x	...	...	...	...	...	x
Coleura	...	...	...	...	...	x	x
Lorisidae	x	x	...	...	...	x	x
Lorisinae	x	x	...	...	...	...	x
†*Mioeuoticus*	x	...	...	...	...	...	...
Galaginae	x	...	...	...	...	x	x
†*Komba*	x	...	...	...	...	...	...
†*Progalago*	x	...	...	...	...	...	...
Galago	...	...	...	...	...	x	x
Cercopithecidae	x	x	x	x	x	x	x
†*Victoriapithecus*	cf.	x	...	...	...	...	...
Cercopithecinae	...	...	...	x	x	x	x
Cercopithicini	...	...	...	...	x	x	x
Cercopithecus	...	...	...	...	x	x	x
Papionini	...	...	...	x	x	x	x
Parapapio	...	...	...	cf.	x	cf.	...
Papio	...	...	...	...	x	x	x
Theropithecus	...	...	...	...	x	x	x
Cercocebus	...	...	...	...	x	x	x
Colobinae	...	...	?	...	x	x	x
Cercopithecoides	...	...	...	...	...	x	...
†*Paracolobus*	...	...	...	...	x	cf.	...
Colobus	...	...	...	...	x	x	x
*Hylobatidae	x	x	...	...	...	...	x
†*Dendropithecus*	x	...	...	...	...	...	...
Pongidae	x	x	...	...	...	...	x
Limnopithecus	x	x	...	...	...	...	...
Dryopithecus	x	x	...	...	...	...	...
Hominidae	...	x	x	x	x	x	x
Ramapithecus	...	x	?	?	...	...	...
Australopithecus	...	...	...	cf.	cf.	x	...
Homo	...	...	...	...	...	x	x
*Ochotonidae	x	x	...	...	...	...	...
†*Kenyalagomys*	x	x	...	...	...	...	...

Table 15.3—Continued

	Kis	Tin	Ngo	Lot-Luk	Kan-Che	Pli-Plei	Rec
Leporidae	...	...	...	...	x	x	x
Lepus	...	...	...	...	x	x	x
†Phiomyidae	x	x	x	...	...	...	...
†*Phiomys*	x	...	...	...	...	...	...
†*Paraphiomys*	x	...	...	...	...	...	...
†*Epiphiomys*	x	...	...	...	...	...	...
†*Diamantomys*	x	...	...	...	...	...	...
†*Kenyamys*	x	...	...	...	...	...	...
†*Simonomys*	x	...	...	...	...	...	...
†*Myophiomys*	x	...	...	...	...	...	...
†*Elmerimys*	x	...	...	...	...	...	...
Thryonomyidae	...	...	...	x	...	x	x
Thryonomys	...	...	...	...	...	x	x
Hystricidae	...	...	...	x	x	x	x
Hystrix	...	...	...	...	x	x	x
†*Xenohystrix*	...	...	...	...	x	x	...
†Bathyergoididae	x	...	...	...	...	...	...
†*Bathyergoides*	x	...	...	...	...	...	...
Bathyergidae	x	...	...	...	...	x	x
†*Proheliophobus*	x	...	...	...	...	...	...
Heterocephalus	...	...	...	...	...	x	x
Anomaluridae	x	x	...	x	...	...	x
†*Paranomalurus*	x	...	...	...	...	...	...
Zenkerella	x	x	...	...	...	...	x
Pedetidae	x	x	x	...	...	...	x
†*Megapedetes*	x	x	...	...	...	...	...
Cricetodentidae	x	x	x	...	...	...	x
†*Afrocricetodon*	x	...	...	...	...	...	...
†*Protarsomys*	x	...	...	...	...	...	...
†*Notocricetodon*	x	...	...	...	...	...	...
†*Leakeymys*	...	x	...	...	...	...	...
†*Cricetodon*	...	x	...	...	...	...	...
Dendromurinae	...	...	x	...	...	x	x
Dendromus	...	...	...	...	...	x	x
Cricetidae	...	...	x	...	x	x	x
Tatera	...	...	...	...	x	x	x
Gerbillurus	...	...	...	...	...	x	x
Muridae	...	...	...	...	...	x	x
Mastomys	...	...	...	...	...	x	x
Arvicanthus	...	...	...	...	...	x	x
Pelomys	...	...	...	...	...	x	x
Aethomys	...	...	...	...	...	x	x
Thallomys	...	...	...	...	...	x	x
Grammomys	...	...	...	...	...	x	x
Mus	...	...	...	...	...	x	x
Lemniscomys	...	...	...	...	...	x	x
Oenomys	...	...	...	...	...	cf.	x
*Dipodidae	...	...	...	...	...	x	x
Jaculus	...	...	...	...	...	x	x
Sciuridae	x	x	x	...	...	...	x
†*Vulcanisciurus*	x	...	...	...	...	...	...
Paraxerus	...	...	...	...	...	x	x
Xerus	...	...	...	...	...	x	x
†Arctocyonidae	x	...	...	...	...	...	...
†*Kelba*	x	...	...	...	...	...	...
†Teratodontidae	x	...	...	...	...	...	...
†*Teratodon*	x	...	...	...	...	...	...
†Hyaenodontidae	x	x	...	...	...	...	...
†*Anasinopa*	x	x	...	...	...	...	...
†*Metasinopa*	x	...	...	...	...	...	...

Table 15.3—*Continued*

	Kis	Tin	Ngo	Lot-Luk	Kan-Che	Pli-Plei	Rec
†*Dissopsalis*	x	x	...	...	...	...	...
†*Metapterodon*	x	...	...	...	...	...	...
†*Pterodon*	x	...	...	...	...	...	...
†*Leakitherium*	x	...	...	...	...	...	..
†*Hyaenodon*	x	...	...	...	...	...	...
†*Megistotherium*	x	x	...	...	...	...	...
†Amphicyonidae	x	...	?	...	...	...	...
†*Hecubides*	x	...	...	...	...	...	...
Viverridae	x	?	x	x	...	x	x
†*Kichechia*	x	...	...	...	...	...	...
Ichneumia	...	...	x	...	...	...	x
Civettictis	...	...	...	x	...	x	x
Helogale	...	...	...	...	...	x	x
Herpestes	...	...	...	...	...	x	x
cf. *Mungos*	...	...	...	...	...	x	x
Viverra	...	...	...	...	...	x	x
Genetta	...	...	...	...	...	x	x
†*Pseudocivetta*	...	...	...	...	...	x	...
Felidae	x	...	...	x	...	x	x
†Nimravinae	x	...	...	...	...	...	...
.†*Metailurus*	x	...	...	...	...	...	...
Felinae	...	...	...	x	...	x	x
†*Dinofelis*	...	...	...	...	...	x	...
Panthera	...	...	...	...	...	x	x
Acinonyx	...	...	...	...	...	x	x
Felis	...	...	...	...	...	x	x
†Machairodontinae	...	...	...	x	x	x	....
†*Megantereon*	...	...	...	...	...	x	...
†*Homotherium*	...	...	...	...	...	x	...
†*Therailurus*	...	...	...	...	...	x	...
†*Machairodus*	...	...	...	...	...	x	...
Mustelidae	...	...	x	x	x	x	x
Mellivora	...	...	x	...	...	...	x
†*Enhydriodon*	...	...	...	x	x	x	...
Lutra	...	...	...	...	...	x	x
Canidae	...	...	?	x	...	x	x
Canis	...	...	...	...	...	x	x
Lycaon	...	...	...	...	...	x	x
Hyaenidae	...	...	x	x	x	x	x
†*Percrocuta*	...	...	x	...	...	x	...
Crocuta	...	...	...	...	...	x	x
Hyaena	...	...	...	...	x	x	x
†*Euryboas*	...	...	...	aff.	...	x	...
Orycteropodidae	x	x	x	x	...	x	x
†*Myorycteropus*	x	x	...	...	...	...	...
†*Leptorycteropus*	...	...	...	x	...	...	...
Orycteropus	...	...	x	...	...	x	x
†Deinotheriidae	x	x	x	x	x	x	...
†*Prodeinotherium*	x	x	...	...	...	...	...
†*Deinotherium*	...	...	x	x	x	x	...
†Gomphotheriidae	x	x	x	...	...	...	...
†*Platybelodon*	x	x	...	...	...	...	...
†*Gomphotherium*	?	?	?	...	...	...	...
†Mammutidae	x	x	...	x	x	x	...
†*Zygolophodon*	x	x	...	...	...	...	...
†*Anancus*	...	...	...	...	x	x	...
†*Stegodon*	...	...	...	...	x	x	...
Elephantidae	...	x	...	x	x	x	x
+*Primelephas*	...	x	...	x	...	...	...
+*Stegotetrabelodon*	...	...	...	x	x	...	...
+*Mammuthus*	...	...	...	...	x	...	...
*Elephas	...	...	...	...	x	x	x
Loxodonta	...	...	...	...	x	x	x

Table 15.3—Continued

	Kis	Tin	Ngo	Lot-Luk	Kan-Che	Pli-Plei	Rec
†Geniohyidae	x	x	...	...	...	x	...
†*Pachyhyrax*	x	x	...	...	...	x	...
Procaviidae	x	...	x	...	...	x	x
†*Meroehyrax*	x	...	...	...	...	...	...
†*Prohyrax*	cf.	...	...	...	...	...	...
†*Parapliohyrax*	...	...	x	...	...	...	...
Heterohyrax	...	...	...	...	...	x	x
†*Gigantohyrax*	...	...	...	...	...	x	...
†Chalicotheriidae	x	...	...	x	...	x	...
†*Chalicotherium*	x	...	...	...	...	x	...
†*Ancylotherium*	...	...	...	...	...	x	...
Rhinocerotidae	x	x	x	x	x	x	x
†*Aceratherium*	x	...	?	...	...	...	...
†*Chilotheridium*	x	...	x	...	...	...	...
Dicerorhinus	x	...	?	...	...	...	x
†*Brachypotherium*	x	x	x	x	...	...	...
†*Paradiceros*	...	x	?	...	...	...	...
Ceratotherium	...	...	...	x	x	x	x
Diceros	...	...	...	...	x	x	x
Equidae	...	...	x	x	x	x	x
Hipparion	...	...	x	x	x	x	...
Equus	...	...	...	...	...	x	x
†Anthracotheriidae	x	x	...	...	...	...	...
†*Masritherium*	x	x	...	...	...	...	...
†*Gelasmodon*	x	...	...	...	...	...	...
†*Brachyodus*	x	x	...	...	...	...	...
Suidae	x	x	x	x	x	x	x
†*Hyotherium*	x	...	...	...	...	...	...
†*Bunolistriodon*	x	x	...	...	...	...	...
†aff. *Propalaeochoerus*	x	...	...	...	...	...	...
†*Listriodon*	x	...	x	...	...	...	...
†*Xenochoerus*	x	x	...	...	...	...	...
†*Sanitherium*	x	...	...	...	...	...	...
†*Nyanzochoerus*	...	...	...	x	x	x	...
†*Mesochoerus*	...	...	...	...	x	x	...
Potamochoerus	...	...	...	...	x	x	x
†*Notochoerus*	...	...	...	...	x	x	...
†*Pronotochoerus*	...	...	...	...	...	x	...
†*Metridiochoerus*	...	...	...	...	...	x	...
Phacochoerus	...	...	...	...	...	x	x
†*Stylochoerus*	...	...	...	...	...	x	...
Hippopotamidae	?	?	x	x	x	x	x
Hippopotamus	...	...	...	x	x	x	x
Tragulidae	x	x	x	...	...	...	x
†*Dorcatherium*	x	x	...	...	...	...	...
†Gelocidae	x	...	...	...	...	...	...
†*Gelocus*	x	...	...	...	...	...	...
†Palaeomerycidae	x	x	x	...	...	...	...
†*Propalaeoryx*	x	...	...	...	...	...	...
†*Canthumeryx*	x	...	...	...	...	...	...
†*Climacoceras*	...	x	x	...	...	...	...
Giraffidae	x	x	x	...	x	x	x
Palaeotraginae	x	x	x	...	x	x	x
†*Palaeotragus*	x	x	x	...	x	x	...
Okapia	...	...	...	...	x	x	x
Giraffinae	...	x	x	x	x	x	x
Giraffa	...	...	...	?	x	x	x
†*Samotherium*	...	x	x	...	...	...	...
†Sivatheriinae	...	...	...	...	x	x	...
†*Sivatherium*	...	...	...	...	?	x	...
Camelidae	...	...	...	...	...	x	x
Camelus	...	...	...	...	...	x	x

Table 15.3—Continued

	Kis	Tin	Ngo	Lot-Luk	Kan-Che	Pli-Plei	Rec
Bovidae	x	x	x	x	x	x	x
†Walangania	x	...	...	...	...	...	...
†Menelikia	...	...	...	...	...	x	...
†Megalotragus	...	...	...	...	...	x	...
†Rabaticeros	...	...	...	...	...	x	...
†Pelerovis	...	...	...	...	...	x	...
Tragelaphini	...	...	x	x	x	x	x
Tragelaphus	...	...	...	x	x	x	x
Boselaphini	?	x	x	...	...	x	x
†Eotragus	?	?	...	...	...	...	...
†Protragoceras	...	x	...	...	...	...	...
†Selenoportax	...	...	...	...	...	x	...
†Pseudotragus	...	x	x	aff.	...	...	...
†Miotragoceros	...	...	...	x	...	...	...
Bovini	...	...	...	x	x	x	x
Syncerus	...	...	...	...	x	x	x
†Hemibos	...	...	...	...	...	?	...
Cephalophini	...	...	x	...	...	...	...
Reduncini	...	...	x	x	x	x	x
Redunca	...	...	...	aff.	...	x	x
Kobus	...	...	...	aff.	...	x	x
Hippotragini	...	...	...	aff.	...	x	x
Hippotragus	...	...	...	...	...	x	x
Oryx	...	...	...	...	...	x	x
Alcelaphini	...	...	...	...	x	x	x
Damaliscus	...	...	...	aff.	...	x	x
Alcelaphus	...	...	...	...	...	x	x
†Parmularius	...	...	...	...	...	x	...
Beatragus	...	...	...	...	...	x	x
Connochaetes	...	...	...	...	...	x	x
Neotragini	...	...	x	x	...	x	x
Madoqua	...	...	...	...	...	x	x
Rhynchotragus	...	...	...	x	...	...	x
Antilopini	...	x	x	x	x	x	x
Gazella	...	x	x	x	x	x	x
Aepyceros	...	...	...	aff.	x	x	x
Antidorcas	...	...	...	...	...	x	x
Antilope	...	...	...	x	...	x	x
*Caprini	...	x	x	x	x	x	x
†Oiocerus	...	x	...	...	...	...	...
†Pultiphagonides	...	...	...	...	...	x	...
†Tossunnoria	...	...	...	...	...	x	...

Note: This is a modification and expansion of a similar chart compiled by Maglio, 1974. x and x both mean taxon present; ... means taxon absent. Kis: Early Miocene (23-17 my), species list from Van Couvering and Van Couvering (1976); Tin: Middle Miocene (16-14 my), species list from Andrews and Walker (1976) and personal field notes; Ngo: early Late Miocene (12-9 my), species list from Bishop and Pickford (1975); Lot-Luk: middle Late Miocene (7-5 my), species list from Pickford (1975) and Smart (1976); Kan-Che: late Late Miocene (5-3 my), species list from Bishop et al. (1971) and Behrensmeyer (1976); Pli-Plei: Plio-Pleistocene (3-1 my), species list from Maglio (1972) and Coppens and Howell (1976).

†Genus extinct.

*Genus extinct in Africa.

possible that in the forest there was a relative time lag in the post-Early Miocene faunal turnover. The "new" (post-Early Miocene) groups which occur in the forest today either (1) adapted to forest conditions secondarily, because they evolved initially in more open habitats (such as bovids and giraffids); (2) immigrated to East Africa from Eurasia and other parts of Africa through a primarily forested corridor (such as many of the cricetid and murid rodents); or (3) evolved in situ from preexisting forms (such as tragulids, lorisids, and tenrecoids).

Although it was mentioned above that there is no true forest fauna known from the

record after 17 my, it should be pointed out that there is a strong forest component in
the faunas from Lothagam I, Kanapoi, and Chemeron (fig. 15.1). Thus there were either
important riverine forests or local districts of true rain forest habitat in the region
southwest of Lake Turkana (= Lake Rudolf) in the Late Miocene. Members of these forest
faunas include forest-adapted representatives of cercopithecids, advanced bovids,
elephantids, and fissiped carnivores. This evidence suggests that by the late Miocene,
at least, these animals had already replaced their Early Miocene trophic counterparts in
the central forest.

The Early Miocene equatorial rain forest community (in East Africa) existed for
at least six millions years (from 23 to 17 Ma) with little apparent change even at the
species level; the faunas from Bukwa (23 Ma) and Rusinga (18 Ma) have an FRI (faunal
resemblance index) of 94 and those from Karungu (?22 Ma) and Rusinga have an FRI of 100
(see Van Couvering and Van Couvering 1976 for further discussion). Unfortunately, there
is no superpositional sequence of faunas which can be examined in detail throughout
this period, although there are various time levels well represented at different localities
in the Kisingiri complex (Karungu, Lower Mfwangano, Rusinga-Hiwegi, and Rusinga-Kulu, in
order of decreasing age) which could be integrated in a close approximation of such a
study.

I have suggested above that rain forest conditions, including the population
dynamics and socioecological relationships (relationship between social organization and
ecology; see Crook 1972), have remained similar throughout the Neogene, except perhaps
at the eastern edge of the forest block. If this is true we have a puzzling situation:
there has been a major taxonomic turnover in the face of very few climatic or structural
changes in the habitat or coexisting biota (note that the plants, gastropods, and perhaps
some lower vertebrates [Estes 1962] are generically identical to modern forms). However,
without more information about Middle and Late Miocene forest forms we cannot evaluate
the cause of the taxonomic turnover.

Modern Savanna-Mosaic

I have used the term *savanna-mosaic* here to describe the mosaic of woodland,
wooded grassland, grassland, and bushland which occur together over large parts of Africa.
Because of the integration of these different habitats, the coherent biogeographical and
evolutionary picture that they make when viewed together, and the difficulty of separating
them in the fossil record at this time, I have considered them here as a single community
type.

Savanna-mosaic exists in areas of highly seasonal (monsoonal, in the tropics)
rainfall. Dry conditions, and in many cases fire (which is the direct result of these dry
conditions), are important in the maintenance of this community type (De Vos 1969).
Seasonality is extremely important in the population dynamics and structure of the overall
community type (cf. Lamprey 1964) can be characterized, in the most extreme conditions,
by those factors listed on the right of table 15.2. The mammals found in two modern
savanna-mosaic communities are listed in table 15.4.

Each of the specific habitats which make up the savanna-mosaic community type is
developed to a different degree, depending on local conditions of rainfall, soil, and
groundwater. The physiognomy of these component habitats and their special attributes
are discussed below.

Woodland

This plant community type is made up of deciduous trees with an open or continuous,

Table 15.4

Mammalian Genera of Recent and Fossil African Savanna-Mosaic Communities

	Kafue	Mkomazi	Amboseli	Shungura-G
Lipotyphla				
Soricidae				
Crocidura	6	2	1	2
Suncus	-	-	-	1
Erinaceidae				
Erinaceus	-	-	1	-
Macroscelidea				
Elephantulus	-	1	1	-
Petrodromus	1	-	-	-
Rhynchocyon	-	1	1	-
Primates				
Lorisidae				
Galago	2	1	1	3
Cercopithecidae				
Cercopithecus	1	2	2	1
Papio	1	1	1	1
Theropithecus	-	-	-	2
Colobus	-	-	-	1
Other				4
Hominidae				
†*Australopithecus*	-	-	-	1
?*Homo*	1	1	1	1
Lagomorpha				
Lepus	1	1	1	1
Rodentia				
Sciuridae				
Paraxerus	1	1	1	1
Xerus	-	1	2	1
Cricetodontidae				
Dendromus	1	-	-	1
Cricetomys	1	-	-	-
Steatomys	1	-	-	-
Cricetidae				
Otomys	1	-	-	-
Tatera	2	1	-	1
Taterillus	-	1	-	-
Gerbillurus	-	-	-	1
Muridae				
Lemniscomys	1	2	-	1
Mastomys	1	1	-	1
Dasymys	1	-	-	-
Pelomys	1	-	-	1
Aethomys	1	-	-	1
Saccostomus	1	-	-	-
Arvicanthus	1	-	-	1
Mus	1	-	-	1
Oenomys	-	-	-	1
Thallomys	-	-	-	1
Grammomys	-	-	-	1
Gliridae				
Graphiurus	1	1	1	-
Dipodidae				
Jaculus	-	-	-	1
Bathyergidae				
Tachyoryctes	-	-	1	-
Heliophobius	-	1	-	-
Cryptomys	1	-	-	-
Other	-	-	-	1
Thryonomyidae				
Thryonomys	1	1	1	1
Pedetidae				
Pedetes	1	1	1	-

Table 15.4—Continued

	Kafue	Mkomazi	Amboseli	Shungura-G
Hystricidae				
Hystrix	1	1	1	2
†*Heterohystrix*	-	-	-	1
†*Xenohystrix*	-	-	-	1
Carnivora				
Viverridae				
Civettictis	1	1	1	1
Viverra	-	-	-	2
Genetta	2	1	2	2
Herpestes	2	2	2	-
Helogale	1	1	1	1
Atilax	1	-	1	1
Mungos	1	1	1	1
Ichneumia	1	1	1	-
†*Pseudocivetta*	-	-	-	1
Felidae				
Felis	3	3	3	1
Panthera	2	2	2	3
Acinonyx	1	1	1	1
†*Homotherium*	-	-	-	1
†*Megantereon*	-	-	-	1
†*Dinofelis*	-	-	-	1
Hyaenidae				
†*Euryboas*	-	-	-	1
Crocuta	1	1	1	1
Hyaena	-	1	1	2
Proteles	1	1	1	-
†*Percrocuta*	-	-	-	1
Pholidota				
Manis	1	1	-	-
Hyracoidea				
Dendrohyrax	1	-	1	-
Procavia	1	1	-	-
Heterohyrax	1	-	-	1
†*Gigantohyrax*	-	-	-	1
Proboscidea				
†*Deinotherium*	-	-	-	1
†*Anancus*	-	-	-	1
†*Stegodon*	-	-	-	1
†*Elephas*	-	-	-	1
Loxodonta	1	1	1	1
Perissodactyla				
Rhinocerotidae				
Diceros	1	1	1	1
Ceratotherium	-	-	-	1
Equidae				
Equus	1	1	1	3
†*Hipparion*	-	-	-	2
Chalicotheriidae				
†*Ancylotherium*	-	-	-	1
Artiodactyla				
Suidae				
Potamochoerus	1	1	-	-
Phacochoerus	1	1	1	2
†*Nyanzochoerus*	-	-	-	3
†*Notochoerus*	-	-	-	3
†*Mesochoerus*	-	-	-	1
†*Metridiochoerus*	-	-	-	2
Hippopotamidae				
Hippopotamus	1	-	1	4
Camelidae				
Camelus	-	-	-	1

Table 15.4—Continued

	Kafue	Mkomazi	Amboseli	Shungura-G
Giraffidae				
Giraffa	-	1	1	3
†*Sivatherium*	-	-	-	1
Bovidae	17	15	20	17

Note: Numbers in columns refer to numbers of species present; dashed line means genus absent at that site. Kafue: Kafue River Flood Plain, Zambia; species list from Sheppe and Osborne (1971). Mkomazi: Mkomazi Game Reserve, Tanzania; species list from Harris (1972). Amboseli: Amboseli National Park, Kenya; species list from Williams (1968). Shungura-G: Shungura Formation through Bed G, Omo River Basin, Ethiopia; species list from Coppens and Howell (1976).

†Genus extinct.

*Genus extinct in Africa (excludes bats).

but not interlaced, canopy, with an undergrowth of shrubs, herbs, and sparse grasses (Pratt et al. 1966; Langdale-Brown et al. 1964; for a more detailed description of all of the savanna-mosaic habitats see Kingdon 1971 and Lind and Morrison 1974). It is found throughout the savanna-mosaic in areas of strong seasonality, with dry subhumid to semiarid climate, and is limited within these areas to districts of locally higher moisture content or water table. The trees fruit seasonally and thus do not provide a constant food supply, in contrast to the forest trees. This habitat is significant for mammals because it provides an area of cover and shade. Few mammals are restricted to the woodlands but many use it as a primary habitat and many others use it as a marginal habitat (cf. Western 1973; Lamprey 1963). Many of the mammals which are restricted to, or prefer, woodlands year-round rather than seasonally are small and live alone or in small family groups much like the mammals of the forest, but breed seasonally (data from Dorst and Dandelot 1970). In addition, the woodland community type has the following characteristics: (1) only 30% of the mammalian species are less than 1 kg, 50% are less than 10 kg and about 25% are greater than 45 kg; (2) 60% of the mammalian species are large ground-dwellers, and less than 15% are arboreal, scansorial, aerial, and aquatic combined; (3) insectivores and frugivores make up about 40% of the mammalian species and grazing herbivores are also important (20%). The structure of the woodland is quite different from that of the forest, especially in having relatively more large ground-dwelling mammals, particularly grazers, and fewer insectivores and frugivores (see p. 280 and fig. 15.2).

Woodland today does not cover extensive tracts of land, as forest does, and its mammalian members in general range between woodland and other savanna-mosaic habitats (Lamprey 1963; Harris 1972). It is thus an integral part of the savanna-mosaic community type.

The first woodland-dominated, rather than forest-dominated, habitat appears in the East African fossil record in the Middle Miocene (see fig. 15.1) and is dominated by browsing herbivores. This marks the first step in the *local* shift towards the drier, seasonal, savanna-mosaic community type which exists in East Africa today. However, this Middle Miocene woodland community is structurally different from modern woodland. These ecological differences and the habitat spectra (fig. 15.1) suggest that there was a strong forest component mixed with the woodland. It should be remembered here that the Early Miocene rain forest may have had a minor woodland component. By the Middle Miocene this woodland component was more pronounced and the ecological structure had changed accordingly.

Predictions about the population dynamics and socioecological features of this community are not yet in order.

Wooded Grassland

This is made up of scattered or grouped trees in which the canopy cover is less than 20% and the dominant ground cover is grass, mixed with some herbs (Pratt et al. 1966). This is *savanna* in the restricted sense of the word. It exists in highly seasonal climates, and it is the seasonally dry conditions that govern its structure. Frequent dry season fires are important in recycling nutrients, stimulating growth, and maintaining the grassy cover (De Vos 1969). This highly seasonal climate also affects plant growth and reproduction and thus the growth, reproduction, behavior, and adaptive strategies of the associated animals (Bourlière and Hadley 1970).

The characteristic mammals of the wooded grassland habitat are grazing and mixed browsing-grazing ungulates and their associated predators, although herbivorous rodents and their predators are also important (De Vos 1969). Extreme adaptations for grazing and cursorial locomotion are found in the larger herbivores, and for burrowing, grazing, and seed-eating in the small herbivores. True browsers are not numerous in this habitat (data from Harris 1972, fig. 17) nor are insectivores and omnivores (Harris 1972; Sheppe and Osborne 1971).

Wooded grassland seems to appear first as a dominant habitat in East Africa in the Late Miocene, and continues its dominance throughout the rest of the Cenozoic in the localities from which the fossil fauna has been collected (fig. 15.1).

Grassland

This is a community type which is dominated by grasses mixed with some other herbaceous plants in which the tree canopy cover does not exceed 2% (Pratt et al. 1966). Seasonality and fire are important in the maintenance of the plant community, although grassland does occur edaphically as well. No extensive stands of grassland occur in East Africa today, although it is the dominant community type in places such as the Serengeti Plains. Grasslands have expanded in temperate regions to form huge tracts of prairie or steppe. These steppe grasslands are species-poor, and carry a much less diverse mammalian fauna than the savanna grasslands, and are really part of a completely different biome (see Gregory 1971).

Savanna-mosaic grasslands are dominated by large herding, grazing, and mixed grazing-browsing ungulates and their predators as well as small herbivorous rodents and their predators (see fig. 15.1). These mammals are similar to those found in wooded grasslands. In many places wooded grassland grades subtly into grassland, and few mammals are completely restricted to the grasslands part of the spectrum. It is thus a difficult community to characterize. The mammalian components of savanna grasslands tend, in general, to have a structure like those listed in the right-hand column in table 15.2.

Grasslands rarely form a dominant community type in the fossil sites sampled but are of great importance within the savanna-mosaic from the Late Miocene to the present (fig. 15.1). The Olduvai fauna from Bed I most closely approximates a grassland-dominated community (fig. 15.2), but differs in the presence of more scansorial and aquatic forms. This is probably due to the presence of waterside communities (cf. Hay 1976), although the higher representation of large browsers and grazers and fewer carnivores may be due to human selectivity or other taphonomic factors.

Bushland and Bushed Grassland

These are similar to woodland and wooded grassland except that the trees are

replaced by shrubby, woody plants less than 6 m high. The bushes occur in relatively
drier environments than trees, and support browsers which are small, solitary, and more
or less water-independent. Bushland also occurs along river banks (Pratt et al. 1966).

This habitat is important from the Late Miocene on (fig. 15.1), but is difficult
to separate from woodland on the basis of mammals. Thus the high levels of bushland seen
in fig. 15.1 may be spurious.

Aquatic Margin

There is always a wet, at least partly plant-covered area along the margin of
any watercourse or lake. This provides a habitat for many animals not directly associated
with those from the surrounding regional habitat (cf. Sheppe and Osborne 1971). The
animals living in this environment generally make up an important part of the fossil record
and may dominate certain samples (see, for example, Kanapoi and Kafue River, fig. 15.1).

Desert

True desert conditions are very minor in East Africa today, although arid,
sparsely vegetated, semidesert areas occur in the north and may have been important in
the periphery of the Omo Basin in the Plio-Pleistocene (fig. 15.1). Many of the mammals
which have been assigned to desert habitat for the purposes of constructing the habitat
spectra may actually have lived in grassland or wooded grassland. Thus the high values
for desert habitat in some of the habitat spectra may be spurious.

Savanna-Mosaic Paleocommunity Types

The earliest record of a woodland-dominated habitat is from Fort Ternan. As
discussed above, this may have been a forest-woodland complex with no modern analogue
except perhaps in some montane areas. This fauna has been interpreted as documenting
the first appearance of local seasonality in East Africa (Van Couvering and Van Couvering
1976; Walker, personal communication, 1975). From the Late Miocene onward, wooded
grassland became the dominant component at most of the known sites (fig. 15.1), and the
forest component became less important overall.

The true savanna-mosaic community type, of similar structure to that of today,
was not present until about 3 Ma. Plio-Pleistocene faunas from the Omo region, the
eastern shore of Lake Turkana, Kaiso, and Olduvai all show variations of a savanna-
mosaic spectrum and compare favorably with that from the modern Kafue River flood plain.
Lukeino (ca. 6.5 Ma) has a higher forest and bush component than the later spectra; the
5 Ma sites (Kanapoi and Chemeron) are poorly known and somewhat peculiar, possibly because
they are biased towards aquatic margin samples. The savanna-mosaic community of the
Pliocene was faunally similar to that of today, but faunal changes within that sequence
and at the end of the Pleistocene have modified the details (table 15.4; Maglio 1974).

Development of the Savanna-Mosaic Biome in East Africa

If, as has been suggested elsewhere (see p. 280 and Andrews and Van Couvering
1975), a lowland equatorial rainforest biome may have existed in most of East Africa in
the Early Miocene, what caused it to be displaced by the savanna-mosaic biome, and what
was the nature of that change? To answer these questions we must briefly examine the
nature of the worldwide development of the savanna-mosaic biome.

The World Record

The mammalian fossil record, together with palynological evidence, suggests that
throughout the world savanna-mosaic biomes progressively replaced forest biomes in the

higher latitudes and locally in low latitudes, from the Early Oligocene onwards. The following sequence of events in the development of grass-dominated communities is suggested:
(1) grasses evolve at least by the Cenomanian (Penny 1969); (2) grasses commonly appear in
waterlogged or shallow-soil forest glades in the Late Eocene and Oligocene, as reflected
by their minor but constant appearance in palynological spectra (Penny 1969); (3) non-
edaphic, wide-ranging grass species and associated herbs (e.g., Compositae) evolve in
Late Oligocene and Early Miocene time (Leopold 1969); (4) mammals adapted to living in
grass-dominated communities appear in the Oligocene of both South America (hypsodont
ungulates in Deseadan faunas; Patterson and Pascual 1968) and Asia (Hsanda Gol fauna in
Mongolia with a hypsodont "bovoid" and numerous species of seemingly grass-adapted
rodents and ochotonids; Mellet 1968) and somewhat later in the Early Miocene of North
America (Hemingfordian faunas with hypsodont ungulates; Wood et al. 1941) and Africa
(Namibian faunas with early bovoids and open-country rodents; Stromer 1926); (5) grasses
and associated herbs become dominant in many world floras by the Middle Miocene (Leopold
1969), as reflected by their proportions in palynological spectra; (6) savanna-mosaic
faunas reach their maximum extent in the Late Miocene, covering much of North America,
Asia, parts of Africa, South America, southeast Europe, the Middle East, and sub-
Himalayan Asia; (7) many high-latitude savanna-mosaic communities are replaced by steppe
grassland communities in the Plio-Pleistocene.

There have been several separate and parallel developments of savanna-mosaic
community types. Certainly, the South American savanna-mosaic mammalian fauna developed
endemically (Patterson and Pascual 1968). The North American savanna-mosaic mammalian
fauna also developed independently but was influenced by intermittent immigration from the
Old World (Simpson 1947). The relationship between the Eurasian and African savanna-
mosaic faunas is, however, uncertain. Evidence presented below indicates that the Old
World savanna-mosaic biome developed from the same ancestral community, and that there was
a great amount of interchange between Eurasia and Africa throughout the Late Tertiary.
There was, however, enough isolation that distinct lineages developed in several different mammalian groups.

The earliest appearance of an open-habitat fauna in the Old World is that from the
Middle Oligocene of Hsanda Gol, central Asia (Mellett 1968). The earliest hypsodont
ruminant is known from this fauna, which was dominated by ctenodactylid rodents and
ochotonid lagomorphs. On the other hand, all of the known Early Miocene faunas from
Eurasia were dominated by aquatic-margin and temperate-forest mammals. Ancestral
members of the later savanna-mosaic faunas are absent from this Asian Oligocene open-
habitat community and, in addition, this Oligocene fauna is not chronologically continuous
with the later savanna-mosaic biome. Thus we do not consider it as ancestral to the
modern Old World savanna-mosaic communities.

It is probable that, instead, the early evolution of the savanna-mosaic mammalian
fauna took place, in part, in Africa. This is indicated by (1) the early occurrence of
groups which were later important in savanna-mosaic communities (giraffoids, bovids, suids,
elephantid ancestors, and hippopotamid ancestors) in at least partially open-habitat
faunas in Africa while it was still isolated from Eurasia (e.g., Namibian fauna, Stromer
1926); (2) the fact that these groups remained more advanced and diverse in Africa, even
after the continents were connected, up until the latest Miocene (perhaps due to the
ameliorating influence of the Paratethys Sea on the Eurasian climates). However, by the
latest Miocene (Turolian-equivalent) distinct lines of bovids, rhinos, pigs, and hyenids
had developed in both Africa and Eurasia and remained essentially restricted to their
respective continents.

The East African Record

Unfortunately the East African mammalian faunas that are known from the Middle and Late Miocene, the time crucial to the development of the savanna-mosaic community type, are either known from poor localities or are still incompletely studied. However, certain important faunal changes which can be associated with the shift in community types (cf. fig. 15.1) can be demonstrated.

1. Maboko and Fort Ternan (16-14 my): appearance of a greater number of bovid and giraffid species; first appearance of cercopithecids in numbers (*Victoriapithecus*); first appearance of *Struthio*, the ostrich; first appearance of the modern Africa rhinocerotid line, *Paradiceros*, and possibly of hominids *(Ramapithecus)*; woodland-forest habitat.

2. Ngorora (?12-9 my): bovid species numerous; hippopotamids, dendromurines, modern-type hyrax (*Parapliohyrax*), and the extant genus of aardvark, *Orycteropus*, evolve in situ; immigration of equids, hyenids (?), and mustelids from Eurasia; woodland-forest habitat.

3. Lothagam I, Lukeino (7-5 my): bovid species much more numerous; evolution in situ of modern viverrids, suids, and elephantids; immigration of hystricids, felids, and machairodontids; only remaining major Early Miocene lineages of those that are now extinct are rhinocerotids, chalicotheres, and deinotheres.

4. Kanapoi-Chemeron (5-3 my): many first appearances: *Papio*, Colobini, *Australopithecus*, *Anancus*, *Mammuthus*, *Loxodonta*, *Elephas*, *Ceratotherium*, *Potamochoerus*, *Okapia* (evolutionary); *Lepus*, *Tatera*, *Giraffa*, and *Syncerus* (?immigrant); "savanna-mosaic" and waterside habitats.

5. Plio-Pleistocene (3-1 my): many modern genera appear both as evolutionary products and as immigrants; true savanna-mosaic habitat.

This sequence of faunal events and changes in habitat spectra (fig. 15.1) suggest that there was a slow shift from forest-adapted mammals in the Early Miocene to savanna-mosaic adapted mammals in the Late Miocene due to both in situ evolutionary change and immigration from Eurasia. By the Plio-Pleistocene there was a true savanna-mosaic community type, very similar to that of today, in which most modern genera were represented.

The Succession

The change from an equatorial rain forest community type to a savanna-mosaic community type is basically one of evolutionary succession (see p. 16). That is, it happened gradually, but in the end a completely different community type with very different dynamics and structure resulted. This gradualness is suggested by fig. 15.1. As the climate became drier and more seasonal, first a forest-woodland complex developed, then a woodland-wooded grassland, and so on until true savanna-mosaic in the modern sense developed.

The causes of this increased seasonality in East Africa are probably worldwide climate change, change in oceanic circulation, and especially rain-shadow effects due to the rising margins of the Western and Eastern Rifts (see Andrews and Van Couvering 1975).

The essential difference between the two community types, in terms of structure and dynamics, is caused by seasonality. In an equatorial rain forest community, seasonality is of negligible importance (Bourlière 1973), while in a savanna-mosaic community the seasonality and unpredictability of climate and resources essentially govern the community structure (Sinclair 1975). Table 15.2 lists many of the differences between the equatorial

rain forest and savanna-mosaic community types. In general, forest mammals are small and
sedentary, have a small home range, live alone or in small family groups, pair bond,
and breed year-round (Bourlière 1973). Savanna-mosaic mammals, especially those from
wooded-grassland and grassland habitats, are for the most part medium to large, migratory
or dispersal-prone (in the sense of Western 1973), have a large home range, live in herds,
are polygynous, and breed seasonally. Mammals restricted to woodland are similar to forest
mammals but breed seasonally.

In addition, savanna-mosaic communities have a variable number of species
present throughout the year, due to the migratory and dispersal habits of the ungulates.
Although the savanna is extremely rich in numbers of species and supports a very large
total biomass (Bourlière 1963), the species diversity is not high. Harris (1972), for
example, points out that four species of large herbivores make up 80-90% of the total
numbers and biomass of the mammalian faunas in the Mkomazi Reserve, and many other studies
of East African savanna-mosaic communities show the same thing (see density/km^2 tables 3-6
in Bourlière 1963).

Trophic and faunal differences are also important. There are fewer species of
insectivores and frugivores, more species of herbivores, especially large grazers and mixed
grazer-browsers, and more species of carnivores in a savanna-mosaic community than in a
forest community (fig. 15.1; also Van Couvering and Van Couvering 1976, table 2).

The changes in dynamics and structure, like the faunal changes, have probably
developed slowly. Indeed, species which have populations living in both forest and
savanna-mosaic communities today often have different reproductive and behavioral patterns in each of these habitats. For example, a particular species which is solitary
or lives in small family groups, has small clutches, breeds year-round, and pair bonds in
the forest may live in larger groups, have larger clutches and more seasonal breeding,
and be relatively polygynous in the savanna-mosaic (Bourlière and Hadley 1970). If these
kinds of differences exist within a species, certainly it is reasonable to assume that
evolutionary change of this sort was not genetically difficult.

In order to study the paleontological record from an ecological perspective
the following information is needed: (1) detailed species distributions; (2) type of
sample bias; (3) correction for this bias; (4) body size of species; (5) guild of species;
and (6) relative abundances of species. Future work should be directed towards obtaining
this information from existing collections as well as from field work designed specifically for these purposes. Some of this information is already available for a few
sites. Detailed records of species distributions have been kept, in some cases, throughout the history of paleontology (e.g., J. B. Hatcher, B. Brown, field notes), and much
taphonomic fieldwork of the last decade has been directed toward determining the type
of sample bias (e.g., Voorhies 1969; Brain, this volume). However, not until recently
has any work been aimed at correcting sample bias (cf. Behrensmeyer and Dechant Boaz, this
volume). Species body size and guild type can be reconstructed for most species by
studying museum collections, but determining the relative abundances of species remains
a major stumbling block.

I have used only presence-absence data in this paper, not because they are
preferable, but because in most cases relative abundance data are either unavailable or un-
reliable. One of the major unresolved problems at the conference was *how* to determine
relative abundances. Discussions at the conference, and with Robert Bakker since that
time, have crystallized my views on this subject as follows: Determination of relative
abundances for any particular site should be made in an *equivalent* manner; that is, only

bones which are taphonomically and sedimentologically similar and equally identifiable should be used. Although other bones can also be used, the mandible is preferable, and it has the additional qualification of being often preserved. Relative abundances determined on bones which are different in their taphonomic and sedimentological properties and which are not equally identifiable will by their nature give *spurious* results. In addition, the bones which are chosen should be merely *counted* (either in total, or as rights or lefts), not studied for eruption or states of wear in the case of tooth-bearing bones, or epiphyseal closure in the case of long bones. Relative abundances obtained in this manner will thus not be automatically more numerous for groups which are well known to the researchers than for groups which are not as well known.

Using the kinds of information discussed above, we hope to be able to evaluate the place of hominoids in their communities and how the evolution of these communities affected hominoid evolution.

Summary and Conclusions

By lumping the known fauna from sites of a single age and a single depositional basin, an overall habitat spectrum and ecological diversity spectra for that particular basin can be constructed. These spectra give a general picture of the habitats represented in that basin and their relative importance. Each species included in a habitat spectrum was weighted on the basis of its ecological range and similarity to living species. The ecological diversity spectra were constructed for several modern as well as fossil sites. These two techniques are important in discerning the differences between fossil and modern sites.

An examination of habitat spectra and ecological diversity spectra derived in this manner and other lines of evidence explored elsewhere (Andrews and Van Couvering 1975) suggest the following:

1. The early Miocene equatorial rain forest community of East Africa remained virtually unchanged for six millions years (23-17 Ma) and then underwent a major taxonomic overturn among the mammals. This community differed from its modern-day counterpart in the presence of relatively more species of large, ground-dwelling, browsing mammals.

2, There was a slow, unidirectional shift in the overall environment in East African from a nonseasonal climate in which equatorial rain forest was the dominant community type (Early Miocene) to a seasonal climate in which savanna-mosaic was the dominant community type (Plio-Pleistocene). The fossil record demonstrates that major faunal changes occurred in the Middle and Late Miocene and again in the Pleistocene, although in general the changes were slow and additive. Hominids do not appear consistently as members of the savanna-mosaic community until the Plio-Pleistocene, when these community types apparently became structurally identical to modern savanna-mosaic communities.

One of the ultimate goals of paleoecological analysis of the East African Cenozoic sites is to be able to evaluate the place of the hominoids in their communities and to detect how the evolution of these communities affected the evolution of the hominoids. This cannot be achieved without analysis of the faunas in terms of taphonomic and paleoecological information as discussed on p. 296. In addition, we need a better understanding of the makeup, structure, and dynamics of modern community types and a better understanding of modern burial conditions in different environments.

Acknowledgments

I appreciate the help and fruitful discussion which I have had with P. Andrews, R.

T. Bakker, P. Robinson, N. Solounias, and A. Walker. P. Andrews has been very generous in providing unpublished data. I am also grateful for the opportunity to attend the Burg Wartenstein conference and thank all of the participants for thought-provoking discussions. In addition, I thank John Van Couvering for drafting the figures and Lorna Serber for final editing and typing.

Many of the data used here were collected with the support of grants from the Wenner-Gren Foundation and the L. S. B. Leakey Foundation.

CONCLUSION

Vertebrate paleoecology encompasses much more than can be covered by the papers in this book, yet these serve to exemplify issues of major importance to the field and its future. In this concluding chapter, we will attempt to distill some central themes in vertebrate paleoecology from the preceding papers, assisted by discussions that occurred among participants at the Burg Wartenstein conference and drawing also upon our own perspective on the current status of paleoecological studies.

Three primary themes can be found in all the contributions, although in some cases they are not specifically discussed by the authors. These are (1) a theoretical framework for the study of paleoecology, (2) the use of modern analogues and experimentation, and (3) the problems of sampling and sampling biases, particularly with respect to the fossil record. Most research in paleoecology falls under one or another of these general headings, but as the field seeks a firmer basis in scientifically established fact, researchers will probably have to take all three into account for any particular project or problem. These major themes in paleoecological research apply to vertebrates from fish to mammals (and, in fact, all fossil organisms), to all time periods relevant to vertebrate history, and to fossil localities of every continent.

The paleoecological problems of the later Cenozoic in sub-Saharan Africa have served as subject material for many of the contributions to this volume. However, we feel that generalizations drawn from this background are applicable to vertebrate paleoecology as a whole. The papers serve a dual purpose; they are important as examples of current approaches to paleoecological problems, and as such they are open to critical appraisal and refinement as ideas progress. They also represent some significant pioneering contributions to African paleoecology and the role of human ancestors in the evolving paleocommunities.

Theoretical Frameworks

The future progress of vertebrate paleoecology depends on a greatly expanded sample of the fossil record, but equally important is the need for refinement of the questions we seek to answer about paleoecological relationships among organisms. Our theoretical approach to the data determines the nature and scope of the questions, and success ultimately depends on whether data and questions are suited to each other.

The process of paleoecological investigation can proceed in two ways: we can pursue the kinds of data that will answer questions of paleoecological significance, as dictated by present theory, or we can alter these questions to accommodate the limitations of the data. A close system of feedback between empirical evidence and hypothesis seems the most productive strategy for advancing the science until broader syntheses are possible. Theory gives us what we *want* to know, the data tell us what we *can* know. Thus, our theoretical approach, unless tied closely to accepted principles of physics or chemistry (as exemplified in the papers by Hanson, Hare, and Parker and Toots), for now must remain flexible. This applies particularly to theoretical approaches based on modern analogues and recent ecology. Paleoecologists involved in data collection and analysis, even for

299

restricted areas or time periods, thus have the freedom to develop new and better questions and to reassess the theoretical background for what they are doing. Such opportunities have not been fully appreciated by many workers in the field, particularly students.

As several of the authors have commented, there are no "laws" in taphonomy, and
as yet neither are there "laws" in paleoecology or paleocommunity evolution. This can be
lamented as a deficiency in paleoecological knowledge; it can also be regarded as a
challenge to the present fields of natural science and all their varied methods for
scientific investigation. There may or may not be discernible "laws" or patterns in
paleoecology, but it can be fairly said that we do not know enough yet to say one way or
the other. An important step will be to define the limits of paleoecological knowledge—
what we can and cannot know about past ecosystems. Once research effort can be focused
on what we can expect to know (e.g., body-size distributions in fossil assemblages), then
testable hypotheses can be formulated which may eventually lead to "laws" in vertebrate
paleoecology. Although there are many acknowledged difficulties, most paleoecologists
are optimistic that patterns will emerge.

The theoretical approach to a paleoecological problem is bound to vary according
to the nature of the fossil evidence. Whether or not we choose to rely heavily on modern
analogues, to resort to experimentation, or to extract mutually supporting lines of
evidence from the fossils themselves, depends on the context and age of our fossil sample.
Olson, in the beginning chapter, sets out two major approaches to paleoecology: (1) working
from recent analogues to the past, and (2) "working back" from the fossil record to
paleoecology. He strongly supports the latter approach, commenting that the use of recent
analogues is "too conservative," but this perspective can be related to the fact that
Olson has worked primarily on Permian vertebrate communities which have no modern analogues.
Most of the other authors in this book deal with fossil faunas of the later Cenozoic,
a time when modern analogues are more relevant to the preserved communities. Thus, the
level at which the "modern analogue approach" can be used as a theoretical basis for
interpretation changes as we go back in time, in that we must use it in broader, less
specific ways to suggest expected taphonomic processes or community structures in progressively older samples.

Hanson has provided a theoretical approach to taphonomy in his "segment model,"
which divides the transfer of bones from the biosphere to the lithosphere into segments
(death, destruction, import-export, and burial), each of which can be studied through
modern analogues, experimentation, or by examination of fossils and their context. This
represents the next stage of refinement in approach beyond Olson's, and Hanson feels that
it permits important taphonomic problems to be studied in a testable theoretical framework.
The theoretical framework for taphonomy, in this respect, appears to be on firmer ground
than the framework for paleoecology in general.

There is little doubt that paleoecology could benefit from increased exchange
with mathematics and statistics in constructing theoretical frameworks. There have been
some pioneering efforts in the use of mathematical models to separate time and ecology
in correlating fossil faunas, particularly due to the efforts of Ralph Shuey, a participant
in the Burg Wartenstein conference. His contributions to the discussions were much appreciated, and it was clear that most present fossil data and taphonomic variables were
not yet well suited to the precision required by the mathematical models. This is a case
in which the questions are greatly in need of modification based on the limitations of the
data, and this will require the attention of workers who are both skilled in appropriate

mathematical methods and well acquainted with the data base. Such people could doubtless
make some fundamental contributions.

Modern Analogues and Experiments

Our knowledge of recent ecosystems and processes, whether specified or not as
the "modern analogue" approach in paleoecological research, forms our basis for understanding the past. At the level of physical and chemical processes this approach can
scarcely be questioned. However, in terms of biological processes, modes of evolution, and
ecological relationships, there is the danger of conservatism, as pointed out by Olson.
He has expressed the problem in the phrase, "me-too ecology." In using the present as
a model, we run the risk of seeing only those paleoecological patterns that agree with
our analogue, and missing those that were different.

The manner in which modern analogues are used can help to alleviate (although
not eliminate) the problems of conservatism. Instead of applying them rigidly as models,
we can use them more cautiously, as a basis of comparison, paying equal attention to the
differences from and similarities to the fossil examples. Modern analogues generally are
used to interpret either the structure of past ecological relationships or processes
which affect and bias bone samples. Analogues and experiments for taphonomic processes,
particularly if tied to physical and chemical principles, can be used with greater confidence than analogues for ecological structures. Common sense should also be reliable in
suggesting which present biological processes (e.g., trampling) are likely to serve as
valid analogues for the past.

Even given the problems inherent in using modern analogues, studies of
recent community structures and interactions between faunas and environments provide
essential guidelines for interpreting the past. However, it is well to remember that the
study of modern vertebrate communities is difficult enough in itself, although there is
an apparent wealth of evidence for the dynamics of living, interacting organisms of
recent ecosystems. Papers by Western and Coe point out that even some of the most basic
components of land ecosystems are not yet well understood. Questions of interest to the
paleoecologist, such as the meaning of body-size distributions or the meaning of the
relative abundances of higher taxonomic categories (e.g., genera and tribes), cannot be
answered as yet by modern ecologists.

Ecologists familiar with the problems of describing community structures and
cause-effect interactions in modern ecosystems tend to be skeptical about the potential
for reconstructing valid paleoecological relationships. This skepticism is justified if
we attempt to reconstruct the configuration of a paleocommunity as a modern ecologist sees
it—at a single point in time. However, paleoecology can tap one component of natural
systems which is essentially unavailable to the recent ecologist: *time depth*. Ideally,
paleoecologists should be able to show which aspects of vertebrate communities are persistent through long periods of time, and such information could be of great importance to
modern ecology. There is potential for beneficial exchange between these perspectives on
the past and the present which has yet to be fully appreciated by either field.

Most of the papers in this volume make use of modern analogues, particularly with
respect to taphonomic processes. Much recent research in taphonomy has been focused on
modern processes which affect carcasses and bones, in an effort to build up a body of
comparative data, and hopefully some general rules, which can be applied to the fossil
record. The range of taphonomic variables in modern systems is enormous, however, and it

is necessary for recent studies to maintain close contact with taphonomic problems in the fossil record in order to determine which of the recent phenomena are worth the greatest expenditures of research effort.

Experimentation in taphonomy is beginning to grow in importance, and its potential goes far beyond what has been demonstrated up to now. Of the papers in this volume, only those by Hanson, Hare, and Walker make extensive use of experimental data on modern bones, although most of the contributions dealing with modern analogues include ideas which could be tested using relatively simple experiments. The few published works on taphonomic experiments (e.g., Voorhies 1969; Dodson 1973) have served to generate many new ideas and have been used extensively in support of taphonomic interpretations and paleoecological hypotheses. Growth in this realm of taphonomy would doubtless lead to considerable strengthening of our basis for reconstructing the taphonomic history of bone assemblages. However, it will be important for laboratory experimentation to be closely tied to problems of the fossil record, and to natural recent situations, so as not to become too far removed from the "real world."

The uses of modern analogues for interpreting community structures and paleo-habitats are demonstrated in the papers by Vrba and Van Couvering. They assume, with some qualifications, that present ecological characteristics of mammal faunas provide the key to interpreting the environmental history of their selected study areas. Rather than interpreting the paleoecology of a fauna from multiple lines of evidence, they use the faunas, supported by their modern analogues, to draw broad conclusions about paleoenvironments. Environments of the present are thus projected, through fossil faunas, to environments of the past. Faunas have traditionally been the prime evidence for such paleo-ecology, but it is all too easy for this to become circular, "me-too ecology." It seems that the obvious way to strengthen such lines of reasoning will be to spend more effort on the supporting lines of evidence—geology, palynology, geochemistry—to build a set of converging hypotheses which are not so narrowly dependent on modern ecological analogues.

Sampling and Sampling Biases

There is a great need to increase our paleoecological sample of the fossil record, but there is an even greater need to establish first what and how to sample. Because of the importance of contextual and taphonomic evidence in paleoecology, considerable fore-thought is required in planning a sampling strategy. This contrasts with paleontological collecting, which has typically required only general stratigraphic and areal control in the recovery of specimens suitable for evolutionary studies.

Papers in this volume have referred to samples and sampling biases of various kinds, at different levels of resolution. These can be organized into two major categories which we will term primary and secondary samples. Primary samples are the fossil as-semblages, as they occur in their original burial contexts, representing a sample of the original living community of organisms. Secondary samples are the collections we subjec-to paleoecological analysis; they are subsets (usually very small subsets) of the primary samples. Hence, sampling biases which affect the primary samples of the paleocommunity can be called primary biases. These form the subject material for taphonomy. Sampling biases affecting our collections due to sampling procedures can be referred to as secondary biases. These also, of necessity, must be studied as taphonomic phenomena since it is usually difficult to separate primary and secondary biases in initial analysis.

Thus, the data upon which we must rest our paleoecological hypotheses usually

consist of a small secondary sample of a relatively small primary sample of the original
community. The need to eliminate as many of the collecting biases as possible is obvious.
Also apparent is the need to apply statistical methods appropriate for small samples
which cannot be assumed to represent normal distributions (in a parametric sense) in the
original populations. Such problems in vertebrate paleoecology have yet to be tackled by
competent statisticians. At present, since many primary and secondary biases in bone assemblages can be recognized, if not corrected for, researchers have attempted to minimize
the problem by comparing only those samples with equivalent, or nearly equivalent, biases.
Thus, assemblages from river channel deposits can be more validly compared with each other
than with those from overbank deposits, since we know that taphonomic processes affecting
bones in channels and in floodplains are widely different. Establishing criteria for the
paleoecological comparability of samples, whether through statistical methods or through
the "equivalent biases" approach, is of primary importance to the future of paleoecology.

What and how we sample for paleoecological evidence depends to a great extent
on the questions we are trying to answer. The papers in this book provide many ideas for
important evidence to recover from fossil deposits. However, the exact formulae for
sampling, the standardized methodologies, have yet to be worked out. Given the uniqueness
of each fossil occurrence, at least based on present knowledge, perhaps it is too early
for standard sampling methodologies. Instead it is easier to establish some standardized
goals for what we know to be important taphonomic and paleoecological evidence, and then
leave to the individual researcher the decisions about how best to sample a particular
deposit.

Based on the work in this volume, we can outline some important "taphonomic indicators" to look for and, if possible, to establish quantitatively in vertebrate fossil
deposits:

1. Skeletal part representation, particularly the ratio of teeth to vertebrae,
the number of limb elements, and the presence of easily destroyed elements, as evidence
for biological and/or geological processes responsible for primary biases in the bone
sample.

2. Surface characteristics, completeness and articulation of bones, as evidence
for original surface weathering and agents such as carnivores that influenced the original
bone sample prior to burial.

3. Spatial relationships and orientation of bones, in excavated samples, as
evidence for geological and biological biasing processes.

4. Microstratigraphic context of bones in excavations, as geological evidence of
such processes as trampling or water flow which affected bones during burial, and as
evidence for sedimentary environment and geological contemporaneity of fossils deposited
together.

5. Representation (presence/absence or relative abundance) of animals of differing body size, including juveniles and adults of the same species, as an indication of
taphonomic biases in representation of the original community structure.

6. Representation of animals of different ecological preferences or taxonomic
category, as evidence for the original community structure and paleoenvironments, with
emphasis on animals of similar body size that are likely to have more or less equivalent
taphonomic biases.

These lines of evidence concern biasing processes affecting the primary bone
sample between death and burial. They are most effectively studied in the field, and at
least require careful initial field observation. Most previous palentological collec-

tions are thus not suited for paleoecological purposes that require knowledge of taphonomic
history. However, such collections may be of use if original field documentation was careful (as in many quarry excavations) or if the original localities can be revisited and
collecting biases assessed. Museum collections can be of value for studies of bone micro-
surface characteristics and geochemistry, as illustrated by Hare, Parker and Toots, and
Walker. Faunal lists such as used by Van Couvering in her paleocommunity analysis are
based on museum collections with minimal taphonomic information, and their usefulness is
limited to broad-scale but nevertheless interesting paleoecological interpretations.

New field sampling programs designed to retrieve information outlined above are
essential to the growth of paleoecology. There is a need for more large, carefully documented fossil samples from single localities, and there is also a need for multiple
smaller samples that can reveal persistent patterns of skeletal part and taxonomic occurrences. Quarriable bone concentrations often result from unusual circumstances of
death and burial, and thus may be limited in their representation of original paleocommunity characteristics, although they provide a great deal of taphonomic information.
Smaller samples which tap the "background" of fossil occurrences in a deposit can provide
a different kind of paleoecological sample, perhaps containing more reliable information
on the original overall paleocommunity. Ideally, a sampling program would include both
sources of paleoecological and taphonomic evidence.

It is interesting to note that, except for Bishop, none of the contributors to this
volume discuss the taphonomy and paleoecology of a specific fossil quarry, although much
of the published work on vertebrate paleoecology is based on such research. In fact, this
reflects our initial desire to steer away from site-specific examples in an effort to
broaden the scope of contributions to this volume. However, in discussions at the Burg
Wartenstein conference, we found ourselves returning often to the best-documented quarry
samples to illustrate our more general points concerning the importance of contextual
evidence. There is no substitute for this kind of basic data in taphonomy and paleoecology.

The major practical problem in recovering large quarry samples lies in the time
needed for careful excavation and documentation. Although the resulting information is
often of great value, samples may inevitably be too small for many paleoecological purposes,
particularly those requiring large taxonomic samples. In some cases, the answer to the
need for a large faunal sample that is not the result of unusual bone-concentrating
processes may lie in the eroded fossils that occur scattered over outcrop surfaces. Surface bones have their own set of complicating secondary biases since they may be combined
from different eroded fossil-bearing horizons. However, under certain circumstances
limits can be set on their provenance, and they can provide another source of paleoecological evidence where increased sample size may balance out the lack of specific contextual
information.

In our usual frame of reference, we regard "sampling biases" as detrimental to
our search for paleoecological facts and testable hypotheses. However, as pointed out
in the paper by Hill, the same processes that destroy or alter some kinds of information
leave other, positive evidence for their own roles in creating the fossil record. Taking
this perspective, primary biases represented in fossil samples can provide a new source of
information on geological and biological processes through time.

The foregoing discussion of three major themes in paleoecology demonstrates that
paleoecological "evidence" derives from a wide variety of approaches applied to a remarkable spectrum of samples from the fossil record. As yet there are no rules about how

this evidence can be put together to establish the relationships of past organisms and environments. Current interpretations are based on converging lines of evidence regarding fossils and context which indicate that one explanation is more probable than another. This leads to paleoecological reconstructions whose merits are often difficult to judge. However, if interpretations can be formulated as testable hypotheses with specified probabilities, subject to experimentation and to repeated sampling of the fossil record, then this could lead to a firmer basis for the field. Such a basis is necessary if paleoecology is to be given the attention it deserves as a contributor to scientific knowledge.

The wise use of modern analogues, established physical and chemical principles, and experimentation can provide us with the essential framework we need to tie fossil evidence in to testable hypotheses. Inevitably, the fossil evidence itself will require that hypotheses built upon previously established theories be modified in ways that will lead to a new understanding of paleoecology. Once we learn how to use the information contained in the fossil record, it may also lead us to unique concepts and theories concerning the evolution of vertebrate communities and ecosystems through time.

REFERENCES

Abel, O. 1912. *Grundzüge der Palaeobiologie der Wirbeltiere*. Stuttgart: E. Schweizer-bartsche Verlagsbuchhandlung Nägele und Dr. Sproesser.

_____. 1914. *Paläontologie und Paläozoolgie*. Leipzig: B. G. Teubner.

_____. 1919. *Die Stamme der Wirbeltier*. Berlin: W. de Gruyter.

Abelson, P. H. 1956. Paleobiochemistry. *Sci. Am.* 195:83-92.

Alexander, A. J. 1956. Bone carrying by a porcupine. *S. Afr. J. Sci.* 52:257-58.

Allsop, R. 1971. The population dynamics and social biology of the bushbuck (*Tragelaphus scriptus* Pallas). M.Sc. thesis, University of Nairobi.

Almagor, U. 1971. The social organization of the Dassanetch of the lower Omo. Ph.D. dissertation, University of Manchester.

_____. 1972a. Name-oxen and ox-names among the Dassanetch of southwest Ethiopia. *Paedeuma* 23:79-96.

_____. 1972b. Tribal sections, territory and myth: Dassanetch responses to variable ecological conditions. *Asian and African Studies* 8(2):185-206.

_____. In press. *Pastoral partners—affinity and bond partnership among the Dassanetch of south-west Ethiopia*. Manchester: University of Manchester Press.

Andere, D. 1975. The dynamics of pasture use by wildebeest and cow in the Amboseli Basin. M.Sc. dissertation, University of Nairobi.

Anderson, R. van V. 1927. Tertiary stratigraphy and orogeny of the northern Punjab. *Bull. Geol. Soc. Am.* 38:665-720.

Andrews, P. J., and Van Couvering, J.A.H. 1975. Paleoenvironments of the East African Miocene. In *Approaches to primate paleobiology,* ed. F. S. Szalay. Basel: Karger.

Andrews, P. J., and Walker, A. 1976. The primate and other fauna from Fort Ternan, Kenya. In *Human origins,* ed. G. Isaac and E. McCown, pp. 279-304. Palo Alto: Benjamin.

Ansell, W. F. H. 1971. Artiodactyla (excluding the genus *Gazella*). In *The mammals of Africa: an identification manual,* ed. J. Meester and H. W. Setzer, Pt. 5, pp. 11-84. Washington: Smithsonian Institution Press.

Ascenzi, A. and Bell, G. H. 1972. Bone as a mechanical engineering problem. In *The biochemistry and physiology of bone,* 2nd ed., ed. G. H. Bourne., pp. 311-46. New York: Academic Press.

Ascher, R. 1961. Analogy in archaeological interpretation. *Southwestern Journal of Anthropology* 17(4):317-25.

Axelrod, D. I. 1950. Evolution of desert vegetation in Western North America. *Contrib. Paleontol., Carnegie Inst. Wash.*, 590:215-306.

Axelsson, V. 1967. The Laitaure delta, a study of deltaic morphology and processes. *Geogr. Annaler, Ser. A.*, 49:1-127.

Bada, J. L., and Deems, L. 1975. Accuracy of dates beyond the ^{14}C dating limit using the aspartic acid racemization technique. *Nature* 255:218-19.

Bada, J. L., and Helfman, P. M. 1975. Amino acid racemization dating of fossil bones. *World Archeology* 7:160-75.

306

Bada, J. L.; Schroeder, R. A.; and Carter, G. F.   1974.   New evidence for the antiquity of man in North America deduced from aspartic acid racemization. *Science* 184:791-93.

Bagnold, R. A.   1954.   Experiments on a gravity-free dispersion of large solid spheres in a Newtonian fluid under shear. *Royal Soc. London Proc., Ser. A*, 225:49-63.

Bakker, R. T.   1971a.   Dinosaur physiology and the origin of mammals. *Evol.* 25:636-58.

_____.   1971b.   Ecology of the Brontosaurs. *Nature* 229 (5281):172-74.

_____.   1975a.   Dinosaur renaissance. *Sci. Am.* 232 (4):58-78.

_____.   1975b.   Experimental and fossil evidence for the evolution of tetrapod bioenergetics. In *Ecological studies,* vol. 12, *Perspectives of biophysical ecology,* ed. D. M. Gates and R. B. Schmerl, pp. 365-99. New York: Springer-Verlag.   .

Barkham, J. P., and Ridding, C. J.   1974.   *The large animal populations and vegetation of the Samburu-Isiolo Game Reserve, Kenya. A preliminary survey.*   School of Environmental Sciences, University of East Anglia, Norwich.

Barrell, J.   1916.   Dominantly fluviatile origin under seasonal rainfall of the Old Red Sandstone. *Bull. Geol. Soc. Am.* 27:345-86.

Beaumont, P. B.   1973.   Border Cave—a progress report. *S. Afr. J. Sci.* 69:41-46.

Beaumont, P. B., and Boshier, A. K.   1972.   Some comments on recent findings at Border Cave, northern Natal. *S. Afr. J. Sci.* 68:22-44.

Beerbower, J. R.   1963.   Morphology, paleoecology and phylogeny of the Permo-Pennsylvanian amphibian *Diploceraspis. Bull. Mus. Comp. Zool.* 130:31-108.

Beerbower, J. R., and McDowell, F. W.   1960.   The Centerfield biostrome: An approach to a paleoecological problem. *Proc. Penn. Acad. Sci.* 34:84-91.

Beerbower, J. R., and Jordan, D.   1969.   Application of informative theory to paleontological problems: Taxonomic diversity. *Jour. Paleontol.* 43:1184-98.

Behrensmeyer, A. K.   1975.   The taphonomy and paleoecology of Plio-Pleistocene vertebrate assemblages of Lake Rudolf, Kenya. *Bull. Mus. Comp. Zool.* 146 (10):473-578.

_____.   1976.   Lothagam, Kanapoi and Ekora: A general summary of stratigraphy and fauna. In *Earliest man and environments in the Lake Rudolf Basin,* ed. Y. Coppens et al., pp. 163-70. Chicago: University of Chicago Press.

_____.   1978.   Taphonomic and ecologic information from bone weathering. *Paleobiology* 4 (2):150-62.

Behrensmeyer, Anna K.; Western, D.; Dechant Boaz, D. E.   1979.   New perspectives in vertebrate paleoecology from a recent bone assemblage. *Paleobiology* 5 (1):12-21.

Bell, R. H. V.   1970.   The use of the herb layer by grazing ungulates in Serengeti.   In *Animal populations in relation to their food resources*, ed. A. Watson, pp. 111-24. Oxford: Blackwell Scientific.

Benzécri, J. P.   1973.   *L'Analyse des correspondences.*   L'Analyse des donnees, vol. 2. Paris: Dunod.

Binford, L. R.   1964.   A consideration of archaeological research design. *American Antiquity* 29 (4):425-41.

_____.   1967.   Reply to Chang.   In Major aspects of the interrelationship between archaeology and ethnology. *Current Anthropology* 8 (3):227-43.

_____.   1968.   Methodological considerations of the archaeological use of ethnographic data.   In *Man the hunter,* ed. R. B. Lee and I. DeVore, pp. 263-73. Chicago: Aldine.

_____.   1973.   Interassemblage variability—the Mousterian and the "functional" argument.   In *The explanation of culture change: Models in prehistory,* pp. 227-56. Pittsburgh: University of Pittsburgh Press.

_____. 1975. Sampling, judgement, and the archaeological record. In J. W. Mueller, ed., *Sampling in archaeology*, pp. 251-57. Tucson: University of Arizona Press.

Bishop, W. W. 1963. The later Tertiary and Pleistocene in eastern Equatorial Africa. In *African ecology and human evolution*, ed. F. C. Howell and F. Bourlière, pp. 246-75. Viking Fund Publications in Anthropology, No. 36.

Bishop, W. W.; Chapman, G. R.; Hill, A. P.; and Miller, J. A. 1971. Succession of Cainozoic vertebrate assemblages from the northern Kenya Rift Valley. *Nature* 233: 389-94.

Bishop, W. W., and Pickford, M. H. L. 1975. Geology, fauna and palaeoenvironments of the Ngorora Formation, Kenya Rift Valley. *Nature* 254:185-92.

Boaz, N. T., and Behrensmeyer, A. K. 1976. Hominid taphonomy: Transport of human skeletal parts in an artificial fluviatile environment. *Am. J. Phys. Anth.* 45 (1):53-60.

Boné, E. L. 1955. Une Clavicule et un nouveau fragment mandibulaire d'*Australopithecus prometheus*. *Palaeontologica Africana* 3:87.

Boné, E. L., and Singer, R. 1965. *Hipparion* from Langebaanweg, Cape Province, and a revision of the genus in Africa. *Ann. S. Afr. Mus.* 48:273-397.

Boshier, A. K., and Beaumont, P. B. 1972. Mining in southern Africa and the emergence of modern man. *Optima* 22 (1):2-12.

Boucot, A. J. 1953. Life and death assemblages among fossils. *Am. Jour. Sci.* 251:25-40.

Bourlière, F. 1963. Observations on the ecology of some large African mammals. In *African ecology and human evolution*, ed. F. C. Howell and F. Bourlière. Chicago: Aldine.

_____. 1973. The comparative ecology of rainforest mammals in Africa and tropical America: Some introductory remarks. In *Tropical forest ecosystems in Africa and South America: A comparative review*, ed. B. J. Meggers, E. S. Ayensu, and W. D. Duckworth, pp. 279-92. Washington: Smithsonian Institution Press.

Bourlière, F., and Hadley, M. 1970. The ecology of tropical savannas. *Ann. Rev. Ecol. Syst.* 1:125-52.

Bourlière, F., and Verschuren, J. 1960. *Introduction à l'ecologie des ongules du Parc National Albert*. Brussels: Institut des Parcs Nationaux du Congo Belge.

Bourn, D. M., and Coe, M. J. 1978. The size, structure and distribution of the giant tortoise population of Aldabra. *Phil. Trans. Roy. Soc. Lond. B.*, 282:139-75.

_____. 1977. Features of tortoise mortality and decomposition on Aldabra. In *Royal Society Discussion Meeting*, ed. D. Stoddart and T. Westoll.

Bourquin, O. 1974. Utilisation and aspects of management of the Willem Pretorious Game Reserve. *J. S. Afr. Mgmt. Asso.* 3 (2):65-73.

Bourquin, O.; Vincent, J.; and Hitchins, P. M. 1971. The vertebrates of the Hluhluwe Game Reserve—Corridor (State Land)—Umfolozi Game Reserve Complex. *The Lammergeyer* 14:5-58.

Brain, C. K. 1958. The Transvaal Ape-Man-Bearing Cave Deposits. *Transv. Mus. Mem.* 13: 1-125.

_____. 1967. Hottentot food remains and their meaning in the interpretation of fossil bone assemblages. *Scientific Papers Namib Desert Research Station* 32:1-11.

_____. 1968. Who killed the Swartkrans Ape-Men? *South African Museums Association Bulletin* 9:127-39.

_____. 1969a. The contribution of Namib Desert Hottentots to an understanding of Australopithecine bone accumulations. *Scientific Papers of the Namib Desert Research Station* 39:13-22.

_____. 1969*b*. New evidence for climatic change during Middle and Late Stone Age times in Rhodesia. *South African Archaeological Bulletin* 24:127-43.

_____. 1969*c*. Faunal remains from the Bushman Rock Shelter, Eastern Transvaal. *South African Archaeological Bulletin* 24:52-55.

_____. 1970. New finds at the Swartkrans australopithecine site. *Nature* 225:1112-19.

_____. 1974. Some suggested procedures in the analysis of bone accumulations from southern African Quaternary sites. *Ann. Transv. Mus.* 29 (1):1-8.

_____. 1976*a*. Some principles in the interpretation of bone accumulations associated with man. In *Human origins, perspectives in human evolution*, vol. 3, ed. G. L. Isaac and E. McCown, pp. 97-116. Menlo Park, Ca.: W. A. Benjamin.

_____. 1976*b*. A re-interpretation of the Swartkrans site and its remains. *S. Afr. J. Sci.* 72:141-46.

Brain, C. K., and Cooke, C. K. 1967. A preliminary account of the Redcliff stone age cave site in Rhodesia. *South African Archaeological Bulletin* 21:171-182.

Brain, C. K.; Vrba, E. S.; and Robinson, J. T. 1974. A new hominid innominate bone from Swartkrans. *Ann. Transv. Mus.* 29:55-63.

Bretsky, P. W. 1968. Evolution of Paleozoic marine invertebrate communities. *Science* 159:1231-33.

_____. 1969. Evolution of Paleozoic benthic marine invertebrate communities. *Paleogeog., Paleoclimat., Paleoecol.* 6:45-69.

Bretsky, P. W., and Lorenz, P. M. 1970. Adaptive response to environmental stability: A unifying concept in paleontology. In *Proc. N. Amer. Paleontol. Congr.*, ed. E. Yochelson, 1:522-50.

Briggs, L. I., and Middleton, G. V. 1965. Hydromechanical principles of sediment structure formation. In *Primary sedimentary structures and their hydrodynamic interpretation*, ed. G. V. Middleton. Soc. Econ. Paleontol. and Mineral., Spec. Pub. No. 12:5-16.

Brooks, P. M. 1975. The sexual structure of an impala population and its relationship to an intensive game removal programme. *The Lammergeyer* 22:1-8.

Brown, A. B. 1974. Bone strontium as a dietary indicator in human skeletal populations. *Contr. to Geol.* 13:1-2.

Brown, F. H.; Howell, F. C.; and Eck, G. G. 1978. Observations on problems of correlation of late Cenozoic hominid-bearing formations in the North Lake Turkana Basin. In *Geological background to fossil man,* ed. W. W. Bishop. Edinburgh: Scottish Academic Press.

Browne, M. W., and Vrba, E. S. In prep. Correspondence analysis applied to biological census data.

Brudevold, F., and Söremark, R. 1967. Chemistry of the mineral phase of enamel. In *Structural and chemical organization of teeth,* ed. E. W. Miles, 2:257-77. New York: Academic Press.

Butzer, K. W. 1971*a*. Another look at the Australopithecine cave breccias of the Transvaal. *Am. Anthrop.* 73:1197-1201.

_____. 1971*b*. Recent history of an Ethiopian delta. Univ. Chicago Dept. Geog. Res. Paper, 136:1-184.

_____. 1973*a*. On the geology of a late Pleistocene *Mammuthus* site, Virginia, Orange Free State. *Navors Nas. Mus. Bloemfontein* 2:386-93.

_____. 1973*b*. Geology of Nelson Bay Cave, Robberg, South Africa. *South African Archaeological Bulletin* 28:97-110.

_____. 1973*c*. On the sedimentary sequence from the "E" quarry, Langebaanweg, Cape Province. *Quaternaria* 17:237-43.

Butzer, K. W., and Helgren, D. M. 1972. Late Cenozoic evolution of the Cape coast between Knysna and Cape St. Francis, South Africa. *Quaternary Research* 2:143-69.

Carr, C. J. 1972. A system of societal/environmental interaction: The Dassanetch of Southwest Ethiopia. Ph.D. dissertation, University of Chicago.

Carrington, A. J., and Kensley, B. F. 1969. Pleistocene molluscs from Namaqualand coast. *Ann. S. Afr. Mus.* 52:189-223.

Case, E. C. 1919. The environment of vertebrate life in the late Paleozoic in North America: A paleogeographic study. *Publ. Carnegie Inst. Wash.* 283:1-273.

_____. 1926. Environment of tetrapod life in the late Paleozoic of regions other than North America. *Publ. Carnegie Inst. Wash.* 375:1-211.

Casebeer, R. L., and Bmai, H. T. M. 1974. Animal mortality 1973/74, Kajiado district. UNDP/FAO Wildlife Management Project, Nairobi, Kenya. Project Working Document No. 5, pp. 1-32.

Caughley, G. 1966. Mortality patterns in mammals *Ecology* 47 (6):906-18.

Chaney, R. W. 1933. A Tertiary flora from Uganda. *J. Geol.* 41:702-9.

Chang, K. C. 1967. Major aspects of the interrelationship of archaeology and ethnology. *Current Anthropology* 8(3):227-43.

Chaplin, R. E. 1971. The study of animal bones from archaeological sites. New York: Seminar Press.

Charig, A. J. 1963. Stratigraphical nomenclature in the Songea series of Tanganyika. *Geol. Surv. Tanganyika, Records* 10:47-53.

Chesters, K. I. M. 1957. The Miocene flora of Rusinga Island, Lake Victoria, Kenya. *Palaeontographica* 101(B):30-67.

Clapham, W. D., Jr. 1970. Evolution of upper Permian terrestrial floras in Oklahoma as determined from pollen and spores. In *Proc. N. Amer. Paleontol. Congr.*, ed. E. Yochelson, 6:411-26. Lawrence, Kansas: Allen Press.

Clark, J.; Beerbower, J. R.; and Kietzke, K. K. 1967. Oligocene sedimentation, stratigraphy and paleoclimatology in the Big Badlands of South Dakota. *Fieldiana Geol.* 5:1-158.

Clark, J. D. 1967. Atlas of African prehistory. Chicago: University of Chicago Press.

_____. 1971. Human behavioral differences in southern Africa during the later Pleistocene. *Am. Anthrop.* 73:1211-36.

Clark, J. G. D. 1936. *The Mesolithic settlement of Northern Europe.* Cambridge: Cambridge University Press.

Cloud, P. E., Jr. 1959. Paleoecology: Retrospect and prospect. *Jour. Paleontol.* 33: 926-62.

Cobb, S. 1976. The distribution and abundance of the large herbivore community of Tsavo National Park, Kenya. D. Phil. thesis, University of Oxford, England.

Cochran, W. G. 1963. *Sampling techniques.* New York: Wiley.

Cody, M. L. 1974. *Bird communities.* Princeton: Princeton University Press.

Coe, M. J. 1976. The decomposition of elephant carcases in the Tsavo (East) National Park, Kenya. *J. Arid Environm.* 1:71-86.

Coe, M. J.; Cumming, D. H.; and Phillipson, J. 1976. Biomass and production of large African herbivores in relation to rainfall and primary production. *Oecologia* 22: 341-54.

Cooke, C. K. 1963. Report of excavations at Pomongwe and Tshangula Daves, Matopo Hills, Southern Rhodesia. *S. Afr. Archaeol. Bull.* 18 (71):73-151.

Cooke, H. B. S. 1962. Notes on faunal material from the Cave of Hearths and Kalkban. In *Prehistory of the Transvaal*, ed. R. J. Mason, pp. 447-53. Johannesberg: Witwatersrand Univ. Press.

_____. 1963. Pleistocene mammal faunas of Africa, with particular reference to southern Africa. In *African ecology and human evolution*, ed. F. C. Howell and F. Bourlière, pp. 65-116. Chicago: Aldine.

Cooke, H. B. S.; Malan, B. D.; and Wells, L. H. 1945. Fossil man in the Lebombo Mountains, South Africa: The "Border Cave," Ingwavuma District, Zululand. *Man* 45 (3):6-13.

Cooke, H. B. S., and Wells, L. H. 1951. Fossil remains from Chelmer near Bulawayo, Southern Rhodesia. *S. Afr. J. Sci.* 47:205-9.

Cope, E. D. 1885. On the evolution of the vertebrata. *Amer. Natur.* 19:140-48.

Coppens, Y., and Howell, F. C. 1976. Mammalian faunas of the Omo Group: Distributional and biostratigraphical aspects. In *Earliest man and environments in the Lake Rudolf Basin*, ed. Y. Coppens et al., pp. 177-92. Chicago: University of Chicago Press.

Coppens, Y.; Howell, F. C.; Isaac, G. Ll.; and Leakey, F. E. F., eds. 1976. *Earliest man and environments in the Lake Rudolf Basin*. Chicago: University of Chicago Press.

Corfield, T. F. 1973. Elephant mortality in Tsavo National Park, Kenya. *E. Afr. Wildl. J.* 11:339-68.

Cotter, G. 1933. The geology of part of the Attock District west of longitude 72°45' E. *Mem. Geol. Surv. India* 55:63-161.

Craig, G. Y. 1953. Fossil communities and assemblages. *Am. J. Sci.* 251:547-48.

Craig, G. Y., and Hallam, H. 1963. Size-frequency and growth-ring analysis of *Mytilus edulis* and *Cardinus edule*, and their paleoecological significance. *Paleontol.* 6:731-50.

Craig, G. Y., and Oertel, G. 1966. Deterministic models of living and fossil populations of animals. *Quart. Jour. Geol. Soc. London* 122:315-55.

Croizat, L.; Nelson, G.; and Rosen, D. E. 1974. Centers of origin and related concepts. *Syst. Zool.* 23:265-87.

Crook, J. H. 1972. The socio-ecology of primates. In *Perspectives on human evolution*, ed. S. L. Washburn and P. Dolhinow, pp. 281-347. New York: Holt, Rinehart and Winston.

Dart, R. A. 1957. The Osteodontokeratic culture of *Australopithecus prometheus*. *Mem. Trans. Mus.* 10:1-105.

Dasmann, R. and Mossman, A. S. 1962. Road strips counts for estimating numbers of African ungulates. *J. Wildl. Mgmt.* 26:101-4.

Dasmann, R. F., and Mossman, A. S. 1961. Commercial utilisation of game mammals on a Rhodesian ranch. Paper presented at the annual meeting of the Wildlife Society, California Section, Jan. 1961, and at the meeting of the National Affairs Association, Bulawayo, Southern Rhodesia, March 1961. Mimeogr.

Dawson, J. B. 1962. The geology of Oldoinyo Lengai. *Bull. Volc.* 24:349-87.

_____. 1964a. Carbonatite volcanic ashes in northern Tanganyika. *Bull. Volc.* 27: 81-92.

_____. 1964b. Carbonatite tuff cones in northern Tanganyika. *Geol. Mag.* 101:129-37.

Day, M. H. 1975. Hominid post-cranial remains from Bed I, Olduvai Gorge. In *Human origins*, ed. G. L. Isaac and E. McCown. Reading, Mass.: W. A. Benjamin.

Day, M. H.; Leakey, R. E. F.; Walker, A.; and Wood, B. A. 1974. New hominids from East Rudolf I. *Am. Jour. Phys. Anthrop.* 42:641-76.

Day, M. H., and Molleson, T. I. 1976. The puzzle from JK2—a femur and a tibial
    fragment (O.H. 34) from Olduvai Gorge, Tanzania. *Jour. Hum. Evol.* 5:455-65.

De Graaf, G.; Bothma, J. du P.; and Moolman, E. Unpublished. Preliminary results of
    six aerial game censuses in the Kalahari Gemsbok National Park and the adjoining
    Kalahari Gemsbok Park (Republic of Botswana) for the period March 1973-April 1974.

De Vos, A. 1969. Ecological conditions affecting the production of wild herbivorous
    mammals on grasslands. *Advances in Ecological Research* 6:137-83.

Deacon, H. J. 1967. Two radiocarbon dates from Scott's Cave, Gamtoos Valley. *S. Afr.
    Archaeol. Bull.* 22:51-52.

_____. 1972. A review of the post-Pleistocene in South Africa. *South African
    Archaeological Society Goodwin Series* 1:26-45.

_____. 1974. An archaeological study of the Eastern Cape in the Post-Pleistocene
    Period. Ph.D. dissertation, University of Cape Town.

Deacon, H. J., and Brooker, M. 1976. The Holocene and Upper Pleistocene sequence in the
    southern Cape. *Ann. S. Afr. Mus.* 71:203-14.

Deacon, H. J., and Deacon, J. 1963. Scott's Cave: A Late Stone Age Site in the Gamtoos
    Valley, Southeastern Cape. *S. Afr. J. Sci.* 70:186-87.

Deacon, J. 1972. Wilton, an assessment after fifty years. *S. Afr. Archaeol. Bull.*
    27:10-48.

_____. 1974. Patterning in the radiocarbon dates for the Wilton/Smithfield
    complex in southern Africa. *S. Afr. Archaeol. Bull.* 29:3-18.

_____. 1976. Report on stone artifacts from Duinefontein 2, Melkbosstrand. *S.
    Afr. Archaeol. Bull.* 31:21-25.

Deans, T., and Powell, J. L. 1968. Trace elements and strontium isotopes in carbonotites,
    fluorites and limestones from India and Pakistan. *Nature* 218:750-52.

Deevey, E. S. 1947. Life tables for natural populations of animals. *Quart. Rev. Biol.*
    22:283-314.

Dingle, R. V. 1971. Tertiary sedimentary history of the continental shelf off southern
    Cape Province, South Africa. *Trans. Geol. Soc. S. Afr.* 74:173-86.

_____. 1973. The geology of the continental shelf between Luderitz and Cape Town
    (Southwest Africa). With special reference to tertiary strata. *Jour. Geol. Soc.
    London* 129:337-63.

Dodson, P. 1971. Sedimentology and taphonomy of the Oldman formation (Campanian)
    Dinosaur Provincial Park, Alberta (Canada). *Paleogeog., Paleoclimat., Paleoecol.*
    10:21-74.

_____. 1973. The significance of small bones in paleoecological interpretation.
    Univ. Wyoming Contrib. Geol. 12(1):15-19.

Dorst, J., and Dandelot, P. 1970. *A field guide to the larger mammals of Africa.* London:
    Collins.

Dowsett, R. J. 1966. Wet season game populations and biomass in the Ngoma area of the
    Kafue National Park. *The Puku* 4:135-45.

Dougall, H. W.; Drysdale, V. M.; and Glover, P. E. 1964. The chemical composition of
    Kenya browse and pasture herbage. *E. Afr. Wildl. J.* 2:86-121.

Douglas-Hamilton, I. 1972. On the ecology of the African elephant. D. Phil. thesis,
    University of Oxford.

Dreyer, T. F. 1938. The archaeology of the Florisbad deposits. *Argeologese Navorsing
    van die Nasionale Museum* 1(8):65-77.

Ducos, P. 1969. Methodology and results of the study of the earliest domesticated
    animals in the Near East (Palestine). In *The domestication and exploitation of*

plants and animals, ed. P. J. Ucko and G. W. Dimbleby, pp. 265-75. Chicago: Aldine.

Durham, J. U. 1971. Evolution of communities. _Geotimes_ 16:34-35.

DuToit, A. L. 1917. The phosphate of Saldanha Bay. _Mem. Geol. Surv. Univ. S. Afr._ 10: 1-31.

Edwards, D. 1974. Survey to determine the adequacy of existing conserved areas in relation to vegetation types. A preliminary report. _Koedoe_ 17:1-38.

Efremov, I. A. 1940. Taphonomy: A new branch of paleontology. _Pan-American Geologist_ 74:81-93.

_____. 1950. Taphonomy and the geological record. _Tr. Paleontol. Inst. Acad. Sci. USSR_ 24:1-177 (in Russian).

_____. 1953. Taphonome et annales géologiques (trans. S. Ketchian and J. Roger). _Ann. du Centre d'Etud. et de Doc. Paléontol._ No. 4, pp. 1-164.

_____. 1957. On the taphonomy of fossil faunas of terrestrial vertebrates of Mongolia. _Vertebrata Asiatica_ 1:83-102.

_____. 1958. Some considerations on the biological bases of paleozoology. _Vert. Palasiatica_ 2:83-99.

_____. 1961. Some considerations on the biological bases of paleontology. _Tr._ IV Session, All Soviet Paleontol. Congr. 4:198-220. (In Russian)

Einstein, H. A. 1950. The bed-load function for sediment transportation in open channel flows. _U. S. Dept. Agriculture Tech. Bull._ 1026, pp. 1-70.

Eldredge, N., and Gould, S. J. 1972. Punctuated equilibria: An alternative to phyletic gradualism. In _Models in paleoecology,_ ed. T. J. M. Schopf, pp. 82-115. San Francisco: Freeman, Cooper and Co.

Elton, C. S. 1958. _The ecology of invasion by animals and plants._ London: Methuen Lanscape.

Estes, R. 1962. A fossil gerrhosaur from the Miocene of Kenya (Reptilia: Cordylidae). _Breviora_ 158:1-10.

Estes, R. D. 1974. Social organization in African Bovidae. In _The behaviour of ungulates and its relation to management_, ed. V. Geist and F. Walther. I.U.C.N. Publ. (n.s.), 24, 1:166-205.

Ewer, R. F. 1955. The fossil carnivores of the Transvaal caves: Machairondontinae. _Proc. Zool. Soc. London_ 125:587-615.

_____. 1956. The fossil carnivores of the Transvaal caves: Felinae. _Proc. Zool. Soc. London_ 126:83-95.

Fagan, B. M., and van Noten, F. L. 1971. The hunter-gatherers of Gwisho. _Musée Royal de l'Afrique Centrale (Tervuren, Belgium) Annales_ 74:1-227.

Fager, E. W. 1957. Determination and analysis of recurrent groups. _Ecol._ 38:586-95.

Fagerstrom, J. A. 1964. Fossil communities in paleoecology: Their recognition and significance. _Bull. Geol. Soc. Am._ 75:1197-1216.

Fatmi, A. N. 1973. Lithostratigraphic units of the Kohat-Potwar Province, Indus Basin, Pakistan. _Mem. Geol. Surv. Pakistan_ 10:1-80.

Fenchel, T. 1974. Intrinsic rate of natural increase: the relationship with body size. _Oecologia_ (Berl.) 14:317-26.

Field, C. R., and Laws, R. M. 1970. The distribution of the larger herbivores in the Queen Elizabeth National Park, Uganda. _J. Appl. Ecol._ 7:237-94.

Fleming, T. H. 1973. The number of mammalian species in north-central American forest communities. _Ecol._ 54:555-63.

Flessa, K. 1975. Area, continental drift and mammalian diversity. *Paleobiology* 1:189-94.

Foin, T. C.; Valentine, J. W.; and Ayala, F. J. 1975. Extinction of taxa and Van Valen's law. *Nature* 257:514-15.

Foster, J. B., and Coe, M. J. 1968. The biomass of game animals in Nairobi National Park. *J. Zool. London* 155:413-25.

Foster, J. B., and Kearney, D. 1967. Nairobi National Park Game Census, 1966. *E. Afr. Wildl. J.* 5:112-20.

Freeman, L. G. 1968. A theoretical framework for interpreting archaeological materials. In *Man the hunter,* pp. 262-67, ed. R. B. Lee and I. DeVore. Chicago: Aldine.

_____. 1973. The significance of mammalian faunas from Paleolithic occupations in Cantabrian Spain. *American Antiquity* 38:3-44.

_____. In press. The analysis of some occupation floor distributions from earlier and middle Paleolithic sites in Spain. In *Views of the past,* ed. L. G. Freeman. Chicago: Aldine.

Fison, G. C. 1974. *The Casper site.* New York: Academic Press.

Garrod, D. A. E.; Buxton, L. H. D.; Smith, G. E.; and Bate, D. M. A. 1928. Excavation of a Mousterian rock shelter at Devil's Tower, Gibraltar. *Journal of the Royal Anthropological Institute* 58:33-113.

Gentry, A. W. 1968. Historical zoogeography of antelopes. *Nature* 217:874-75.

_____. 1970a. The Bovidae (Mammalia) of the Fort Ternan fossil fauna. In *Fossil vertebrates of Africa,* ed. L. S. B. Leakey and R. J. G. Savage, vol. 2. London and New York: Academic Press.

_____. 1970b. The Langebaanweg Bovidae. Appendix in Q. B. Hendey, A review of the geology and paleontology of the Plio/Pleistocene deposit at Langebaanweg, Cape Province. *Ann. S. Afr. Mus.* 56:114-17.

_____. 1971. The earliest goats and other antelopes from the Samos Hipparion fauna. *Bull. Br. Mus. Nat. Hist. (Geol.)* 20:231-96.

Gifford, D. P. In press. Ethnoarchaeological observations on natural processes affecting cultural materials. In *New Directions in Ethnoarchaeology,* ed. R. A. Gould. Albuquerque: University of New Mexico Press.

Gifford, D. P., and Behrensmeyer, A. K. 1977. Observed formation and burial of a recent human occupation site in Kenya. *Quat. Res.* 8:245-66.

Gill, W. D. 1952a. The stratigraphy of the Siwalik Series in the northern Potwar, Punjab, Pakistan. *Quart. J. Geol. Soc. London* 107:375-94.

_____. 1952b. The tectonics of the sub-Himalayan fault zone in the northern Potwar region and in the Kangra District of the Punjab. *Quart. J. Geol. Soc. London* 107: 395-421.

Gingerich, P. D. 1976. Paleontology and phylogeny: Patterns of evolution at the species level in early Tertiary mammals. *Amer. Jour. Sci.* 276:1-28.

Goddard, J. 1970. Age criteria and vital statistics of a black rhinoceros population. *E. Afr. Wildl. J.* 8:105-21.

Gosling, L. M. 1975. The ecological significance of territorial behaviour in the male hartebeest. Ph.D. dissertation, University of Nairobi.

Gould, J. S. 1970. Land snail communities and Pleistocene climates in Bermuda: A multivariate analysis of microgastropod diversity. In *Proc. N. Amer. Paleontol. Congr.,* ed. E. Yochelson, pp. 486-521.

Gould, R. A. In press. Beyond analogy in ethnoarchaeology. In *New directions in ethno-archaeology,* ed. R. A. Gould. Albuquerque: University of New Mexico Press.

Goulden, C. E. 1966. La Aquada de Santa Ana Vieja; An interpretive study of the cladoceran microfossils. *Archiv. Hydrobiol.* 62:373-404.

Graham, A., and Bell, R. 1969. Factors influencing the countability of animals. *E. Afr. Agric. For. J.* 34:38-43.

Gregory, J. T. 1971. Speculations on the significance of fossil vertebrates for the antiquity of the Great Plains of North America. Abh. hess. L.-Amt. Bodenforsch. 60, Heinz Tobien Festschrift, pp. 64-72.

Hallam, A. 1976. The Red Queen dethroned. *Nature* 259:12-13.

Halstead, L. B. 1974. *Vertebrate hard tissues*. London: Wykeham Press.

Hare, P. E. 1969. Geochemistry of proteins, peptides and amino acids. In *Organic geochemistry,* ed. G. Eglinton and M. T. J. Murphy. Berlin: Springer-Verlag.

Hare, P. E. 1974. Amino acid dating of bone—the influence of water. *Carnegie Institution of Washington Year Book* 73:576-81.

Hare, P. E.; Miller, G. H.; and Tuross, N. C. 1975. Simulation of natural hydrolysis of proteins in fossils. *Carnegie Institution of Washington Year Book* 74:609-12.

Hare, P. E., and Mitterer, R. M. 1969. Laboratory simulation of amino acid diagenesis in fossils. *Carnegie Institution of Washington Year Book* 67:205-8.

Harris, L. D. 1972. An ecological description of a semi-arid East African ecosystem. *Range Sci. Dept. Sci. Ser.* No. 11, Colorado State University.

Haughton, S. H. 1932. On the phosphate deposits near Langebaan Road, Cape Province. *Trans. Geol. Soc. S. Afr.* 35:119-24.

Hay, R. L. 1973. Lithofacies and environments of Bed 1, Olduvai Gorge, Tanzania. *Quaternary Research* 3(4):541-60.

_____. 1976. *Geology of Olduvai Gorge*. Berkeley: University of California Press.

Hecker, R. F. 1965. *Introduction to paleoecology*. New York: Elsevier.

Hendey, Q. B. 1969. Quaternary vertebrate fossil sites in the South-western Cape Province. *S. Afr. Archaeol. Bull.* 24:96-105.

_____. 1970a. A review of the geology and paleontology of the Plio/Pleistocene deposits at Langebaanweg, Cape Province. *Ann. S. Afr. Mus.* 56:75-117.

_____. 1970b. The age of the fossiliferous deposits at Langebaanweg, Cape Province. *Ann. S. Afr. Mus.* 56:119-31.

_____. 1974. The Late Cenozoic Carnivora of the South-western Cape Province. *Ann. S. Afr. Mus.* 63:1-369.

Hendey, Q. B., and Hendey, H. 1968. New Quaternary fossil sites near Swartklip, Cape Province. *Ann. S. Afr. Mus.* 52:43-73.

Hendey, Q. B., and Singer, R. 1965. The faunal assemblages from the Gamtoos Valley shelters. *S. Afr. Archaeol. Bull.* 20(80):206-13.

Hendrichs, H. 1970. Schätzongen der Huftier-Biomasse in der Dornbusch-Savanne nördlich und westlich der Serengeti-Steppe in Ostafrika nach einem neuen Verfahren und Bemerzungen zur Biomasse der anderen pflanzenfressenden Tierarten. *Säugetierk. Mitt.* 18:237-55.

Herbert, H. J. 1972. *The population dynamics of the Waterbuck Kobus ellipsiprymnus (Ogilby, 1833) in the Sabi-sand Wildtuin*. Hamburg: Paul Parey.

Herring, G. M. 1972. The organic matrix of bone. In *The biochemistry and physiology of bone,* 2nd ed., ed. G. H. Bourne, pp. 128-84. New York: Academic Press.

Hershfield, D. M. 1962. A note on the variability of annual precipitation. *J. Appl. Meteor.* 1:575-78.

Hershkovitz, P. 1966. Mice, land bridges and Latin America faunal interchange. In *Ectoparasites of Panama,* ed. R. Wenzel and J. Tipton, pp. 725-51. Chicago: Field Museum of Natural History.

Hill, A. 1975. Taphonomy of contemporary and late Cenozoic East African vertebrates. Ph.D. dissertation, University of London.

_____. 1976. On carnivore and weathering damage to bone. *Current Anthropology* 17(2):335-36.

_____. In prep.*a*. Taphonomy of some contemporary East African vertebrate assemblages.

_____. In prep.*b*. Butchery and natural disarticulation; an investigatory technique.

Hill, J. N. 1968. Broken K Pueblo: Patterns in form and function. In *New perspectives in archaeology,* ed. S. R. Binford and L. R. Binford, pp. 103-42. Chicago: Aldine.

_____. 1970. Prehistoric social organization in the American Southwest: Theory and method. In *Reconstructing prehistoric pueblo societies,* ed. W. A. Longacre, pp. 11-58. Albuquerque: University of New Mexico Press.

Hillman, J. C. 1974. Ecology and behavior of the wild eland. *African Wildlife Leadership Foundation News* 9(3):6-9.

Hillman, J. C., and Hillman, A. K. K. 1977. Mortality of wildlife in Nairobi National Park, during the drought of 1973-1974. *E. Afr. Wildl. J.* 15:1-18.

Hoeck, H. N. 1975. Differential feeding behavior of the sympatric hyrax *Procavia johnstoni* and *Heterohyrax brucei*. *Oecologia* 22:15-47.

Hoffman, A. C. 1955. Important contributions of the Orange Free State to our knowledge of primitive man. *S. Afr. J. Sci.* 51:163-69.

Höhnel, L. von. 1894. *Discovery of Lakes Rudolf and Stephanie*. London: Longmans, Green. (Repr. ed. Frank Cass Press, 1968.)

Hole, F., and Flannery, K. V. 1967. The prehistory of Southwestern Iran: A preliminary report. *Proceedings of the Prehistoric Society* 33:147-206.

Hopwood, A. T. 1933. Miocene primates from Kenya. *J. Linn. Soc. (Zool.)* 38:437-64.

Huffaker, C. B. 1958. Experimental studies on predation dispersion factors and predator-prey oscillations. *Hilgardia* 27:343-83.

Hughes, A. R. 1954. Hyaenas versus australopithecines as agents of bone accumulation. *Am. J. Phys. Anthrop.* 12(4):467-86.

Hussain, S. T. 1971. Revision of *Hipparion* (Equidae, Mammalia) from the Siwalik Hills of Pakistan and India. *Bayer Akad. Wiss., Math.-Naturwiss. Kl. Abh.* 147:1-68.

Hutchinson, G. E. 1957. Concluding remarks. *Population Studies: Animal Ecology and Demography,* Cold Spring Harbor Symposium on Quantitative Biology 22.

Hvidberg-Hansen, H., and de Vos, A. 1971. Reproduction, population and herd structure of two Thomson's Gazelle (*Gazella thomsoni* Günther) populations. *Mammalia* 35:1-16.

Imbrie, J., and Newell, N. eds. 1964. *Approaches to paleoecology*. New York: John Wiley & Sons.

Isaac, G. Ll. 1967. Towards the interpretation of occupation debris: Some experiments and observations. *Kroeber Anthropological Society Papers* 37:31-57.

Isaac, G. Ll.; Pilbeam, D.; and Walker, A. 1972. Fossil hominids with potassium-argon age control. In *Calibration of hominid evolution,* ed. W. W. Bishop and J. A. Miller, appendix I, chart 1. Edinburgh: Scottish Academic Press.

Isaac, Ll., and McCown, E., eds. 1976. *Human origins: Louis Leakey and the East African evidence*. Menlo Park, Ca.: W. A. Benjamin.

Jarman, P. J. 1974. The social organisation of antelope in relation to their behaviour. *Behaviour* 58(3,4):215-67.

Johnson, R. G. 1957. Experiments on the burial of shells. *J. Geol.* 65:527-35.

_____. 1960. Models and methods for analysis of the mode of formation of fossil assemblages. *Bull. Geol. Soc. Am.* 71:1075-86.

_____. 1962. Mode of formation of marine fossil assemblages of the Pleistocene, Millerton Formation of California. *Bull. Geol. Soc. Am.* 73:113-30.

_____. 1970. Variations in diversity within marine communities. *Amer. Natur.* 104:285-300.

_____. 1972. Conceptual models of benthic marine communities. In *Models in paleobiology*, ed. J. M. Schopf, pp. 148-59. San Francisco: Freeman, Cooper & Co.

Joubert, E. 1971. Observations on the habitat-preferences and population dynamics of the black-faced impala *Aepyceros petersi* Bocáge 1875 in South West Africa. *Madoqua* 3(1):55-56.

Kauffman, E., and Scott, R. W. 1976. Basic concepts of community ecology and paleo-ecology. In *Structure and classification of paleocommunities*, ed. R. W. Scott and R. R. West, pp. 1-28. Stroudsburg, Pa.: Dowden, Hutchinson & Ross.

Kent, P. E. 1944. The age and tectonic relationships of East African volcanic rocks. *Geol. Mag.* 81:15-27.

Kingdon, J. 1971. *East African mammals, an atlas of evolution in Africa*, vol. 1. London: Academic Press.

Kleiber, M. 1961. *The fire of life*. New York: Wiley.

Klein, R. G. 1972a. A preliminary report on the June through September 1970 excavations at Nelson Bay Cave, Cape Province, South Africa. *Paleoecology of Africa* 6: 177-206.

_____. 1972b. The late Quaternary mammalian fauna of Nelson Bay Cave (Cape Province, South Africa): Its implications for megafaunal extinctions and for cultural and environmental change. *Quaternary Research* 2:135-42.

_____. 1973. Geological antiquity of Rhodesian Man. *Nature* 244:311-12.

_____. 1974a. Environment and subsistence of prehistoric man in the southern Cape Province, South Africa. *World Archaeology* 5:249-84.

_____. 1974b. On the taxonomic status, distribution, and ecology of the blue antelope, *Hippotragus leucophaeus* (Pallas, 1766). *Ann. S. Afr. Mus.* 65(4):99-143.

_____. 1974c. A provisional statement on terminal Pleistocene mammalian extinctions in the Cape Biotic Zone (southern Cape Province, South Africa). *South African Archaeological Society Goodwin Series* 2:39-45.

_____. 1975a. Middle Stone Age man-animal relationships in southern Africa: Evidence from Klasies River Mouth and Die Kelders. *Science* 190:365-67.

_____. 1975b. Paleoanthropological implications of the non-archeological bone assemblage from Swartklip 1, South-Western Cape Province, South Africa. *Quarternary Research* 5:275-88.

_____. 1976a. A preliminary report on the Middle Stone Age open-air site of Duinefontein 2 (Melkbosstrand, South-Western Cape Province, South Africa). *S. Afr. Archaeol. Bull.* 31:12-20.

_____. 1976b. The mammalian fauna of the Klasies River Mouth sites. *S. Afr. Archaeol. Bull.* 31:12-18.

Klein, R. G., and Scott, K. 1974. The fauna of Scott's Cave, Gamtoos Valley, South-eastern Cape. *S. Afr. J. Sci.* 70:186-87.

Knight, B., and Lauder, I. 1969. Methods of dating skeletal remains. *Human Biology* 41: 322-41.

Kolata, G. B. 1975. Paleobiology: Random events over geological time. *Science* 189:625-26, 660.

Konerman, H. 1971. Quantitative Bestimmung der Material verteilung nach Rontgenbilbern des knochens miteines neun fotografishen methods. *Z. Anat.* 134:13.

Kravtchenko, K. N. 1964. Soan Formation—upper unit of Siwalik group in Potwar. *Science and Industry* 2:230-33.

Krynine, P. D. 1937. Petrography and genesis of the Siwalik series. *Am. J. Sci.* 34: 422-46.

Kurtén, B. 1952. The Chinese Hipparion fauna. *Commentat. Biol.* 13(4):1-82.

_____. 1953. On the variation and population dynamics of fossil and recent mammal populations. *Acta Zool. Fennica* 76:1-122.

_____. 1954. Population dynamics—a new method in paleontology. *Jour. Paleontol.* 28:286-92.

_____. 1960. Chronology and faunal evolution of the earlier European glaciations. *Soc. Sci., Fennica; Comment. Biol.* 21:1-62.

_____. 1963. Villafranchian faunal evolution. *Soc. Sci., Fennica; Comment. Biol.* 26:3-18.

_____. 1968. *Pleistocene mammals of Europe*. Chicago: Aldine.

Kutilek, M. J. 1974. The density and biomass of large mammals in Lake Nakuru National Park. *E. Afr. Wildl. J.* 12:201-12.

Lamprey, H. F. 1963. Ecological separation of the large mammal species in the Tarangire Game Reserve, Tanganyika. *E. Afr. Wildl. J.* 1:63-92.

_____. 1964. Estimation of the large mammals densities, biomass and energy exchange in the Tarangire Game Reserve and the Masai Steppe in Tanganyika. *E. Afr. Wildl. J.* 2:1-46.

Langbein, W. B., and Schumm, S. A. 1958. Yield of sediments in relation to mean annual precipitation. *Trans. Am. Geophys. Union* 39:1076-84.

Langdale-Brown, I.; Osmaston, H. A.; and Wilson, J.G. 1964. *The vegetation of Uganda and its bearing on land use*. Kampala: Government of Uganda.

Lawrence, D. R. 1968. Taphonomy and information losses in fossil communities. *Bull. Geol. Soc. Am.* 79:1315-30.

_____. 1971. The nature and structure of paleoecology. *Jour. Paleontol.* 45(4): 593-607.

Laws, R. M. 1966. Age criteria for the African elephant, *Loxodonta a. africana*. *E. Afr. Wildl. J.* 4:1-37.

_____. 1968. Dentition and ageing of the hippopotamus. *E. Afr. Wildl. J.* 6:19-52.

Laws, Richard M.; Parker, I. S. C.; and Johnstone, R. C. B. 1975. *Elephants and their habitats*. Oxford: Clarendon Press.

Leakey, M. D. 1970. Stone artifacts from Swartkrans. *Nature* 255:1112-19.

_____. 1971. *Excavations in Beds I and II, 1960-1963. Olduvai Gorge*, vol. 3. Cambridge: Cambridge University Press.

Lee, R. B. 1965. Subsistence ecology of the !Kung Bushmen. Ph.D. dissertation, University of California, Berkeley.

Le Gros Clark, W. E., and Leakey, L. S. B. 1951. *The Miocene Hominoidea of East Africa*. Fossil Mammals of Africa, vol. 1. London: British Museum of Natural History.

Lent, P. C. 1974. Mother-infant relationships in ungulates. In *The behavior of ungulates and its relation to management*, ed. V. Geist and F. Walther. I.C.U.N. Publ. (N.S.), no. 24, 1:14-55.

Leopold, E. B. 1969. Late Cenozoic palynology. In *Aspects of palynology,* ed. R. H. Tsudy and R. A. Scott. New York: Wiley-Interscience.

Leopold, L. B., and Miller, J. P. 1956. Ephemeral streams—hydraulic factors and their relation to the drainage net. U. S. Geol. Surv. Prof. Paper 282-A, pp. 1-37.

Leopold, L. B.; Wolman, M. G.; and Miller, J. P. 1964. *Fluvial processes in geomorphology.* San Francisco: Freeman.

Leroi-Gourhan, A. and Brezillon, M. 1966. L'Habitation magdalenienne No. 1 de Pincevent pres Montereau (Seine-et-Marne). *Gallia Prehistoire* 9(2):263-385.

Leuthold, W. and Leuthold, B. M. 1973. *Ecological studies of ungulates in Tsavo (East) National Park, Kenya.* Tsavo Research Project, Kenya National Parks.

Levins, R. 1968. *Evolution in changing environments.* Monogrs. in Pop. Biol. no. 2. Princeton: Princeton University Press.

Lewis, G. E. 1937. A new Siwalik correlation. *Am. J. Sci.* 33:191-204.

Lind, E. M. and Morrison, M. E. S. 1974. *East African vegetation.* London: Longman.

Lisle, T. E. 1976. Components of flow resistence in an alluvial channel. Ph.D. dissertation, University of California, Berkeley.

Livingstone, D. A. 1971. Speculations on the climatic history of mankind. *Am. Scientist* 59(3):332-37.

Longacre, W. A. 1970. *Archaeology as anthropology: A case study.* Anthropological Papers of the University of Arizona, no. 17.

_____. 1974. Kalinga pottery-making: The evolution of a research design. In *Frontiers in anthropology,* ed. M. J. Leaf, pp. 51-67. New York: Van Nostrand.

Louv, A. W. 1969. Bushman Rock Shelter, Ohrigstad, eastern Transvaal: A preliminary investigation. *S. Afr. Archaeol. Bull.* 24:39-51.

Lull, R. 1915. Triassic life of the Connecticut Basin. *Bull. Connecticut Geol. Soc.* 24:1-285.

MacArthur, R. O., and Wilson, E. O. 1967. *The theory of island biology.* Monogrs. in Pop. Biol. no. 1. Princeton: Princeton University Press.

McBurney, C. B. M. 1967. *The Haua Fteah (Cyrenaica) and the Stone Age of the Southeast Mediterranean.* Cambridge: Cambridge University Press.

McDonald, P.; Edward, R. A.; and Greenhalgh, J. F. D. 1966. *Animal nutrition* (2d. ed.). Edinburgh: Oliver and Boyd.

McNab, B. K. 1963. Bioenergetics and the determination of home range size. *Amer. Nat.* 97:133-40.

Maglio, V. J. 1972. Vertebrate faunas and chronology of hominid-bearing sediments east of Lake Rudolf, Kenya. *Nature* 239:379-85.

_____. 1974. Late Tertiary fossil vertebrate successions in the northern Gregory Rift, East Africa. *Ann. Geol. Surv. Egypt* 4:269-86.

Makacha, S., and Schaller, G. B. 1969. Observations on lions in the Lake Manyara National Park, Tanzania. *E. Afr. Wildl. J.* 7:99-103.

Malan, B. D., and Cooke, H. B. S. 1941. A preliminary account of the Wonderwerk Cave, Kuruman District. *S. Afr. J. Sci.* 37:300-312.

Malan, B. D., and Wells, L. H. 1943. A further report on the Wonderwerk Cave, Kuruman. *S. Afr. J. Sci.* 40:258-70.

Marchiafava, V.; Bonucci, E.; and Ascenzi, A. 1974. Fungal Osteoclasia: A model of dead bone resorption. *Calc. Tiss. Res.* 14:195-210.

Margalef, R. 1968. *Perspectives in ecological theory.* Chicago: University of Chicago Press.

Martin, P. S. 1966. Africa and the Pleistocene overkill. *Nature* 212(5060):339-42.

_____. 1967. Prehistoric overkill. In *Pleistocene extinctions: The search for a cause,* ed. P. S. Martin and H. E. Wright, pp. 75-102. New Haven: Yale University Press.

_____. 1973. The discovery of America. *Science* 179:969-74.

_____. 1975. Palaeolithic players on the American stage: Man's impact on the Late Pleistocene megafauna. In *Arctic and alpine environments,* ed. J. D. Ives and R. G. Barry, pp. 669-700. London: Methuen.

Mason, R. J. 1962. *Prehistory of the Transvaal.* Johannesburg: Witwatersrand University Press.

_____. 1967. Prehistory as a science of change—new research in the South African interior. *Occasional Papers of the Archaeological Research Unit of the University of Witwatersrand* 1:1-19.

Matthews, W. D. 1915. Climate and evolution. *Ann. New York Acad. Sci.* 24:171-318.

Mayr, E. 1970. *Populations, species and evolution.* Cambridge: Harvard University Press.

Meiring, A. J. D. 1956. The acrolithic culture of Florisbad. *Navorsinge van die Nasionale Museum (Bloemfontein)* 1(9):205-37.

Meland, N., and Norrman, J. O. 1966. Transport velocities of single particles in bedload motion. *Geogr. Annaler, Ser. A.* 48(4):165-82.

Mellett, J. 1968. The Oligocene Hsanda Gol Formation, Mongolia: A revised faunal list. *Amer. Mus. Novit.* 2318.

_____. 1974. Scatological origin of microvertebrate fossil accumulations. *Science* 185:349-50.

Mendard, H. W., and Boucot, A. J. 1951. Experiments in the movement of shells by water. *Am. J. Sci.* 249:131-51.

Menzel, R. G., and Heald, W. R. 1959. Strontium and calcium contents of crop plants in relation to exchangeable strontium and calcium of the soil. *Soil Sci. Soc. America Proc.* 23:110-12.

Middleton, J. 1844. On the fluorine in bones, its source and its application to the determination of the geological age of fossil bones. *Geol. Soc. London Proc.* 4: 431-33.

Mills, M. G. L. Unpublished. The food ecology of the brown hyaena in the Kalahari Gemsbok National Park.

Modha, K. L. and Eltringham, S. K. 1976. Population ecology of the Uganda kob *(Adenota kob thomasi* Neumann) in relation to the territorial system in the Rwenzori National Park, Uganda. *J. Appl. Ecol.* 13:453-73.

Moen, A. N. 1973. *Wildlife ecology.* San Francisco: Freeman.

Mohr, E. 1967. *Der Blaubock, Hippotragus leucophaeus* (Pallus, 1766). *Eine Dokumentation. Mammalia Depicta.* Hamburg: Paul Parey Verlag.

Mukherjee, A. K. 1974. Some examples of recent faunal impoverishment and regression. In *Ecology and biogeography in India,* ed. M. S. Mani, pp. 330-70. The Hague: Junk.

Müller, A. H. 1951. Grundlagen der Biostratonomie. *Abhandl. Deut. Akad. Wiss., Berlin, KL. Math. Allgem. Naturw.* 1950(3):1-146.

_____. 1963. *Lehrbuch der Paläozoologie,* vol. 1, *Allgemeine Grundlagen.* Jena: Fischer.

Muller, C. H. 1958. Science and philosophy of the community concept. *Amer. Sci.* 46: 249-308.

Napier, J. R., and Davis, P. R.   1959.   The forelimb and associated remains of *Proconsul africanus*. *Fossil Mammals of Africa* 16.   British Museum of Natural History.

Newell, N. D.   1949.   Phyletic size increase, an important trend illustrated by fossil invertebrates.   *Evolution* 3(2):103-24.

Nordin, C. F., Jr., and Dempster, G. R., Jr.   1963.   *Vertical distribution of velocity and suspended sediment, Middle Rio Grande, New Mexico.*   U. S. Geol. Surv. Prof. Paper 462-B.

Norton-Griffiths, M.   1975.   The numbers and distribution of large mammals in Ruaha National Park, Tanzania.   *E. Afr. Wildl. J.* 13(2):121-40.

Odum, E. P.   1969.   The strategy of ecosystem development.   *Science* 164:262-70.

Odum, H. T.   1957.   Biogeochemical deposition of strontium.   *Inst. Marine Sci. Pub.* 4: 38-114.

Olson, E. C.   1952a.   The evolution of a Permian vertebrate chronofauna.   *Evol.* 6:181-96.

_____.   1952b.   Fauna of the Vale and Choza: 6.   Diplocaulus.   *Fieldiana Geol.* 10: 147-66.

_____.   1957.   Size-frequency distributions in samples of extinct organisms.   *J. Geol.* 65:309-33.

_____.   1958.   Fauna of the Vale and Choza: 14.   Summary, review, and integration of the geology and the faunas.   *Fieldiana Geol.* 10:397-448.

_____.   1961.   Food chains and the origin of mammals.   Intern. Colloq. on the evolution of lower and unspecialized mammals.   *Kon. Vlaamse Acad., Wentensch, Lett. Schone Kunsten Belg.*, pt. 1, pp. 97-116.

_____.   1962.   Late Permian terrestrial vertebrates, USA and USSR.   *Trans. Amer. Philos. Soc.* 52(2):3-224.

_____.   1966.   Community evolution and the origin of mammals.   *Ecol.* 47:291-302.

_____.   1971.   *Vertebrate paleozoology.*   New York: Wiley-Interscience.

_____.   1974.   On the source of therapsids.   *Ann. S. Afr. Mus.* 64:27-46.

_____.   1975.   Permo-Carboniferous paleoecology and morphotypic series.   *Am. Zool.* 15:371-89.

_____.   1976.   The exploitation of land by early tetrapods. In *Morphology and biology of reptiles*.   Linnean Soc., Symposium series no. 3, pp. 1-30.

Olson, E. C., and Beerbower, J. R.   1953.   The San Angelo Formation, Permian of Texas, and its vertebrates.   *J. Geol.* 61:389-423.

Olson, E. C., and Miller, R. L.   1958.   *Morphological integration.*   Chicago: University of Chicago Press, 317 pp.

Olson, E. C., and Vaughn, P. P.   1970.   The changes of terrestrial vertebrates and climates during the Permian of North America.   *Forma et Functio* 3:113-38.

Opperman, H.   1978.   Excavations in the Buffelskloof Rock Shelter near Calitzdorp, Southern Cape.   *S. Afr. Archaeol. Bull.* 33:18-34.

Ortner, D. J.   1976.   *Microscopic and molecular biology of human compact bones: An anthropological perspective.*   *Yearbook of Physical Anthropology* 20:35-44.

Ortner, D. J.; Von Endt, D. W.; and Robinson, M. S.   1972.   The effect of temperature on protein decay in bone: Its significance in nitrogen dating of archeological specimens.   *American Antiquity* 37:514-20.

Oxnard, C. E.   1973.   *Form and pattern in human evolution.*   Chicago: University of Chicago Press.

_____.   1975.   *Uniqueness and diversity in human evolution.*   Chicago: University of Chicago Press.

Oyen, O. J., and Walker, A. C.   1977.   Stereometric craniometry.   *Am. J. Phys. Anth.* 46(1): 177-82.

Parker, R. B.   1967.   Electron microprobe analyses of fossil bones and teeth.   *Geol. Soc. Amer. Spec. Paper* 101:415.

Parker, Ronald B.; Murphy, J. W.; and Toots, H.   1974.   Fluorine in fossilized bone and tooth, distribution among skeletal tissues.   *Archaeometry* 16:98-102.

Parker, Ronald B., and Toots, Heinrich.   1970.   Minor elements in fossil bones. *Bull. Geol. Soc. Am.* 81:925-932.

_____.   1972.   Hollandite-coronadite in fossil bone.   *Am. Mineral.* 57:1527-30.

_____.   1974.   Minor elements in fossil bone—application to Quaternary samples. *Geol. Survey Wyoming, Rept. Inves.* 10:74-77.

_____.   1976a.   A final kick at the fluorine dating method.   *Arizona Acad. Sci. Jour., Proc.* 11:9-10.

_____.   1976b.   Sodium in fossil vertebrates as a paleobiological tool.   *Arizona Acad. Sci. Jour., Proc.* 11:87.

_____.   1976c.   Minor element distribution among vertebrate calcified tissues. *Arizona Acad. Sci. Jour., Proc.* 11:158.

Parker, Ronald B.; Toots, Heinrich; and Murphy, J. W.   1974.   Leaching of sodium from skeletal parts during fossilization.   *Geochem. et Cosmochem. Acta* 38:1317-21.

Parkington, J. E.   1972.   Seasonal mobility in the Late Stone Age.   *African Studies* 31: 223-44.

_____.   1976.   *Follow the San: An analysis of prehistoric seasonality in the South Western Cape, South Africa.*   Ph.D. dissertation, University of Cambridge.

Partridge, T. C.   1978.   Re-appraisal of lithostratigraphy of Sterkfontein hominid site. *Nature* 275:282-87.

Patterson, B., and Pascual, R.   1968.   Evolution of mammals on southern continents.   V. The fossil mammal fauna of South America.   *Quart. Rev. Biol.* 43:409-51.

Payne, S.   1972.   On the interpretation of bone samples from archaeological sites.   In *Papers in economic prehistory,* ed. E. S. Higgs, pp. 65-81. Cambridge: Cambridge University Press.

Penny, J. S.   1969.   Late Cretaceous and Early Tertiary palynology.   *Aspects of palynology,* ed. R. H. Tschudy and R. A. Scott.   New York: Wiley-Interscience.

Pennycuick, C. J.   1972.   *Animal flight.*   London: Arnold.

Pennycuick, C. J., and Western, D.   1972.   An investigation of some sources of bias in aerial transect sampling of large mammal populations.   *E. Afr. Wildl. J.* 10:175-92.

Perkins, D.   1964.   Prehistoric fauna from Shanidar, Iraq.   *Science* 144:1565-66.

_____.   1969.   Fauna of Çatal Hüyük: Evidence for early cattle domestication in Anatolia.   *Science* 164:177-79.

Perkins, D., and Daly, P.   1968.   A hunters' village in Neolithic Turkey.   *Sci. Am.* 219(11): 97-106.

Peterson, J. C. B., and Casebeer, R. L.   1972.   *Distribution, population status, and group composition of wildebeest and zebra on the Athi-Kapiti Plains, Kenya.*   UNDP/FAO Management Project, Nairobi Project Working Document.

Pflug, J. D., and Strübel, G.   1967.   Umwandlungen im Wirbeltierknochen während der Fossilisation.   *Oberhessischen Gesell. Natur. u. Heklkunde Giessen Ber., Neue Folge, Naturw. Abt.* 35:5-22.

Phillipson, J.   1973.   The biological efficiency of protein production by grazing and other land based systems.   In *The biological efficiency of protein production,* ed. J. G. W.

Jones, pp. 217-35. Cambridge: Cambridge University Press.

_____. 1975. Rainfall, primary production and 'carrying capacity' of Tsavo National Park, Kenya. *E. Afr. Wildl. J.* 13(3-4):171-201.

Pianka, E. R. 1970. On r- and K-selection. *Amer. Natur.* 104:592-97.

_____. 1974. *Evolutionary ecology.* New York: Harper and Row.

Pickford, M. 1975. Late Miocene sediments and fossils from the Northern Kenya Rift Valley. *Nature* 256:279-84.

Pienaar, U. de V. 1969. Predator-prey relationships amongst the larger mammals of the Kruger National Park. *Koedoe* 12:108-76.

_____. 1974. Habitat-preference in South African antelope species and its significance in natural and artificial distribution patterns. *Koedoe* 17:185-95.

Pienaar, U. de V.; Van Wyk, P.; and Fairall, N. 1966. An aerial census of elephant and buffalo in the Kruger National Park and the implications thereof on intended management schemes. *Koedoe* 9:40-107.

Piez, A., and Likens, R. C. 1960. The nature of collagen. II. Vertebrate collagens. In *Calcification in biological systems,* ed. R. F. Soghnaes, pp. 411-20. American Association for the Advancement of Science, Publication No. 64.

Pilbeam, D. 1969. Tertiary pongidae of East Africa: Evolutionary relationships and taxonomy. *Bull. Peabody Mus.* 31:1-185.

Pilgrim, G. E. 1910. Preliminary note on a revised classification of the Tertiary freshwater deposits of India. *Rec. Geol. Surv. India* 40:185-205.

_____. 1913. The correlation of the Siwaliks with mammal horizons of Europe. *Rec. Geol. Surv. India* 43:264-325.

_____. 1937. Siwalik antelopes and oxen in the American Museum of Natural History. *Bull. Am. Mus. Nat. Hist.* 72:729-874.

_____. 1939. The Fossil Bovidae of India. *Mem. Geol. Surv. India, Palaeont. Indica (n.s.)* 26:1-356.

Pilgrim, G. E., and Hopwood, A. T. 1928. *Catalogue of the Pontian Bovidae of Europe in the Department of Geology.* British Museum (Natural History), London.

Prasad, K. W. 1971. Ecology of the fossil Hominidae from the Siwaliks of India. *Nature* 232:413-14.

Pratt, D. J.; Greenway, P. J.; and Gwynne, M. D. 1966. A classification of East African rangeland, with an appendix on terminology. *J. Appl. Ecol.* 3:369-82.

Pritchard, J. J. 1972. General histology of bone. In *The biochemistry and physiology of bone,* 2d. ed., vol. 1, ed. G. H. Bourne, pp. 1-20. New York: Academic Press.

Pugh, J. W.; Rose, R. M.; and Radin, E. L. 1973. A possible mechanism of Wolff's law— trabecular microfractures. *Arch. Internat. Physiol. Bioch.* 81:27.

Rand, R. W. 1956. *The Cape Fur Seal, Arctocephalus pusillus (Schreber), its general characteristics and moult.* Union of South Africa, Department of Commerce and Industries, Division of Fisheries Investigational Report no. 21.

Raudviki, A. J. 1967. *Loose boundary hydraulics.* Oxford: Pergamon.

Rautenback, I. L. 1970. Ageing criteria in the springbok, *Antidorcas marsupialis* (Zimmerman, 1780) (Artiodactyla: Bovidae). *Ann. Trans. Mus.* 27:83-133.

Reher, C. A. 1974. Population study of the Casper Site bison. In *The Casper Site,* ed. G. C. Frison, pp. 113-24. New York: Academic Press.

Reid, J. J.,; Schiffer, M. B.; and Neff, J. M. 1975. Archaeological considerations of intrasite sampling. In *Sampling in archaeology,* ed. J. W. Mueller, pp. 209-24. Tucson: University of Arizona Press.

Reyment, R. 1971. *Introduction to quantitative paleoecology.* New York: Elsevier.

Richards, H. G., and Fairbridge, R. W. 1965. Annotated bibliography of Quaternary shorelines (1945-1964). *Acad. Nat. Sci., Philadelphia, Spec. Pub.* 6.

Richter, R. 1928. Aktuopaläontologie und Paläobiologie, eine Abgrenzung. *Senckenbergiana* 10:285-92.

_____. 1929. Gründung und Aufgaben der Forschrugsstelle fur Meeresgeologie "Senkenberg" in Wilhelmshaven. *Frankfurt, G. M. Natur. und Mus.* 59:1-30.

Ricklefs, R. E. 1973. *Ecology.* Newton, Mass.: Chiron Press.

Rigby, J. K. 1958. Frequency curves and death relationships among fossils. *Jour Paleontol.* 32:1007-9.

Robinette, W. L. 1964. *Notes on the biology of the lechwe (Kobus leche).* The Puku Occ. Papers, vol. 2, pp. 84-117. Dept. Game and Fisheries, N. Rhodesia.

Robinson, P. L. 1971. A problem of faunal replacement on Permo-Triassic continents. *Paleontol.* 14:131-42.

Roche, J. 1965. Données récentes sur la stratigraphie et la chronologie des amas coquilliers d'âge mésolithique du muge (Portugal). *Quaternaria* 7:155-63.

Rollins, H. B., and Donahue, J. 1975. Towards a theoretical basis of paleoecology: Concepts of community dynamics. *Lethia* 8:256-69.

Root, R. B. 1967. The niche exploitation pattern of the blue-gray gnatcatcher. *Ecol. Monogr.* 37(4):317-50.

Rosenthal, H. L. 1963. Uptake, turnover and transport of bone-seeking elements in fishes. *Ann. N. Y. Acad. Sci.* 109:278-93.

Rosenzweig, M. L. 1968. Net primary productivity of terrestrial communities: Prediction from climatological data. *Amer. Natur.* 102:67-74.

Rouse, H. 1950. *Engineering hydraulics.* New York: John Wiley and Sons.

Rozovskii, I. L. 1957. *Dvizhenie vodyna povorote otkrytogo rusla.* Institut Gidrologii i Gidrotekhniki, Akad. Nauk Ukr. S.S.R., Kiev. English translation (Flow of water in bends of open channels) by Y. Prushansky, IPST Cat. No. 363. Jerusalem: IPST Press, 1961.

Sabloff, J. A.; Beale, T. W.; and Kurland, A. M., Jr. 1973. Recent developments in archaeology. *Annals of the American Academy of Political and Social Science* 408: 103-18.

Sayer, J. A., and Van Lavieren, L. P. 1975. The ecology of the Kafue lechwe population of Zambia before the operation of hydro-electric dams on the Kafue River. *E. Afr. Wildl. J.* 13:9-37.

Schäfer, W. 1972. *Ecology and paleoecology of marine environments,* ed. G. Craig. Transl. I. Ortel. Chicago: University of Chicago Press.

Schaffer, W. M., and Reed, C. A. 1972. The co-evolution of social behaviour and cranial morphology in sheep and goats (Bovidae, Caprini). *Fieldiana Zoology* 61:1-88.

Schaller, G. B. 1972. *The Serengeti lion.* Chicago: The University of Chicago Press.

Schiffer, M. B. 1972. Archaeological context and systemic context. *American Antiquity* 37(2):156-65.

_____. 1975a. Archaeology as behavioral science. *Am. Anthrop.* 77(4):836-48.

_____. 1975b. Behavioral chain analysis: activities, organization, and the use of space. *Fieldiana Anthropology* 65:103-19.

_____. In press. Methodological issues in ethnoarchaeology. In *New directions in ethnoarchaeology,* ed. R. A. Gould. Albuquerque: University of New Mexico Press.

Schlösser, M. 1903. Die fossilen Säugetiere Chinas, nebst einer Odontographie der recenten Antilopen. *Abh. Bayer. Akad. Wiss.* 22:1-221.

Schopf, T. J. M. ed. 1972. *Models in paleobiology.* San Francisco: Freeman, Cooper and Co.

Schweitzer, F. R. 1970. A preliminary report of excavations of a cave at Die Kelders. *S. Afr. Archaeol. Bull.* 25:136-38.

_____. 1974. Archaeological evidence for sheep at the Cape. *S. Afr. Archaeol. Bull.* 29:75-82.

_____. 1975. *The ecology of post-Pleistocene peoples on the Gansbaai Coast, South-Western Cape.* M.A. thesis, University of Cape Town.

Schweitzer, F. R. and Scott, K. 1973. Early occurrence of domestic sheep in sub-Saharan Africa. *Nature* 241:547-48.

Seber, G. A. F. 1973. *The estimation of animal abundance.* London: Griffin.

Selley, R. C. 1971. *Ancient sedimentary environments; a brief survey.* Ithaca, N. Y.: Cornell University Press.

Sera, G. L. 1935. I caratteri morfologici di *Palaeopropithecus.* *Arch. Ital. Anat. Embriol.*, vol. 229.

Shackleton, R. M. 1951. A contribution to the geology of the Kavirondo Rift Valley. *Quart. J. Geol. Soc. Lond.* 106:345-88.

Shackleton, N. J. Unpublished. Stratigraphy and chronology of the Klasies River Mouth deposits: Oxygen isotope evidence.

Shelford, V. E. 1963. *The ecology of North America.* Urbana: University of Illinois Press.

Sheppe, W., and Osborne, T. 1971. Patterns of use of a flood plain by Zambian mammals. *Ecol. Monog.* 41(3):179-205.

Shortridge, E. C. 1934. *The mammals of South West Africa.* London: W. Heineman.

Shotwell, J. A. 1955. An approach to the paleoecology of mammals. *Ecol.* 36(2):327-37.

_____. 1958. Inter-community relationships in Hemphillian (mid-Pliocene) mammals. *Ecol.* 39(2):271-82.

_____. 1963. The Juntura Basin: Studies in earth history and paleoecology. *Trans. Amer. Philos. Soc.* (ns.) 53(1):3-77.

_____. 1964. Community succession in Mammalia, the Late Tertiary. In *Approaches to paleoecology,* ed. J. Imbrie and N. Newell. New York: Wiley.

Shuey, R. T.; Brown, F. H.; Eck, G. G.; and Howell, F. C. 1978. A statistical approach to temporal biostratigraphy. In *Geological background to fossil man,* ed. W. W. Bishop. Edinburgh: Scottish Academic Press.

Simberloff, D. 1972. Models in biogeography. In *Models in paleobiology,* ed. J. M. Schopf, pp. 160-91. San Francisco: Freeman, Cooper.

Simpson, C. D., and Elder, W. H. 1969. Tooth cementum as an index of age in greater kudu *(Tragelaphus strepsiceros).* *Arnoldia (Rhod.)* 4(20):1-10.

Simpson, G. G. 1944. *Tempo and mode in evolution.* New York: Columbia University Press.

_____. 1945. The principles of classification and a classification of mammals. *Bull. Am. Mus. Nat. Hist.* 85:1-350.

_____. 1947. Holarctic mammalian faunas and continental relationships during the Cenozoic. *Bull. Geol. Soc. Am.* 58:613-88.

_____. 1953. *The major features of evolution.* New York: Columbia University Press.

_____. 1960. Notes on the measurement of faunal resemblance. *Amer. Jour. Sci.*, Bradley vol., 258-A, pp. 300-311.

_____. 1961. Some problems of vertebrate paleontology. *Science* 133:1679-89.

_____. 1963. The historical factor in science. In *This view of life,* chap. 7. New York: Harcourt, Brace and World.

Simpson, G. G., and Roe, A. 1939. *Quantitative zoology.* New York: McGraw Hill.

Sinclair, A. R. E. 1972. Long term monitoring of mammal populations in the Serengeti: census of non-migratory ungulates. *E. Afr. Wildl. J.* 4:287-97.

_____. 1974. Natural regulation of a buffalo population in East Africa. *E. Afr. Wildl. J.* 12:135-54.

_____. 1975. The resource limitation of trophic levels in tropical grassland ecosystems. *J. Anim. Ecol.* 44(2):497-520.

Singer, R. 1961. The new fossil sites at Langebaanweg, South Africa. *Current Anthropology* 2:385-87.

Singer, R. and Hooijer, D. A. 1958. A Stegolophodon from South Africa. *Nature* 182:101-2.

Sloan, R. E. 1970. Cretaceous and Paleocene terrestrial communities of western North America. In *Proc. N. Amer. Paleontol. Congr.* pt. 3:427-53.

Smart, C. 1976. The Lothagam 1 fauna: Its phylogenetic, ecological and biogeographic significance. In *Earliest man and environments in the Lake Rudolf Basin,* ed. Y. Coppens et al., pp. 361-69. Chicago: University of Chicago Press.

Spinage, C. A. 1972. African ungulate life tables. *Ecology* 53(4):645-52.

Spinage, C. A.; Guiness, F.; and Eltringham, S. K. 1972. Estimation of large herbivore numbers in the Akagera National Park and Mutara hunting reserve. *La Terre et la Vie* 4:562-70.

Stanislawski, M. B. 1969. The ethnoarchaeology of Hopi pottery making. *Plateau* 42(1): 26-33.

_____. In press. If pots were mortal. In *New directions in ethnoarchaeology,* ed. R. A. Gould. Albuquerque: University of New Mexico Press.

Stanley-Price, M. R. 1974. *The feeding ecology and energetics of Coke's hartebeest.* D. Phil. dissertation, Oxford University.

Stewart, D. R. M. 1963. Wildlife census—Lake Rudolf. *E. Afr. Wildl. J.* 1:121.

Stewart, D. R. M., and Zaphiro, D. R. P. 1963. Biomass and density of wild herbivores in different East African habitats. *Mammalia* 27:438-96.

Stockley, G. M. 1932. The geology of the Ruhuhu coalfields, Tanganyika Territory. *Quart. J. Geol. Soc. Lond.* 88:610-22.

Stockley, G. M. and Oates, F. 1931a. The Ruhuhu coalfields, Tanganyika Territory. *Min. Mag. Lond.* 45:73-91.

_____. 1931b. Report on the geology of the Ruhuhu coalfields, Njombe-Sangea Districts. *Geol. Surv. Tanganyika Bull.* 2:1-63.

Stromer, E. 1926. Reste land- und süsswasser-bewohnender Wirbeltiere aus den Diamentenfelden Deutsch-Südwestafrikos. In *Die Diamantenwuste Südwestafrikos,* ed. E. Kaiser. Berlin: Verlag von D. Reimer.

Sutcliffe, A. J. 1970. Spotted hyaena: Crusher, gnawer, digestor, and collector of bones. *Nature* 227:1110-13.

Sylvester-Bradley, P. C. 1958. The description of fossil populations. *Jour. Paleontol.* 32:214-35.

Talbot, L. M., and Talbot, M. H.   1963.   The wildebeest in Western Masailand, East Africa.   *Wildl. Monogr.* 12:1-88.

Tankard, A. J.   1974.   Varswater Formation of the Langebaanweg-Saldanha area, Cape Province.   *Trans. Geol. Soc. S. Afr.* 77:265-83.

Tankard, A. J., and Schweitzer, F. R.   1974.   The geology of Die Kelders Cave and environs: A palaeoenvironmental study.   *S. Afr. J. Sci.* 70:360-69.

_____.   1976.   Textural analysis of cave sediments: Die Kelders, Cape Province, South Africa.   In *Geoarchaeology: Earth science and the past*,   ed. D. A. Davidson and M. L. Shackley.   London: Duckworth.

Tattersall, I.   1969*a*.   Ecology of north Indian *Ramapithecus*.   *Nature* 221:451-52.

_____.   1969*b*.   More on the ecology of north Indian *Ramapithecus*.   *Nature* 224:821-22.

Thompson, W. D'Arcy.   1944.   *Growth and form*.   Cambridge: Cambridge University Press.

Thurber, D. L.; Kulp, J. L.; Hodges, E.; Gast, P. W.; and Wampler, J. M.   1958.   Common strontium content of the human skeleton.   *Science* 128:256-57.

Till, R.   1974.   *Statistical methods for the earth scientist*.   New York: Wiley.

Tobias, P. V.   1965.   *Australopithecus, Homo habilis,* tool-using and tool-making.   *S. Afr. Archaeol. Bull.* 20:167-92.

Toots, H.   1963.   The chemistry of fossil bones from Wyoming and adjacent states.   *Contrib. Geol.* 2:69-80.

Toots, Heinrich, and Voorhies, M. R.   1965.   Strontium in fossil bones and the reconstruction of food chains.   *Science* 149:854-55.

Tsukada, M.   1967.   Fossil cladocera in Lake Nojiri and ecological order.   *The Quarternary Research (Japan)* 6:101-10.

Tuggle, H. D.; Townsend, A. H.; and Riley, T. J.   1972.   Laws, systems, and research designs: A discussion of explanation in archaeology.   *American Antiquities* 37(1):3-12.

Turnbull-Kemp, P.   1967.   *The leopard*.   Cape Town: Howard Timmins.

Turner, M. and Watson, R. M.   1964.   A census of game in Ngorongoro Crater.   *E. Afr. Wildl. J.* 2:165-68.

Valentine, J. W.   1972.   Conceptual models in ecosystem evolution.   In *Models in paleobiology*,   ed. J. M. Schopf, pp. 192-215.   San Francisco: Freeman, Cooper.

_____.   1973.   *Evolutionary paleoecology of the marine biosphere*.   Englewood Cliffs, N.J.: Prentice-Hall.

Valentine, J. W., and Mallory, B.   1965.   Recurrent groups of bonded species in mixed death assemblages.   *J. Geol.* 73:683-701.

Valentine, J. W. and Peddicord, R. G.   1967.   Evaluation of fossil assemblages by cluster analysis.   *Jour. Paleontol.* 41:502-7.

Van Couvering, J. A. H., and Van Couvering, J. A.   1976.   Early Miocene mammal fossils from East Africa: Aspects of geology, faunistics and paleoecology.   In *Human origins: Louis Leakey and the East African evidence*,   ed. G. Isaac and E. McCown, pp. 155-207.   Menlo Park, Ca.: W. A. Benjamin.

Vanoni, V. A., and Nomicos, G. N.   1959.   Resistance properties of sediment-laden streams.   *J. Hydraulics Div. Amer. Soc. Civil Engineers* 85(HY5):77-107.

Van Valen, L.   1964.   Relative abundance of species in some fossil mammal faunas.   *Amer. Natur.* 98:109-16.

_____.   1970.   Late Pleistocene extinctions.   *Proc. N. Amer. Paleontol. Congr.*, ed. E. Yochelson, vol. 6, pp. 469-85.   Lawrence, Kansas: Allen Press.

_____.   1973.   Pattern and the balance of nature.   *Evol. Theory* 1:31-49.

Vokes, H. E.   1935.   Unionidae of the Siwalik Series. *Mem. Conn. Acad. Arts and Sci.* 9: 37-48.

Voorhies, M.   1969.   Taphonomy and population dynamics of an early Pliocene vertebrate fauna, Knox County, Nebraska. *Contrib. Geol., Spec. Paper no. 1.* Wyoming: University of Wyoming Press.

_____.   1970.   Sampling difficulties in reconstructing Late Tertiary mammalian communities. In *Proc. N. Amer. Paleontol. Congr.,* ed. E. Yochelson, vol. 6, pp. 454-68. Lawrence, Kansas: Allen Press.

Vrba, E. S.   1970.   Evaluation of Springbok-like fossils: Measurement and statistical treatment of the teeth of the Springbok, *Antidorcas marsupialis marsupialis* Zimmermann (Artiodactyla: Bovidae). *Ann. Transv. Mus.* 26:285-99.

_____.   1973.   Two species of *Antidorcas* Sundevall at Swartkrans (Mammalia: Bovidae). *Ann. Transv. Mus.* 28:287-352.

_____.   1974.   Chronological and ecological implications of the fossil Bovidae at the Sterkfontein Australopithecine site. *Nature* 250(5461):19-23.

_____.   1975.   Some evidence of chronology and palaeoecology of Sterkfontein, Swartkrans and Kromdraai from the fossil Bovidae. *Nature* 254:301-4.

_____.   1976.   The fossil Bovidae of Sterkfontein, Swartkrans and Kromdraai. *Transv. Mus. Mem.* 21:1-166.

Walker, A.   1969.   Fossil mammal locality on Mount Elgon, eastern Uganda. *Nature* 223(5206): 591-96.

_____.   1976.   Remains attributable to *Australopithecus* from the East Rudolf succession. In *Earliest man and environments in the Lake Rudolf Basin,* ed. Y. Coppens et al., p. 484. Chicago: University of Chicago Press.

Walker, K. R., and Alberstadt, L. P.   1975.   Ecological succession as an aspect of structure in fossil communities. *Paleobiol.* 1:238-57.

Walter, H.   1954.   Le facteur eau dans les régions andes et sa signification pour l' organisation de la végétation dans les contrées sub-tropicales. In *Les divisions ecologiques du monde,* pp. 27-39. Centre Nat. de la Res. Scient., Paris.

_____.   1973.   *Vegetation of the earth in relation to climate and eco-physiological conditions.* London: English University Press.

Walther, J.   1930.   *Die methoden der geologie als historische und biologische Wissenshaft,* in E. Abderhalden, ed., Handbuch der biologischen Arbeitsmethoden, Abt. 10. Methoden der Geologie, Mineralogie, Paläontologie und Geographie, pp. 529-658. Berlin: Urban and Schwarzenberg.

Warme, J. E.   1971.   Paleoecological aspects of a modern coastal lagoon. *Univ. Calif. Publ. Geol. Sciences* 87:1-112.

Waston, G. S.   1966.   The statistics of orientation data. *J. Geol.* 74(2):786-97.

Watson, R. M.   1967.   The population ecology of the wildebeest *(Connochaetes taurinus albojubatus* Thomas). Ph.D. dissertation, Cambridge University.

_____.   1972.   *Results of aerial livestock surveys of Kaputei division, Samburu district and Northeastern Province.* Statistics division, Min. of Finance and Planning, Republic of Kenya.

Watson, R. M.; Graham, A. D.; and Parker, I. S. C.   1969.   A census of the large mammals of Loliondo controlled area, Northern Tanzania. *E. Afr. Wildl. J.* 7:45-59.

Watson, R. M., and Turner, M. I. M.   1965.   A count of large mammals of the Lake Manyara National Park; results and discussion. *E. Afr. Wildl. J.* 3:95-98.

Watt, K. E., ed.   1966.   *Systems analysis in ecology.* New York: Academic Press.

Weare, P., and Yalala, A.   1971.   Provisional vegetation map of Botswana. *Botswana Notes and Records* 3:131-47.

Weigelt, J. 1927. *Resente Wirbeltierleichen und ihre Paläobiologische Bedeutung.* Leipzig: Verlag von Max Weg.

Weir, J. S. 1972. Spatial distribution of elephants in a central African national park with relation to environmental sodium. *Oikos* 23:1-13.

Weiss, B. 1975. The new archaeology. *Psychology Today* 8(5):50-58.

Welbourne, R. G. 1971. *A study of prehistoric relationships between environment, animals, and man at one Earlier Stone Age site, one Middle Stone Age site, and eight Iron Age sites in the Vaal-Limpopo Basins.* M.A. thesis, University of Witwatersrand.

Wells, L. J. 1967. Antelopes in the Pleistocene of Southern Africa. In *Background to evolution in Africa,* ed. W. W. Bishop and J. D. Clark. Chicago: University of Chicago Press.

Wells, L. H., and Cooke, H. B. S. 1956. Fossil Bovidae from the Limeworks Quarry, Makapansgat, Potgietersrust. *Palaeont. Afr.* 4:1-55.

Wendorf, F.; Laury, R. L.; Albritton, C. C.; Schild, R.; Haynes, C. V.; Damon, P. E.; Shafquillah, M.; and Scarborough, R. 1975. Dates for the Middle Stone Age of East Africa. *Science* 187:740-42.

Wepfer, E. 1916. Ein wichtiger Grund für die Lückenhaltigkeit paläontologischer Überlieferung. Centralbl. für Min., Geol. u. Paleont. 1916:105-13.

Western, D. 1973. *The structure, dynamics and changes of the Amboseli ecosystem.* Ph.D. dissertation, University of Nairobi.

_____. 1975. Water availability and its influence on the structure and dynamics of a savannah large mammal community. *E. Afr. Wildl. J.* 13:265-86.

_____. 1979. The environment and ecology of pastoralists in arid savannahs. In *The future of hunter-gatherer and nomadic pastoral societies,* ed. J. J. Swift. London: International African Institute.

Western, D., and Van Praet, C. 1973. Cyclical changes in the habitat and climate of an East African ecosystem. *Nature* 241:104-6.

Wheat, J. B. 1972. The Olsen-Chubbock site: A Paleo-Indian bison kill. *Memoirs of the Society for American Archaeology* no. 26.

White, T. E. 1953. Observations on the butchering technique of some aboriginal peoples. No. 2. *American Antiquity* 19:160-64.

_____. 1954a. Observations on the butchering technique of some aboriginal peoples. No. 4. *American Antiquity* 19:257-59.

_____. 1954b. Observations on the butchering technique of some aboriginal peoples. No. 5. *American Antiquity* 19:259-62.

Whittaker, R. H. 1970. *Communities and ecosystems.* London: Collier-Macmillan.

Williams, C. B. 1964. *Patterns in the balance of nature.* London: Academic Press.

Williams, J. G. 1968. *A field guide to the national parks of East Africa.* London: Collins.

Wolff, R. G. 1971. *Paleoecology of a late Pleistocene (Rancholabrean) vertebrate fauna from Rodeo, California.* Ph.D. dissertation, University of California, Berkeley.

_____. 1973. Hydrodynamic sorting and ecology of a Pleistocene mammalian assemblage from California (USA). *Palaeogeog., Palaeoclimat., Palaeoecol.* 13:91-101.

_____. 1975. Sampling and sample size in ecological analysis of fossil mammals. *Paleobiol.* 1(2):195-204.

Wood, B. A. 1976. Remains attributable to *Homo* from the East Rudolf succession. In *Earliest man and environments in the Lake Rudolf Basin,* ed. Y. Coppens et al., p. 490 Chicago: University of Chicago Press.

Wood, H. E., et al. 1941. Nomenclature and correlation of the North American continental Tertiary. *Bull. Geol. Soc. Am.* 52:1-48.

Wyckoff, R. W. G. 1972. *The biochemistry of amino fossils*. Baltimore: Williams and Wilkins Company.

Wymer, J. J., and Singer, R. 1972. Middle Stone Age settlements on the Tzitzikama coast, eastern Cape Province, South Africa. In *Man, settlement and urbanism,* ed. P. J. Ucko, R. Tringham, and G. W. Dimbleby, pp. 207-10. London: Duckworth.

Wysoczanski-Minkowicz, Tadeusz. 1969. An attempt at relative age determination of fossil bones by fluorine-chlorine-apatite method (Polish with English summary). *Studia Geologica Polonica,* vol. 28.

Yellen, J. E. 1974. *The !Kung settlement pattern: An archaeological perspective.* Ph.D. dissertation, Harvard University.

_____. 1975. Cultural patterning in faunal remains: Evidence from the !Kung Bushmen. In *Experimental archaeology,* ed. J. E. Yellen. New York: Columbia University.

Yochelson, E. L. 1970. *Proceedings of the North American paleontological convention.* 2 vols. Lawrence, Kansas: Allen Press.

Zangerl, R., and Richardson, J. 1963. The paleoecological history of two Pennsylvanian black shales. *Fieldiana Geol. Mem.* vol. 4.

Zeuner, F. E. 1959. *The Pleistocene period. Its climate, chronology and faunal succesions.* 2d. ed. London: Hutchinson.

Zimmerman, E. C. 1960. Possible evidence of rapid evolution of Hawaiian moths. *Evol.* 14: 137-38.

Zinderen Bakker, E. M. van. 1957. A pollen analytical investigation of the Florisbad deposits (South Africa). In *Proceedings of the Third Pan-African Congress on Prehistory,* ed. J. D. Clark, pp. 56-67. London: Chatto and Windhus.

_____. 1977. The evolution of Late Quaternary palaeoclimates of Southern Africa. *Palaeoecology of Africa* 9.

Zinderen Bakker, E. M. van, and Butzer, K. W. 1973. Quaternary environmental changes in southern Africa. *Soil Science* 116:236-48.

Zittel, K. A. 1918. *Grundzüge der Paläontologie (Palaozoologie)*, vol. 2, *Vertebrata.* Munich: R. Oldenbourg.

PARTICIPANTS IN THE SYMPOSIUM

Anna K. Behrensmeyer
Department of Geology and Geophysics
Yale University, New Haven,
Connecticut 06520

W. W. Bishop (deceased)
Department of Geology
Queen Mary College
University of London
London El 4NS, England

Dorothy E. Dechant Boaz
Department of Anthropology
University of California
Berkeley, California 94720

Arthur Bourne
Academic Press Inc., Ltd.
24-28 Oval Road
London NW1 7DX, England

C. K. Brain
Transvaal Museum, P. O. Box 413
Pretoria
Republic of South Africa 0001

Malcolm Coe
Department of Zoology
Animal Ecology Research Group
South Parks Road
Oxford OX1 3PS, England

Diane P. Gifford
Department of Anthropology
Merrill College
University of California
Santa Cruz, California 95064

C. Bruce Hanson
Department of Paleontology
University of California
Berkeley, California 94720

P. E. Hare
Geophysical Laboratory
Carnegie Institution
2801 Upton Avenue, N.W.
Washington, D. C. 20008

Andrew P. Hill
The International Louis Leakey Memorial
    Institute for African Prehistory (TILLMIAP)
Box 46727
Nairobi, Kenya

Richard Klein
Department of Anthropology
University of Chicago
Chicago, Illinois 60637

Everett C. Olson
Department of Biology
University of California
Los Angeles, California 90024

Ronald B. Parker
R. R. 1
Henning, Minnesota 56551

Ralph T. Shuey
Gulf Science and Technology Company
P. O. Drawer 2038
Pittsburgh, Pennsylvania 15230

Judith A. H. Van Couvering
University of Colorado Museum
University of Colorado
Boulder, Colorado 80302

Elisabeth S. Vrba
Transvaal Museum, P. O. Box 413
Pretoria
Republic of South Africa

Alan C. Walker
Anatomy Department
Johns Hopkins University
Baltimore, Maryland
21200

David Western
Department of Zoology
University of Nairobi
Box 30197
Nairobi, Kenya